Nature and Psyche

NATURE AND PSYCHE

RADICAL ENVIRONMENTALISM AND THE POLITICS OF SUBJECTIVITY

David W. Kidner

State University of New York Press

Published by
State University of New York Press, Albany

Printed in the United States of America

For information, address the State University of New York Press, 90 State Street, Suite 700, Albany, NY 12207

Production by Kelli Williams
Marketing by Anne M. Valentine

Library of Congress Cataloging-in-Publication Data

Kidner, David W., 1947–
Nature and psyche : radical environmentalism and the politics of subjectivity / David W. Kidner.
p. cm.
Includes bibliographical references and indexes.
ISBN 0-7914-4751-0 (alk. paper). — ISBN 0-7914-4752-9 (pbk. : alk. paper)
1. Deep ecology. 2. Subjectivity. I. Title.

GE195.K53 2000
363.7'001—dc21

00-033894

10 9 8 7 6 5 4 3 2 1

CONTENTS

PREFACE

Over the past several decades awareness of the potentially catastrophic effects of the degradation of the natural world has risen dramatically. Even in the relatively affluent areas, which most readers of this book inhabit, almost every feature of our lives that depends on interaction with the world outside our physical boundaries is increasingly viewed as a matter of environmental "risk" and as a threat to our well-being. Air and water pollution, contamination of food by pesticides, drug-resistant pathogenic organisms, climate change, and so on have all generated a public foreboding regarding technological and economic "progress"—a foreboding that is magnified by an increasingly conscious recognition that what is happening is determined at least partly by factors outside our conscious control. At the same time a range of other, apparently distinct but less tangible problems have emerged as clouds on the horizon of material prosperity: occupational stress and alienation, depression and anxiety, loneliness and a sense of isolation are a few of the more obvious ones. What is more, we are aware that even these problems would be experienced as almost incomprehensible indulgences in those less affluent parts of the world that are in the acute stages of industrialization.

It is typical of the fragmentation of knowledge in the modern world that these problems are generally seen as unconnected, existing within separate realms such as those we label "psychopathology," "environmental issues," "human rights," "social problems," and so on. Our cultural predisposition to view reality as a collection of loosely connected "things" not only determines the connection, or more often lack of connection, between these things, but in many ways defines their character. Ecologists, of course, have tried to correct this fragmentation by pointing to the larger patterns that biological entities inhabit; but even their vision seldom extends into the realms of subjectivity and culture. But is it really plausible that ecology, culture, and individuality—which, after all, have evolved together over millions of years—can be viewed as separate realms obeying unconnected laws and suffering unrelated forms of pathology? Could it be that our experience of the natural world as a collection of "natural resources" is related to the social and psychological fragmentation so characteristic of industrial society? And if we accept that contemporary forms of consciousness are inept at recognizing what Gregory Bateson called "the pattern that connects," then what is it that we are losing sight of through the fragmentariness of our epistemology? In particular, how does our understanding of the diverse range of afflictions from which we suffer—existentially, socially, ecologically—change if we are prepared to question their distinctiveness?

This book attempts to challenge the isolation of academic fields by drawing on a variety of disciplines. However, even this uncomfortable gathering of academic specialisms which usually have little to say to each other runs the risk of continuing the intellectualist bias that is often hostile to lived experience and to the sorts of experiential reality encompassed by certain other cultural systems. Even anthropology, potentially the most humane as well as the most human of disciplines, has only recently been prepared to challenge the invisible barrier between "researchers" and "informants" by admitting that the former experience the same needs, desires, and fallibilities as the rest of us, and that the latter function in ways that, while they may be different in style to our own, may be no less comprehensible and adaptive.[1] For this reason, I am as concerned here with the dissociations and boundaries that define the landscape of modern subjectivity as with the complementary restructuring of nature in the outside world—a complementarity that has been widely misinterpreted in the social sciences, in a remarkable collusion of hubris and solipsism, as indicating that nature is *a product of* human subjectivity. European social theorists seem particularly vulnerable to this profound misreading of our place in the world; and as I argue later, this may well have something to do with the almost complete disappearance of wilderness within Europe. If we lose touch with any authentically wild reality that could confront and correct what Alastair Macintosh has referred to as our "academic autism," then our understanding of nature tends to be patterned either on the degraded ecological world we physically inhabit or, reactively, on an idealistic fantasy of a "pure" nature defined mainly by its opposition to industrialism. Under these circumstances, what we have lost is not only a particular form of wild nature, but also our capacity to envision and work toward *any* realistic nature that could exceed and transcend either of these industrialist inventions.

Under current conditions, in which a real, preexisting order is in the process of being replaced by a manufactured one, and in which language mostly refers to the latter while tacitly abandoning the former, there are peculiar difficulties involved in using conventional terminology to point to the sort of healthy natural world that is increasingly sidelined as an unrealistic fantasy. Terms such as "nature" and "ecology," for example, while they resonate with deeply felt possibilities and relations that have largely been lost from consciousness, are mostly used to refer to aspects of the presently existing world. Usually, there are no recognized terms capable of referring to the natural order that do not simultaneously incorporate the perversion and simplification of this order by consciousness. The reader needs to be aware of these difficulties—reading the book not only with the intellect, but also digging beneath the words toward felt, embodied reactions whose shapes can only dimly be perceived by consciousness.

To a lesser extent, there are related problems in discussing the industrial order that increasingly determines our lives. Industrialism relies on and exploits the narrowness of consciousness in order to conceal its own systemic character, so that the alliance between technology, capitalism, the legal system, education, consumerism, and everyday common sense is concealed beneath the superficial independence of these areas within a democratic political system. This concealment is reflected in the absence of words that could indicate either the character of what is being hidden or the process of concealment itself. I refer to the order that patterns these aspects of our lives as "industrialism"—a deliberately vague term that will, I hope, avoid the fate of assimilation to existing social theories and concepts. This term is intended to refer not simply to overt and presently existing characteristics of industrial society, but also to their roots in patterns, practices, and forms of thinking that may be discernible in societies distanced from us by centuries or even millennia. I have tried here to extend our understanding of this industrialist system beyond the usual references to modes of production and their immediate social implications, exploring the way industrialism reconfigures the world and replaces the natural order as the accepted basis for all life. The character of industrialism is way beyond the power of consciousness to define accurately or completely, not least because it structures consciousness precisely to maintain its own partial invisibility; and we have more to gain by acknowledging this partial invisibility than by the pretence of complete, rational understanding.

This is not to suggest, unfortunately, that industrialism and the natural order can be separated in any simple manner, so that a revolutionary ecological consciousness, drawing on the Romantic tradition, can identify and uproot an invasive technological and economic system that can be rejected in its entirety. Part of industrialism's hold upon us is precisely that it conflates and incorporates the natural and the manufactured, the healthy and the perverted, the authentic and the illusory; and so teasing apart the healthy roots from those that are invasive will require both intellectual awareness and a sure sense of our own groundedness in the earth. Simplistic oppositional viewpoints are not helpful; and pointing out the limitations of science is to reject not science, but *scientism*—the assumption that science frames human life and provides a complete account of the character of the world. The reader will find in these pages criticisms not of science, economics, and technology (although I point to their limitations), but rather of scientism, economism, and technologism. For similar reasons, I have treated "individuality" as at least potentially healthy; while I regard "individualism"—the dogma that human life can be adequately understood simply in terms of individual characteristics—as unhelpful.

Many people have contributed to the ideas I develop in this book. My former colleagues at Fort Lewis College, Durango, shared a sense of academic conviviality, enthusiasm, and passionate engagement with wider realities that has remained with me as a model of how academia can and should function. In particular, I want to thank Hal Mansfield for his support, friendship, and vision, all of which I have drawn on freely. Closer to home, Matt Connell and Conrad Lodziak have been rocks of integrity, demonstrating that ideas and ethics can be rooted in something deeper than academic fashion and economic pressures. I would also like to thank Ian Conolly for his warmth, friendship, and encouragement over many years. Of the many academic writers who have influenced my writing, none has done so more profoundly than Robert Romanyshyn; and my debt to his insights will be obvious to readers of this book. Clifford Geertz, Theodore Roszak, and Frederick Turner have been pioneers, in their different ways, of the sort of interdisciplinary awareness I advocate in this book; and each of them has significantly influenced my thinking. Among the many others who have contributed to the book in a diversity of ways are Annina Kidner, Simon Kidner, and Susie Spencer. I would also like to thank Eugene Hargrove and Linda Kalof for permission to reprint passages from papers originally published in, respectively, *Environmental Ethics* and the *Human Ecology Review*; and Renault (UK) Ltd. and WWAV Rapp Collins for permission to reproduce advertising material. As for the more-than-human world, the inhabitants and natural forms of the Four Corners area of the American southwest influenced my writing in ways that are more difficult to express but just as significant, confirming my deeply sensed but inarticulate intuition that wild nature is not merely a social construct dreamt up by academics, but a powerful and fundamental order in its own right. All these human and nonhuman inhabitants of the world demonstrate something in common, implicitly or explicitly, with the sort of healthy biosphere I envision in these pages.

Part One

Nature and Psyche in an Industrialist Landscape

Chapter 1

INTRODUCTION: IN SEARCH OF THE "NATURAL"

. . . Love moves and spurs the intellect to go before it, like a lantern, to the forests, uncultivated and lonely, very rarely visited and explored.

—Giordano Bruno, *The Heroic Frenzies*

Natural Dilemmas

The rise of environmental awareness over the past few decades, and the emergence of an environmental movement containing various shades of green and transcending the tired polarizations of left- and right-wing, has been one of the most remarkable features of modern political life. Every politician today, whatever their position within the political spectrum, needs to be able to demonstrate some degree of environmental awareness, however superficial and reactive the resulting policies might be. Nevertheless, in spite of some notable successes in heading off attempts by industry or governments to convert what remains of the natural world into material "wealth," the green movement, at a deeper level, shares modern industrial society's puzzlement as to ultimate goals, finding its voice most effectively through *protest against* existing or planned activities rather than putting forward its own lucid agenda for the future. In a sense, this puzzlement is more profound within radical variants of greenery, since these openly admit the *need* for a new vision of our relation to the rest of the natural world, in contrast to reformist varieties that merely try to make existing political and commercial processes less obviously damaging, and so ultimately still believe in, or try to convince themselves that they believe in, the basic soundness of these processes. "Deeper" greens, perceiving the ultimate unworkability and destructiveness of the technological-economic system, are forced to confront the need for a radically different approach. Unfortunately, deep ecology, ecofeminism, and other "radical" approaches, as I will argue below, often incorporate many of the features of existing ideology, and so are bound within the same epistemological, moral, and experiential universe as the structures they attempt to challenge. The result is an environmentalism that is at its most effective when challenging the details of our affluent lifestyles—ozone-damaging refrigerants, mahogany furniture, a road-centered transport policy—but that suffers from an acute sense of impotence and puzzlement over how to alter the direction of industrialism. This is one of the fundamental predicaments addressed by this book: the psychological, social, and epistemological foundations of the green movement's critique of modern industrialism are often the same foundations on which industrialism itself is built. Building on this recognition, I attempt to show how a reconfigured social/environmental theory can transcend the deadening grasp of industrialist assumptions, and point toward a revitalized relation between ourselves and the rest of the natural world that is both realistic and healthy.

The thinly veiled despair that underlies present-day political modernism (a broad category in which I mean to include virtually all mainstream political parties) is associated with a dawning awareness, now common in some form across the whole social spectrum, that concepts such as "progress" or "civilisation," which in the not too distant past were capable of grounding us within a universe that seemed solid and predictable, carry with them a shadow side that becomes more inescapable as these concepts are pushed toward their limits—an awareness that has long been focused on by critical theory.[1] This awareness surfaces differently—and not necessarily consciously—within various contexts: in the intellectualised nihilism of "postmodern" views; in trivially subversive gossip in the local pub; in the subtly frantic behavior of supermarket shoppers; in the casual vandalism that disfigures our cities; or in the narcissistic obsession with appearance and fashion that attempts to paper over the poverty of real relation either to other individuals or, indeed, to anything outside the carefully tended world of egoic consciousness. While one can only respect the courage, ingenuity, and creativity that people show in the face of a context whose character is all the more psychologically debilitating for being difficult to identify or articulate, these adaptive qualities also allow our easy colonisation by ideologies that are ultimately destructive both to their human hosts and to the world we inhabit. Even if the statistics—of population increase, species extinction, exhaustion of "natural resources"—were not enough to convince us, we sense the untenability of our lifestyle in a way that is difficult to articulate but also difficult to deny. But trapped within a universe that seems to be inescapably defined by the laws of industrialist rationality, the only apparent alternative to a bleak and helpless surrender to this reality seems to be a hedonistic, digitally assisted escape that ultimately enmeshes us further within the industrialist process. This understandable tendency to abandon a natural world that often appears as already lost expresses itself in a resigned collaboration with industrialist interests—a collaboration that is exemplified within academia by the increasing domination of research by commercial interests, by the tacit abandonment of any authentically critical stance, by the soothing retreat into theories that deny the existence of what we are abandoning, and by the phoney celebration of this retreat as liberating and inevitable.

Environmental theory and practice exist within this ideological world as surely as their political opponents; and environmentalists are all too often trapped within the same psychological and social confusion as the most traditional industrialist. Both the pretended optimism of an industrial-economic system that continues on the course set in previous centuries largely because the alternatives seem so unthinkable (and so unsellable to voters) and the short-sighted heroism of an environmental movement that protests in a heartfelt and occasionally effective way against what is and what is planned, but is unable to offer anything truly attractive or convincing in the way of a positive vision of the future, are alternative responses within the same ideological universe, albeit responses that demonstrate vastly differing degrees of integrity. Consequently, there is a danger that our alternative visions of the future may be a lot less alternative than we think, and that the sort of future envisioned by "environmental correctness" may be at least as "rationally" determined as the industrialism that it intends to replace. Like the nautical misadventuring of Columbus, our "discovery" of the "new" may often be more soberly viewed as an elaboration of the old; and there is a danger that environmentalists may be the unwitting carriers of the virus of industrialism to previously uninfected areas. Our dawning but still often preconscious awareness that the geographical and ecological limits of the earth are paralleled by our own conceptual limits, and that the environmental futures we propose are rooted within the same intellectual landscape as the rest of industrialism, suggests that we should view these solutions with a good deal of caution. Our recognition of the extent to which we are trapped within industrialist ideology, whether as academic researchers or in our day-to-day lives, is strongly related to the underlying sense of numbness, of fragmentation, of despair, and of ungroundedness that psychotherapists sometimes encapsulate in the term "ontological insecurity." All of these feelings, as we will see, are in some sense authentic, in that they express facets of our lives that are repressed; but since they necessarily emerge into a conceptual, political, and linguistic universe that construes them as mere individual aberrations, they are seldom articulated in ways that allow us to engage with these political realities. Environmentalists are probably more alert to these problems than most, since an unflinching openness to what lies beneath the glossy front of modern industrial society is basic to environmental awareness. We need to find ways of tracing such feelings back to their roots in otherwise taken-for-granted political, social, and technological "realities," so that these realities are challenged by our felt protest against them, and against their concealed inconsistency with our intuited relation to the rest of the natural world.

Such a program of reinterpretation has its precursors. The women's movement of the last several decades, for example, allowed many women who were dissatisfied with conventionally "female" roles, and who might otherwise have taken refuge in dubious solutions such as minor tranquilizers, to reconceptualize the world and their place in it, redefining their earlier "depression" as the consequence of political and social repression. To the extent that this movement was successful, its success lay in the way it redefined not only individual experience or some aspect of the world "outside" the individual, but *both* simultaneously. On a larger scale, if we are to reconfigure the "boundless repressed rage," widespread depression, and other "psychological" symptoms that Christopher Lasch and others have identified as characteristic of industrial society,[2] in addition to the "environmental" problems that are covertly connected to them, then we need to accept that all these symptoms may be bound up with a particular culturally constructed definition of the world and our place within it. Understanding such symptoms as depression *simply* as individual pathology is to perpetuate such definitions; but experiencing them as hints of a frustrated but still vital relationality could provoke a reconfiguring of the "inner" and the "outer" in environmentally revolutionary ways. In Robert Romanyshyn's words, "for a spectator too distant from a world that has become only a spectacle and too distant even from his or her body . . . depression is a cure."[3]

Our entrapment within deep-seated industrialist structures has generally been recognized covertly rather than explicitly, sensed somatically as much as recognized intellectually, almost as if we would prefer not to admit this entrapment or take on board its implications. Just as industrialism generally seems to be unable to confront the enormous changes that a frank environmentalist comprehension of our situation demands, so theory, too, often prefers to shelter in an intellectual backwater of language or logic rather than opening itself to the traumatic realities beyond. If the necessary changes seem too politically and practically inconceivable, the temptation is to take refuge in a more comfortable intellectual world, perhaps adjusting the boundaries of moral considerability here, deconstructing differing environmental discourses there. Pragmatically, too, we have often preferred to focus on modifying industrialism so as to veil the extinction of natural structure, making it appear more humane, reasonable, balanced; and looking for "softer," "greener," more "appropriate" technology. But while such reformist approaches are not without value, their overall direction, and the likely end-point, remain utterly unchanged: the extinction of natural structures, and their complete replacement by those based on technique. This is recognized by "deeper" varieties of environmentalism, for whom the apparent hegemony of industrialism is an immediate problem. Simplistically put, we need a standpoint that is outside the universe of industrialism; but if nothing seems to exist outside this realm—linguistically, conceptually, politically—how can we envision any coherent alternative to it? We are left in the position of protesting against *what is*—through direct action, through our rejection of the products of consumerism, and less articulately, through our own felt anguish—without being able to envision any feasible future. No movement, however, can hope to succeed unless it has a clear sense of what would constitute success. To the extent that it lacks such a sense, environmentalism will continue to define itself largely through its opposition to the industrial-commercial system, ultimately merely reacting to, and so allowing itself to be defined by, this system. As a protest movement, it will be able to slow down industrialism and cosmeticize its effects; but it will be incapable of changing its underlying course in the fundamental way that is needed if the living diversity of the planet is to flourish or, indeed, to survive at all. Like the Titanic, civilization will ultimately either sink or float; and slowing down the rate of sinking is not a particularly appealing option. More energy-efficient types of transport, or less polluting means of electricity generation, or the survival of a rare plant species, are ultimately valuable if, and only if, they are part of a positive and genuinely alternative vision of a future, more healthy natural world. One of the aims of this book is to outline the form that such a vision might take, and to address some of the obstacles that will need to be overcome if we are to secure its eventual realization.

Misconceived Mindscapes

In the absence of such a vision, it is hardly surprising that we turn optimistically to the least unlikely candidate offered by the spectrum of knowledge structures in our search for ways of knowing that might best articulate our feelings, hopes, and fears about the natural world. The term "ecology" is one of those iceberg-like notions that at a conscious level seem to denote something clear and unambiguous—in this case, a natural science that studies the interdependencies of organisms—but that at an unconscious level reverberate profoundly among the unexpressed possibilities and foreclosed relations that find no place within the conscious knowledge of modern industrial society. The use of the term "ecology," as in "deep ecology" or "social ecology," is symptomatic of a fantasy projected onto a biological science, implying a wholeness, a harmony, an integration that are absent from our conscious lives, but whose presence in the unconscious constantly threatens to subvert the claimed completeness of consciousness. But while fantasy, as Harry Guntrip argued,[4] may be healthy if it is a precursor to action, it becomes pathological if it is merely a substitute for it; and unfortunately, much environmental talk of wholeness, ecology, and related terms comes into the latter category.

Although a thorough critique of ecology as a basis of environmentalism[5] is beyond the scope of this chapter, reflecting on some of its limitations throws light on some of the general problems that face any sort of environmentalist praxis. While ecology has often been eulogized by outsiders as representing the holism and integration that are absent from the more traditionally reductionist sciences, these characteristics have often been more imagined than real, and there has in fact been a history of internecine strife within the discipline between those who have sought to align ecology as a biological science among others, incorporating accepted mathematical and physical principles, and those who have argued that ecology necessarily subverts, or at least modifies, these principles, on the other. To the extent that ecology *has* succeeded in establishing itself within the scientific community, this suggests that the former might claim victory. As Robert McIntosh has pointed out, ecology owes much to existing social and scientific ideologies,[6] and its extensive colonization by physics, chemistry, and mathematics undermines its subversive potential, allowing it to become comfortably drawn into the orbit of the technological system that is directly implicated in the destruction of the natural world. For example, the relationships between natural entities are often understood in terms of energy flows, and such understandings can only be seen as reductive.

More pragmatically, Peters has argued that a most serious shortcoming of ecology is that "the problems that ecology should solve are not being solved. They are worsening, growing more imminent, more monstrous."[7] While this criticism seems unfair given that ecology as a science can hardly be expected to reach into the cultural, political, and psychological domains, it does illustrate that even to scientists such as Peters the term "ecology" transcends scientific categories, extending the unspoken promise of a holistic alternative to conventional science. To the extent that ecology has failed in the way that Peters suggests, this failure represents a disappointment of the symbolic and metaphorical connotations of ecology rather than any scientific failure. It is a failure, however, that throws into stark relief the gap between the two meanings of "ecology": ecology the science, and ecology the cipher for those qualities whose poignant absence partly defines the industrialist landscape.

Ecology has undoubtedly generated insights important to the environmental movement; but its main function has been in hinting at and keeping alive these possibilities that are all but suffocated by the technological rationality that pervades the contemporary world. This is one example of how the environmental movement struggles to express what we know at some level, but is denied by virtually all the articulatory structures that are available to us. And in turn, this illustrates the significance of the fundamental ecopsychological insight that environmental problems are simultaneously *psychological*,[8] in that our difficulties in recognizing and articulating them reflect parallel repressions, dissociations, and projections also present within the structure of selfhood. Thus the repression of particular forms of psychological awareness is matched by the extermination in the "outside" world of those species that are inconvenient to industrialism; the dissociation of "wilderness" from areas that are zoned for "development" is simultaneously an ecological and a psychological dissociation; and the zoos, taxonomies, and "management" programs through which we attempt to come to terms with a nature experienced as alien are paralleled by our own internal defences and rationalizations. And, of course, these parallels between the way industrialism orders nature "outside" and "inside" are not coincidences; firstly because the colonization of humanity is only one aspect of a more general colonization; and secondly, because what happens "internally" is often causally related to what happens "externally." Not only does consciousness becomes consistent with a world that is increasingly commodified, rationalized, and devoid of wildness, but in addition, technological power enables us to realize physically a fantasized mechanical world. As Theodore Roszak has written:

> [W]e can read our transactions with the natural environment . . . as projections of unconscious needs and desires, in much the same way we can read dreams or hallucinations to learn about our deep motivations, fears, hatreds. In fact, our wishful, wilful imprint upon the natural

> environment may reveal our collective state of soul more tellingly than the dreams we wake from and shake off. . . . Far more consequential are the dreams that we take with us out into the world each day and maniacally set about making "real"—in steel and concrete, in flesh and blood, out of resources torn from the substance of the planet. Precisely because we have acquired the power to work out our will upon the environment, the planet has become like that blank psychiatric screen on which the neurotic unconscious projects its fantasies. Toxic wastes, the depletion of resources, the annihilation of our fellow species; all these speak to us, if we would hear, of our deep self.[9]

But there is a further ideological twist here; for what is "within" ourselves and what exists "outside" is itself partly defined by the same industrialist patterning out of which the destruction of the natural world arises; and so terms such as "repression" that *assume* this division of the world are themselves problematic, along with the theoretical models to which they belong. In other words, viewing the drama of environmental destruction in terms of such assumed dualisms as those that oppose "consciousness" to Freud's individualistically conceived "unconscious," or "humanity" to "the environment," is as inadequate as viewing it in terms of a fifteenth-century morality play that opposes Riches, Sloth, and Gluttony to Good Deeds and Hope. Rather like those politically imposed boundaries that divide or assimilate peoples who have a strong sense of their own national identity, the boundaries that define the industrial world do not simply need to be relaxed: rather, what is required is ultimately a complete remapping that would explicitly reframe social and political life as part of the natural order. It is not simply, therefore, that "internal" repressions parallel the elimination from the natural world of patterns and species inconsistent with industrialism, implying that solutions would leave intact the boundaries between self and world. Problems in both these areas stem from the historically fabricated opposition between self and world on which industrialism relies; and as Roszak implies, we need a concept of subjectivity that is ecological rather than individual, and so recognizes the continuity and overlap between these two forms of repression. The parallels and synchronicities between "personal," "social" and "ecological" realms challenge their assumed distinctiveness, suggesting structures hidden by our tired reifications and literalizations; and emerging concepts such as the "metaphoric mirroring" suggested by John Rodman imply subjective landscapes that go beyond our geographical grids and physical understandings. In contrast to the European colonialists to whom I refer in the next chapter, it is essential to the radical environmentalist project that we are open to the revolutionary implications of such concepts rather than defensively redefining them in terms of existing categories. What we are dealing with here is a situation in which all language, all concepts, all theory are already suspect, in which "all that is solid melts into air," and in which there are few stable rocks to step on as we attempt to cross the chasm that separates the evolving disaster of industrialism from a way of living that recognizes and respects natural structures. Our awareness of the ongoing dialectic between industrialism's historical transformation of the physical landscape and its parallel transformation of the human person into the autonomous individual is fundamental to understanding environmental issues; and seemingly revolutionary changes that are confined to a single realm are all too easily reassimilated by the industrialist system as a whole.[10]

If the natural sciences, including ecology, are the most obvious intellectual descendants of Descartes' division of the world into that which is capable of thought and that which can be thought about, the social sciences are no less ideologically compromised. Sociology, anthropology, and particularly psychology, when they are not busy trying to imitate the natural sciences, have often taken for granted a biological world that is viewed, at best, as a source of energy and raw materials, and increasingly often, as possessing only a discursive reality, located within the realm of human language—surely the ultimate in anthropocentric hubris. If the natural sciences could be accused of reducing nature to an array of mechanistic fragments, the social sciences—especially given the fashionable influences of postmodernism and literary theory—often seem to portray nature as lacking any substantive existence at all except as the forum for a particular area of debate. It has been widely recognized—intellectually, at least—that the Cartesian separation of "thinking matter" and "extended matter" does violence to the nonhuman world; but we have been slower to acknowledge the complementary violence that this dissociation does to the human person, isolating us from our bodily groundedness, imprisoning subjectivity within us, and, in another turn of the screw, forcing us to engage in psychodynamic gymnastics in order to adjust to this ostensibly "natural" arrangement. Wholeness and integration are not just qualities that we ascribe *to* nonhuman nature: they are also qualities that *prescribe* our place *within* the natural order, implying a reconfiguration of selfhood. This is why the disembodied intellect is inadequate as a starting point for environmental theory: however clever our theories, they exist within a space that is *already* separate from the natural order, and the "nature" they refer to is, all too often, the cultural artifact they theorize about rather than the natural order whose existence they are oblivious to.

Cultural Absences

This brings us to one of the fundamental dissociations on which the whole edifice of industrial society is erected: that which separates the "human" from the "natural," so dividing the previously integrated unity of nature into dominators and dominated. As Val Plumwood has made clear,[11] there is an interlocking system of overlapping dualisms that guide our thought and actions in environmentally significant ways; and these include civilised/wild, modern/primitive, human/animal, conscious/unconscious, rational/irrational, culture/nature, mind/body, and so on. In each case, the first term of each pair represents a preferred state or entity, whereas the second indicates something that we try to distance ourselves from, composing a value system that gives the impression of being based on "factual" distinctions. This value system, as it is articulated through technological power, constructs an industrial world that dominates and consumes nature, and so realizes in material form the original value judgment suggested by the dualism. Every freeway constructed across a previously wild landscape, every tree turned into bags for supermarkets, and every river that is dammed, physically embodies a value structure that makes the arguments for each of these appear almost unassailably rational. The dualistic separations that underlie this rationality therefore oppose and fragment natural structure, reducing nature to the raw material for an alternative, industrialist structure. These conceptual polarizations underpin and legitimate the industrialist domination of the world and the continuing historical fantasy of our emergence from the "primitive" realm of nature. Today, this illusory emergence is nearing completion: many in the more affluent corners of the globe inhabit a society that has established itself at a comfortable distance from the natural order, and the violence of the modern world is remote from us psychologically, geographically, and temporally.

The values implied by this structure that legitimates the domination of the "natural" by the "human" are taken-for-granted parts of our lives to such an extent that it is not easy to question them without appearing absurd. For example, many people take it for granted that the breeding of animals for food, pets, or other uses is acceptable; that we have a right to build on, farm, or otherwise alter the land without much regard for its nonhuman inhabitants; and that "pests" that consume crops intended for human consumption should be "controlled." Real alternatives to such practices will meet opposition for a variety of legal, nutritional, ethical, practical, commercial, and other reasons, illustrating the way that industrialism acts systemically to make alternatives seem bizarre, impractical, and uneconomic. Although nominally distinguishable components of society such as the legal system, academia, industry, social life, agribusiness, and the media are viewed as more or less autonomous components of a "democratic" society, they are rooted in the same values and assumptions, and so act together to form a system that is highly integrated and well defended against fundamental change. In addition, since we are educated to take our place within this system, our thought processes also become consistent with it. In effect, then, the adaptability that is a distinctive feature of the human animal ensures that we become identified with this system to such an extent that alternatives seem unthinkable and pragmatically almost impossible. As we will see in chapter 4, humans, no less than other creatures, are dependent on some form of external structure for our adequate functioning; and so if industrialism apparently represents the only structure available to us we will inevitably become integrated within it.

Industrialism's monopoly on structure, however, is hidden from view by its overt emphases on personal choice, equal opportunities, competition between alternatives, and individual creativity. It is true that as individuals in modern industrial society, we have a certain amount of freedom to arrange our own lives as we please; and this is a form of freedom that we should value and defend. However, it is a freedom that has quite distinct limits. For example, I may choose to drive a Ford, a Volkswagen, or a Toyota; but choosing not to drive at all may at best be impractical and at worst may cost me my job. Similarly, I have some choice as to where I live; but if I choose a more nomadic lifestyle, then I may discover, like the convoy of travelers whose vehicles were impounded and destroyed by Wiltshire police during the Thatcher era, that this is socially and legally unacceptable. And while I am, of course, free to demonstrate my "creativity" in a variety of ways, I am likely to find, like the American painter Frederic Church, whose work I will refer to in chapter 3, that only certain fruits of this creativity are commercially recognized or socially comprehensible. There are powerful but usually invisible forces that maintain the congruence between almost every aspect of our lives and the underlying assumptions of industrialism; and it is a measure of the depth to which we are permeated by these forces that we usually experience our accordance with them as deriving from our individual preferences. The character of the self therefore has to be viewed as an inescapable part of the problem that this book addresses rather than as a starting point for our proposed solutions.

Our allegiance to industrialism, then, is to some extent related to the apparent absence of any alternative structure; and beneath the ostensibly democratic character of industrial society lies a concealed totalitarianism. To most of us, the description of the world offered by disciplines such as physics, economics, and biology has become the *only possible* description, taken for granted as the basis of common sense. This, as we will see later, contrasts with the medieval European world, wherein the social order, whatever its many injustices and oppressions, was modeled on the natural order, and personal identity was derived from an understanding of one's place in this social/natural order. For us, however, the world defined by the sciences, to the extent that it appears to be the *only* world, becomes the sole basis of what R. D. Laing has termed our "ontological security." According to Laing, the ontologically secure person has "a sense of his presence in the world as a real, alive, whole, and, in a temporal sense, a continuous person. As such, he can live out into the world and meet others: a world and others experienced as equally real, alive, whole, and continuous."[12] But now that the "human" world has become so extensively separated from the realm of the "natural," our lives often feel "ungrounded," and ontological insecurity becomes a defining problem for the modern self. The world described by science may be experienced as "real," but to varying degrees it is less than "alive, whole, and . . . continuous." It is, rather, the world described by Edwin Burtt—"a world of quantity, a world of mathematically computable motions in mechanical regularity" in which subjective awareness "of colour and sound, redolent with fragrance, filled with gladness, love, and beauty, speaking everywhere of purposive harmony and creative ideals" has become "just a curious and quite minor effect of that infinite machine beyond."[13] Such an understanding does violence not only to the earth—either conceptually or, through its implementation by technology, physically—but also to subjectivity, which becomes "crowded into minute corners of the brains of scattered organic beings." This relocation of intelligence from the world as a whole into the human mind, leaving the world as mere physical matter, was a crucial step in the separation of the "human" from the "natural" and the domination of the latter by the former; for, as Burtt points out, how otherwise could nature "be reduced to exact mathematical formulae by anybody as long as his geometrical concentration was distracted by the supposition that physical nature is full of colours and sounds and feelings and final causes as well as mathematical units and relations?"[14]

Unsurprisingly, then, the modern person suffers from a distinct sense of existential uncertainty that is strongly related to our having banished so much of the meaning of the world. Ironically, the poverty of our relation to the world that results from our scientism drives us further toward scientistic assumptions: like the survivors of a shipwreck, clinging to whatever pieces of wreckage are still floating, we obstinately perpetuate the illusion of science's monopoly on meaning simply because there seems to be no alternative. This, of course, does not mean that science is to be rejected as illusory, invalid, or meaningless; but only that as a complete handbook for human existence, it has distinct limitations, as we will see later. Even a passing familiarity with other cultural traditions or with our own history over the past several millennia indicate that science is far from being a natural, inevitable basis for comprehending our relation to the rest of the world. It is, rather, a humanly constructed model of the world that is instrumentally powerful and that in certain respects has been of enormous human benefit. However, while its benefits are mostly fairly apparent to those of us who live in the industrialized world, it has indirect costs: and these costs include both the widely acknowledged degradation of the nonhuman world and the covertly related impoverishment of subjectivity—an impoverishment that tends to be invisible to us precisely because we are, *in part*, constituted and constructed by it.

I am not, therefore, driving toward the conclusion that we humans are inevitably and inescapably destructive. We know of too many societies—a few of which I will discuss later—that not only preserved their natural environment but in many cases enhanced it to conclude that humanity is simply lethal to nature. But *industrial* humanity is a different matter altogether; and it is one of my purposes in this book to show why this is so. Although the reasons for the destructiveness of industrialist ideology and practice have been explored before, these explorations have invariably taken industrialism as an object of academic study that is separate from personhood. For example, ideas such as anthropocentrism, objectivity, rationality (by now, the reader can imagine the inverted commas around each of these problematic terms) are often subjected to the processes of intellectual deconstruction and critique as if they did not also apply to the self doing the deconstructing and critiquing; and those academic approaches that *do* recognize the significance of this problem of reflexivity have so far failed to provide any adequate alternative to grounding discourse within some nonconstructed realm.[15] This intellectual externalization is complemented by a temporal distancing: we identify the causes of our problems in the past, focusing on environmental villains such as Descartes in a way that makes our own, current practices appear to be somehow inevitable. The danger here is that "understanding" becomes a rationalization of current practices, and a substitute for and an alternative to *change*, since this understanding *presumes* the split between self and world. It conceals rather than illuminates the way environmental problems are mute expressions of an incompatibility between the social phantasy systems that we inhabit and those characteristics of the natural world that we are not only unaware of, but are unaware that we are unaware of. It is not easy, as we ponder environmental issues, simultaneously to be aware that the forms taken by our own pondering are themselves partly determined by centuries of co-evolution with industrialist and pre-industrialist structures. Our understandings, therefore, often serve to integrate the process of environmental degradation into industrialist realities without seriously challenging these realities; and both the integrity of the world and our own integrity are silently surrendered as a form of selfhood emerges that is "environmentally aware," "liberated," and wholly consistent with industrialism.

But understanding is not *simply* the agent of industrialism, since it has also evolved as part of our embodied selves, despite the post-Enlightenment denial of this embodiedness. Although the influences of feeling, intuition, and spiritual awareness are difficult to articulate, they are the basis of thought's deconstructive and critical potential, enabling us to stand aside, albeit temporarily, from those dominant and rationalistic frameworks with which we normally identify, as well as from the suffocating "normality" of our time. As a result, an enlarged subjective space is opened within which we can begin to sense other, less exclusively intellectual, forms of relation to the world. By analogy, if we are lost in a wilderness area, climbing a convenient hill may enable us to realign ourselves with the lie of the land, to perceive where we are as well as where we might be, even though doing so does not in itself take us nearer to our objective. This metaphor, of course, would be rejected by those who argue that no such hills exist in the purely discursive "wilderness" that they claim to inhabit; but such arguments are flawed in ways that will become apparent during the course of this book. In effect, then, intellectually problematizing our own intellectuality can have an effect on our minds not unlike the use of the *koan* in Zen Buddhism: the mind becomes more still, its own buzzings embarrassed into a self-conscious quietness, and sensed awarenesses that normally would be hidden begin to appear. Thus the intellect, even if it is incapable of generating solutions to the problems that arise from its own dominance, can recognize the forms and qualities that would characterize a whole subjectivity as it transcends the boundaries of the individual mind. Intellect cannot generate or define this subjectivity; but it can, I suggest, glimpse its necessity and the conditions under which it might appear. This book attempts to employ a largely intellectual analysis in this fashion—to provoke and uncover those nonintellectual forms that, together with a reconfigured intellectuality, could define a subjectivity that is not merely *in* the world, but also *of* it.

But this intention, which is simultaneously a psychological and an ecological one, is all too easily subverted if we relax our awareness of the way our thoughts and actions are constantly pulled toward consistency with industrialism. The psychological, commercial, and technological structures within which we move are not simply plucked by historical accident from an inexhaustible array of ideological possibilities, but rather incorporate fateful decisions to recognize one type of ontological structure rather than another. Industrialism, as we have seen, forms a *system* that contains a number of highly interdependent parts; and like any system, it embodies its own homeostatic tendencies, resisting attempts to change its direction in ways that may be either direct and obvious or subtle and beyond the reach of consciousness. For example, the British company Traidcraft was set up to foster fair trade with the indigenous populations of Third World countries, encouraging native crafts and other environmentally and socially constructive types of commerce. However, the "realities of the market" made this relatively benign form of trade untenable, and the company was forced toward more traditional business practices.[16] Similarly, much psychological practice has been formed, however unwittingly, to be consistent with the industrialism it serves and legitimates, as we will see in the next chapter. In effect, industrialism exerts its own gravitational pull; and insufficiently powerful efforts to escape this pull result in our falling back into line with existing structures. Attempts to reform industrialism, therefore, are likely to be easily assimilated or corrected; which may say something about why, in spite of growing public awareness, green technology, and so on, industrialism's rapacious consumption of natural resources grows more obviously unsustainable with each year that passes. The term "natural resources," of course, itself suggests something about why this is, since it denies the systemic character of what is happening: the assimilation and destruction of one system by another. Recognizing this may help us to avoid the pitfalls of quantifying environmental destruction simply in terms of numbers of species lost, percentage of a natural resource remaining unexploited, and so on; for what is being lost is not merely these quantifiable and often reified aspects of the natural world, but more basically, the *system of relations* of which they are a part. Understanding the natural world as consisting of species, individuals, even ecosystems is to deny that what these are derives partly from their interrelatedness with each other—an interrelatedness that cannot be expressed in terms of the characteristics of separate "things." Our attempts to quantify environmental destruction in such terms is therefore itself symptomatic of our colonization by industrialism. The natural order cannot be protected simply by preserving its component parts, as if in an ecological museum, for such measures in effect enlist the alternative system—industrialism—that is destroying nature. Rather, our starting point must be a tenaciously defended relation to the natural order itself, experienced not as a "nature" external to ourselves, which we conceptually or geographically visit from time to time, but as a felt resonance that is basic to our identities as human animals.

Environmental theory has itself often been infected by industrialism's assumptions of a fragmented world. While it has attempted to recognize the systemic quality of the natural world as something external to ourselves, the self has invariably remained a spectator, separate from a world that we are distanced from. We are correspondingly unable to recognize that this spectator-like experience is consistent with the character of the industrial-economic system, and that this configuration of selfhood, far from being "natural," "inevitable," and so neutral with respect to environmental issues, is part of a perniciously destructive system that renders us blind to natural structures and eventually causes us to become unwitting agents in the destruction of these structures. It is not simply a question, then, of humans adopting an anthropocentric position with respect to the rest of nature; but rather that both participants in this fabricated opposition have been infected by the industrialist ideology that domesticates them in parallel ways. The violence that industrialism does to nature, in other words, is not *just* a matter of the violence that it separately inflicts on nature "out there" or to nature "within"; it is *primarily* the violence that separates these two natures in the first place, destroying that resonance between the psychic and material worlds that constitutes the cultural realm. This enforced separation diminishes both selfhood and the nature we experience as "outside," although it is a diminution that we have learned to accept as "natural." Our distorted consciousness, depicting our role as one of guilty domination, conceals the extent to which we are victims as well as agents of the spread of industrialism. John Rodman points out that

> Descartes' depiction of beasts as machines was followed by the proliferation of mechanistic models of man; Marx's indictment of capitalist industrialism for treating human workers as machines is followed by Harrison's and Singer's indictment of factory farming for creating the monstrosity of "animal machines"; the *Natural Resources Journal* is followed by the *Journal of Human Resources*; and Darwin's projection onto nature of a model derived from man's "domestic productions" (plant and animal varieties created by artificial selection) now returns to haunt us as the prospect of the genetic engineering of human beings by human beings, as the literal fulfilment of the metaphor of domestication.[17]

The parallels between these distortions of the human and nonhuman worlds become apparent only if we can envision both of these worlds as potentially entwined with each other—a task that is made easier if we appreciate the historical process that separated them in the first place. But these distortions, and this separation, are not the result of a prising apart of two autonomous, whole, and unchanged entities: rather, the character of selfhood and that of the natural world have been profoundly altered and diminished through the loss of those common structures that were integral to them both. If this process sounds rather abstract and esoteric, it has the most profound and wide-ranging implications both for our lives and for what is happening to the natural world; and I will explore it in more detail in later chapters. But for the moment, let us simply note the difficulty of simplistically attempting to reconnect an individual and a nature that have both been reduced by this historical separation; for without the "common structures" that I referred to above, the only possible solutions appear to involve an abandonment of our separateness—a very unsophisticated and arguably regressive form of relation, as ecofeminists such as Jim Cheney have pointed out.[18] What we need is a form of reconnection that does not involve the abandonment of our sense of individuality, although it *will* necessarily require the abandonment of some of the more egoistic expressions of this individuality. If we mistakenly identify individuality with *isolation*, then it follows that *connection* must involve *relinquishing* individuality; and this is the Achilles heel of the otherwise profoundly important deep ecology viewpoint, as I will argue later. But while individuality as experienced within modern industrial society does indeed imply alienation from the world, there are, as we will see, forms of individuality that manage to retain a clear sense of self while at the same time expressing a profound degree of interconnection with the world beyond individual boundaries.

It is symptomatic of industrialism's denial of structures inconsistent with its own that there are no convenient terms either for the holistic structure of the natural world or for the various intermediary structures that might relate human experience to it, and that could therefore articulate a subjectivity that reaches beyond the solipsistic boundaries of the modern psyche. "Culture" is the term that I will later use to describe these intermediary structures, since symbolic anthropology, at least, has recognised that culture can be understood as playing this integrative role, unifying the world in a way that respects the diversity of its component parts. However, it is—once again—symptomatic of the industrial world that "culture" has been dualistically viewed as the "opposite" of "nature," rather than as a sphere that has the potential to integrate us *into* nature. It is in this sense that Val Plumwood can—quite accurately, given industrialist definitions of culture—discuss the "set of interrelated and mutually reinforcing dualisms which permeate Western culture," forming "a fault line which runs through its entire conceptual system."[19] John Shotter's statement that we are not "beings immersed in nature," but rather "beings in a culture in nature"[20] is therefore less a factual description of reality in the modern world than a plea for an ontology quite distinct from that assumed by industrialism. The absence of structures that could mediate our relation to the world, together with the absence of a suitable term that could point to such (potential) structures, joins with an experience of self to which any such structure seems threatening and almost inconceivable, so reinforcing the hegemony of industrialism.

Culture can be part of such an alternative ontology, then, only if we broaden our horizons beyond the global but still restrictive gaze of industrialism; and this would make possible the radical reconfiguration of self-in-the-world that I will explore later. Under existing conditions, the study of culture is awkward for the biological sciences, since it cannot easily be assimilated to a quantitative and material emphasis, and in any case, is not seen as relevant to the nonhuman world; but it also sits uncomfortably with the social sciences, since it threatens the individualist and empiricist paradigm that is mostly still dominant. For many older civilizations, material explanations such as atomism existed side-by-side with cultural and spiritual paradigms; but industrial society is characterized by the almost complete acceptance of the former and the almost complete rejection of the latter. Culture becomes not only invisible and intangible, but also *unreal* in a world of "things"; and it also threatens the clarity of the distinction between self and world that is basic to virtually every academic discipline. Thus the study of culture is often marginalized outside anthropology, due to its covertly recognized potential to subvert those epistemological structures that have maintained the conceptual separation of the social and natural worlds. This potential for subversion, however, precisely indicates culture's significance for environmental theory; for the epistemological separation of the social and the natural, together with the dualistic oppositions that flow from it, is fundamental to the dismantling of the natural order, and their reintegration is correspondingly fundamental to the survival of this order.

In this respect academia, although it likes to claim a lofty autonomy, faithfully replicates trends in the industrialized world at large; and the separation of the "human" (the arts, humanities, sociology, and so on) from the "natural" (sciences) mirrors both the formal organization of modern society and our day-to-day praxis. Both in academia and in modern life more generally, the cultural sphere that could offer us a sense of integration into the world, as indeed it has done in some eras and in some places, has been overwhelmed to such an extent that we are for the most part no longer even aware of its necessity nor of the chronic consequences of its atrophy. This is no mere triviality, however, simply requiring that we add a cultural dimension to an otherwise adequate situation; for our understanding of terms such as "individual" and "environment" are predicated on a worldview that implicitly denies the cultural realm. Consequently, recognising the significance of culture implies a corresponding problematization of such terms. An effective environmentalism, therefore, must ultimately transcend even such apparently basic buildingblocks as these, recognizing that any reconstituted relation between them must be preceded by the rediscovery of their common overtones.

Our denial of these overtones has something to do with our unwillingness to accord subjectivity to nature. We may empathize with natural entities, or accept that they may be sentient, and we may experience an emotional relation to landscape; but the prejudice is hard to shake that the subjectivity that we clasp so jealously to ourselves can never transcend our own physical boundaries. However sincere our intentions, the configuration of self we assume is usually not negotiable, and we remain somehow aloof from the environment whose fate we debate. This being so, the reality we construct tends to coalesce around us like a crystal growing in a mother liquor, reproducing our unexamined assumptions and doctrines in apparently inevitable processes of physical ordering.

Articulating Nature

Recognizing that the building blocks out of which we attempt to construct a defense of the natural world may have the character of ideological Trojan horses, directing our theories in directions that are ultimately ineffective, does not mean that we should, or can, avoid them altogether. Unless we are to remain silent, then we have to use whatever materials are available to us, even if these are ideologically tainted. But they need to be used in full recognition of their ideological implications so that we minimize the extent to which they covertly determine the form of our theorizing and the conclusions we arrive at—suggesting a provisional, tongue-in-cheek stance that is quick to sense divergence from our intuitions. In this book, I will—initially at least—use inverted commas to signal particularly problematic terms; but the reader will no doubt soon be able to imagine them around many others as well.

Our reliance on language exemplifies the problems that entangle and constrict us as we attempt to construct forms of theory that are consistent with the natural world. Whereas medieval styles of language were taken as expressing a Divine order that was immanent within the cosmos, post-Renaissance language developed toward a more nominal relation to the world as order was increasingly seen as originating within the human mind, and as imposed on an environment experienced as essentially passive.[21] Today, the world is increasingly technicized and rationally ordered: it is, as Fredric Jameson has said,

> A world from which nature as such has been eliminated, a world saturated with messages and information, whose intricate commodity network may be seen as the very

> prototype of a system of signs. There is therefore a profound consonance between linguistics as a method and that systematised and disembodied nightmare which is our culture today.[22]

Environmentalists' struggle toward a more open awareness of the natural order is therefore handicapped by the need to use a language that implicitly imposes a constructed, technocratic order on the world and therefore denies the natural order, illustrating Stanley Aronowitz's insight that "the meaning of 'hegemony' consists precisely in its presence within the discourse of opponents of the dominant ideology."[23] One of the frequently overlooked features of modern languages of European origin, and especially technical and academic discourses, is their tendency toward abstract disconnection from felt relation to the world, which in turn allows the rationality and humanly imposed order of language to take precedence over natural order. In extreme cases, this allows the claim that discourse determines both selfhood and the natural world, that "there is nothing outside the text"—a complete swing of the dualistic pendulum away from the opposite, biologistic, pole. From this perspective, there is *no* natural order, since the only order is that generated by language. This is clearly anthropocentric, and replicates what might be referred to as the fundamental project of modernity: the creation of a human realm that is free from any natural patterning or constraint. Academia has long been a faithful advocate of this project, following the guidelines of a long philosophical tradition stemming from Plato, through Kant, to postmodernism, that sees order as necessarily imposed by human understanding. This applies even to writing that concerns the obviously endangered natural realm, so that instead of a passionate engagement with and defense of this realm, we have talk of "alternative natures" being "socially constituted," as if such "natures" were an artifact of social life rather like bowls clubs or Labor Day parades. I will argue, however, that nature is *prior* to human existence or activity—historically, ontologically, and materially—and is a *condition* of social life rather than a *consequence* of it.

Academic writing about nature frequently reproduces these anthropocentric assumptions. For example, Phil Macnaghten and John Urry, while claiming to "transcend the . . . debate between 'realists' and 'constructivists'," nevertheless locate themselves firmly among the latter group. Few could take exception to claims that "[w]hat is viewed and criticised as unnatural or environmentally damaging in one era or one society is not necessarily viewed as such in another," or that "[n]ature does not simply provide an objective ethics which tells us what to do"; but the slide from such statements into the view that "there is no singular nature as such, only natures [which] . . . are historically, geographically, and socially constituted"[24] is one that moves beyond a recognition that our means of articulating nature are diverse and inadequate, to the assertion that nature "out there," independent of our means of articulating it, doesn't exist at all! Such assertions, which we will examine in more detail in the next chapter, abandon the natural world to whatever social and political fashions are currently accepted, so that an "appropriate politics of nature would be . . . one which stems from how people talk about, use, and conceptualise nature and the environment."[25] Such a politics of nature, then, would be one that fits appropriately and uncontroversially into the political realities of the day, and so would clearly be impotent to challenge these realities.

What is more, if the dualistic ontology that emerges from the Enlightenment tradition is collapsed—as it is by constructionism—into a single, anthropocentric world in which nature is merely an artifact of human activity, then we lose any sense of the natural world as a "ground" from which human life emerges, and nature is viewed with the same "incredulity" as any other socially derived structure. However, while our suspicion of modernistic notions such as "progress" (as in "economic progress") or the attempt to extract values from facts (as in the view that "humans have a God-given right to dominate nature") is fully justified when applied to *social* structures, it is mistaken when generalized to a nature that predates and grounds us. While few today would defend the view of those medieval historians who declared that we can use the natural world directly as a model of virtue by observing, for example, the chastity of the camel or the altruism of the stork[26], it is equally mistaken to interpret the current absence of those cultural forms that could allow nature to appear as morally meaningful as indicating that nature is *intrinsically* devoid of moral meaning. If nature is recognized both as partly separable from us and as embodying an intrinsic order from which we ourselves have emerged, then it provides a context through which all social and psychological life must, however indirectly, be patterned; and it takes little imagination to see that constructionism's denial of social life's necessary alignment with such natural patterns can only be environmentally disastrous.

This is a point which Macnaghten and Urry clearly recognize at some level, since they frequently draw back from the more unpalatable conclusions implied by their argument and instead adopt a more realist tone, claiming that "there is little doubt that some of these patterns of contemporary consumerism have had disastrous consequences for the environment."[27] This retreat into a more embodied, emotionally involved relation to nature is a rare lapse in an otherwise unsoiled discursive edifice; and the detachment of the authors is indicated by their dispassionate discussion of "what appears to be environmentally damaged."[28] Such language becomes a means of distancing us from nature rather than articulating our relation to it; and this is a distancing that, as we will see in later chapters, is entirely consistent with the industrialist paradigm. In contrast, the view adopted here could broadly be described as a "critical realist" one, in the tradition of Roy Bhaskar's philosophy.[29] That is, while recognizing that nature cannot provide any unmediated ethics and that a diversity of understandings of nature are possible, I will nevertheless argue that there is something "out there" that is real not merely by virtue of any social or linguistic process; and that any social, moral, or intellectual system that is not grounded, however indirectly, within this reality is ultimately untenable.

I will have more to say about constructionism's influence on academia in the next chapter; but in the meantime we should note that the disconnection of language from the natural world allows us to impose, or traps us into imposing, implicit properties of language onto the natural world while perceiving these properties as already in the world. For example, we tend to see the world as made up of the "things" that we have learned to isolate from their contexts and to refer to by nouns, rather than in terms of the overall web of relations between these things. As we will see, the ability of language to abstract particular forms or specifics from their context parallels a more general decontextualization and literalization within the modern world, so that we remove item from context in the belief that properties belong only to particular things, and not to the relation between them. In effect, then, those languages that are derived from a European context often deny natural structure, articulating those attributes of the world that are consistent with the fragmentary understanding appropriate to industrialism while denying properties that involve natural relation and process.

Once again, environmental theory is vulnerable to infection by these characteristics of language. Our use of the notion of "value" can serve as an example. "Extrinsic" or "instrumental" value—that is, the value that something derives from its usefulness to us—is an unashamedly anthropocentric idea that clearly accords with industrialist aims. But "intrinsic value" is a term that has been used by environmentalists to justify the preservation of aspects of the world that have no obvious use to humankind. "Value" is usually thought of as an attribute of a "thing"; so our attempts to demonstrate the value of the nonhuman world become instead demonstrations of the values of those independent bits of the world that we can recognize and name, so unwittingly confirming industrialism's understanding of the world as made up of these separate bits. Preserving the natural world by preserving a collection of separate bits may be successful in certain respects: that is, we may be able to demonstrate that our actions have resulted in certain individuals or certain species being no longer endangered. This, of course, is in itself an important outcome; but there is a danger that such successes may encourage us to think of the natural world only in terms of such individuals or species, blinding us to that more elusive wholeness that flows from the *relations between* these individuals and species. Describing the world in terms of nouns and quantities may be convenient; but it is a form of description that omits important characteristics of nature, albeit characteristics that are less available to consciousness. Not only that, but the particular attribute we select as conferring intrinsic value inevitably reflects our particular biases: "sentience" for example, reflects the projection onto the world of our historically constructed preference for mind over body. There is thus a danger that the notion of intrinsic value can lead us into a comprehension of the world that is reductive, fragmentary, and anthropocentric, and so easily integrated with destructive ideologies. Frederick Turner finds the antecedents of this modern attitude in the lust for *things* demonstrated by Elizabethan explorers, noting that

> The native peoples who lived amidst vast, unexploited lodes of these very things often regarded them as mere sparkling parts of an infinitely larger and more beautiful design. Maybe no single aspect of the cultural difference between Christians and natives is more revealing of the difference between a civilisation ruled by a dead mythology and people animated by vibrant ones than this contrast in attitudes toward stones and metals. . . .
>
> . . . Gold, silver, and stones, like technology, are pathetic substitutes for a lost world, a lost spirit life, and to the extent that they rule a culture we may infer its inmost health.[30]

The attempt to quantify, literalize, categorize aspects of the world is bound up with the assumption that the world contains "natural resources" that are more or less valuable or beautiful. This (literal or conceptual) "collecting" of aspects of the world is also motivated by the desire to restore to one's life some of the meaning that has been lost through the belief that the world is devoid of intelligence and spirit—a narcissistic appropriation of salvaged parts of a world that is felt to be already damaged beyond repair. By taking, holding on to something valuable, beautiful, even "natural," within a reality in which these qualities are denied ideologically and destroyed physically, we attempt to incorporate within the private realm of our own, personally composed lives some of what has been lost from the world outside—a poignant environmental survivalism. This, like most destructive behaviours, is a perversion of a basically healthy impulse—to preserve the beautiful; but unfortunately it is an impulse that is effortlessly assimilated to a market economy in which the beautiful has a price. In a world where these properties retained the meaning conferred by their relation to the whole, however, there would be no need to asset-strip and collect those bits of it that we see as valuable. Similarly, in an environmental theory that successfully resonates with the forms of the natural world it seeks to model, there will be no need to divide off those bits of the world that are intrinsically valuable from those that are less so. To do so would be to embody and to collude with the divisions that underlie the exploitation of the world.

As another example of how we can unwittingly be drawn into accepting subtly destructive dichotomies, consider the easy way we can slide toward discussing, say, the survival of wilderness while ignoring its connection with the wildness that permeates the world,[31] an approach that prioritizes the material fact of wilderness while denying those symbolic characteristics that cannot be contained within wilderness areas. The significance of wilderness lies partly in the way it arouses resonances within living beings such as ourselves, and so maintains a wholeness that transcends geographical boundaries. The integrity of the world is not merely ecosystemic or geographical: it is also symbolic, metaphorical, and subjective; and an environmentally healthy world cannot be the result of our selecting certain types of wholeness while rejecting others. While stopgap measures such as isolating and preserving remaining wilderness areas are vital first steps, they do not realize the necessary extent of our vision. As environmentalists, we try to preserve the constituent parts of the natural world not so much for their own sake, but in the hope that at some future time the earth may rediscover the elusive resonances between these parts. Trying to preserve the world by defending its parts in a piecemeal fashion is rather like a physician attempting to protect the health of an individual by preserving, say, the liver, the heart, and the brain, while allowing other vital organs to degenerate.

Fragmentary viewpoints such as this illustrate the extent to which the conceptual and imaginative foundations of the human-as-embodied-being have become eroded, even as we retain within ourselves some felt intuition of our embodiedness. There is a sense in which the world, too, functions as a whole, even if this wholeness, like that of the body, is elusive to consciousness; and just as conventional medicine has been accused of replacing the whole person by a sort of mechanized corpse,[32] perhaps there is a danger that environmentalism, adopting the language of science, may preserve the corpse of the world while allowing its vitality to slip away unawares. The long-term regeneration of the natural world cannot be effected *merely* by the preservation of scattered pockets of wilderness, maintained by a life support system of ecological management; nor can an adequate environmentalism align itself with a spectator-like consciousness that represses its own felt connection with the wild world, preserving a reified wilderness while ceding the rest of the world, including our own personalities, to industrial civilization. Such approaches relieve us of the responsibility for addressing the survival of wildness within the entire world and within our own lives, colluding in the framing of nature within a human scheme of categorization and division. This is not, of course, to deny the value of wilderness preservation as a necessary first step; but only to point out that the environmentalist vision needs to go well beyond this first step, obstinately insisting on a vision of an authentically whole world in which the human and the natural are reintegrated with each other.

Language, however, can mislead by denying distinctions as well as by imposing them. For example, several writers have argued that *all* cultures are "technological," since even the most "primitive" cook foods, or wear clothing made of vegetable matter, or make fire.[33] Along similar lines, it has been suggested that "what we do to the world is as much a part of the 'natural' process of change as is the work of termites, beavers, or the elements."[34] Such arguments extend concepts such as "technological" or "natural" to the point of absurdity, ignoring distinctions that we intuitively sense, even if they are difficult to articulate within available discourses. To argue that Glen Canyon Dam is as environmentally benign as a beaver dam requires a distortion of awareness that would be obvious in a world less colonized by the technological imagination. And the extent to which language can "launder" the destruction of the natural world so as to make it appear reasonable and unremarkable is well illustrated by Gregg Easterbrook's description of the oil contamination resulting from the 1989 Exxon Valdez disaster as the "repositioning" of "a natural contaminant from inside a rock formation to the surface of a water body."[35] Here, the term "reposition" functions in much the same way as the word "rearrange" in sentences such as "the thug offered to rearrange my teeth." In both cases, there is a denial of natural *structure*; and this denial is consistent with industrialism's view of the world as a stockpile of "natural resources" arranged in no particular order. In effect, such arguments are pleas for the priority of industrialist realities over natural realities, and we would do well to listen to our felt unease about them.

There are therefore profound difficulties, which none of us can evade or ignore, in using conventional forms of language in order to express feelings and intuitions that attempt to escape the suffocating gravitational pull of the existing conceptual-commercial system. The linear and logical forms of expression through which we have learned to communicate can be effectively mapped onto certain aspects of our everyday reality—in particular, onto those features of our lives that are humanly constructed and so are ultimately generated within the same historical and psychological contexts as language itself. Such forms of expression, however, are not comfortable in articulating realities that are marginalized or suppressed within our technological world—those shadows of the rational dream that language covertly represses. In order to recognize and communicate such realities, we need a language that is capable of expressing relations and states that are more holistic, associative, intuitive, dreamlike. In Robert Romanyshyn's terms,

> The lines of a book inscribe a logic of linear connections where sequence means consequence, where effect follows cause, and where dispassionate argument is valued over the passion of emotions. Dreams, however, are not at all like that. They have no linear logical lines. On the contrary, they are patterns, webs of interconnections which more often than not follow aesthetic values rather than logical rules, and in listening to a dream, in attending to the story it unfolds, one is, more often than not, taken up by the dream, moved by it.[36]

Other writers, too, such as Susan Griffin and Gary Snyder, have begun this task of generating a language more consistent with the natural order; and the ecopsychologist Robert Greenway has expressed the opinion that the new language of environmental awareness will be music. Poets in particular have been doing this since the dawn of consciousness, for humanity will always produce those blasphemers against conventional conceptual structures whose task it is to speak up for what lies hidden or mangled in the shadows of our technological vision. Bill Devall has expressed this point more concretely:

> When . . . documents speak of "predator control," I see beautiful mountain lions, bears, and wolves. Refusing to use the language of the opponent is as much an act of resistance as spiking trees. . . . We protect the integrity of landscapes by speaking of them in the voice of the creative writer, poet, lover of the land. We can find the voice to speak for the forest.[37]

In the absence of a language that is sufficiently resonant with the natural world, we will have to make do with what we have available; and this requires that we use words in a way that is self-critical, inconsistent, and sometimes ironic. This will not be the postmodernist use of language that problematizes *any* nondiscursive structure; but rather one that uncovers the naturalizing and legitimizing function of words so as to reveal the organic structures that they occlude. For example, in the sentence "the thug offered to rearrange my teeth," the denial of structure is obvious; but in the Easterbrook statement above, the word "reposition" is all too easily accepted as an "objective" description of what happened, or at least as one of many equally valid descriptions. Similarly, words such as "pests," "weeds," or "development" also carry their own particular ideological baggage; and by pointing out their hidden implications we challenge the industrialist structures that they are part of, and so uncover the indigenous forms that lie beneath them.

But in pointing out that language has practical implications for the ecological fate of the world, we should not ignore the other side of this dialectic, for language is itself affected by what frames it ideologically and physically. Just as those characteristics of nature that are difficult to name tend to disappear physically from the world as we restructure it, it is equally true that what has been lost physically tends to disappear linguistically and conceptually. While the first part of this dialectic is accomplished through technological power, academia plays an important role in the second part. It is no coincidence, for example, that claims that nature is socially constructed are usually made by writers who inhabit "overdeveloped" parts of the world such as Britain where wilderness has already been virtually eliminated; and the effect of such claims is to deny the possibility of a nature that transcends its current domesticated state. By making language consistent with this impoverished ecological reality, and by denying the possibility of anything that is "beyond the text," constructionism undermines any possible role of language in pointing to and formulating states of ecosystemic health that are potential rather than actual. In this case, the industrialist worldview becomes the *only possible* worldview; and a major task of environmental theory is to keep alive those ecological scenarios that *do* exceed such industrialized views of nature.

Perceiving "Reality"

Just as the language we use can unwittingly smuggle in ideological features that derail our intentions, the strongly visual emphasis of most Western languages, as exemplified by words such as "viewpoint," "perspective," "focus," and so on, also contain ideological quicksands that will imperil the environmental project if unrecognized. The visual sense is peculiarly compatible with industrialism, and is closely associated with a reductionist view of the world, as Walter Ong has pointed out:

> Sight isolates, sound incorporates. Whereas sight situates the observer outside what he views, at a distance,

> sound pours into the hearer. Vision dissects. . . . Vision comes to a human being from one direction at a time: to look at a room or landscape, I must move my eyes around from one part to another. When I hear, however, I gather sound simultaneously from every direction at once. . . .
>
> By contrast with vision, the dissecting sense, sound is thus a unifying sense. A typical visual ideal is clarity and distinctness, a taking apart. . . . The auditory ideal, by contrast, is harmony, a putting together.[38]

As our distance from the natural order has increased, the world we have substituted—a mechanical, silent world—is one that we interact with in a mostly abstract, conceptual fashion, aided by imagery that is often visual. And in a world in which sound is often synonymous with noise, our insensitivity to the notions of harmony and discord is inscribed in the world we are building. The everyday consciousness that we inhabit is one that has developed dialectically with the technological and economic realities of our time, and so it tends to embody similar assumptions and to reflect the same visual emphasis. Although we will leave a fuller exploration of the significance of this visual emphasis till later, it is worth pointing out the corrosive effect that it has had on alternative ways of construing the world, since many traditional connections to the natural world are of a felt, visceral nature, and these have often been displaced by more immediately striking visual/rational modes of relating.

For example, if we explore a landscape new to us, the visual sense—aided by cameras, binoculars, and postcards—is likely to overwhelm any more holistic, felt—or auditory—connection with the place, since the latter tends to become established only within longer time spans and a more open and subtle form of awareness. Vision, however, is ideally suited to gaining an immediate, atemporal impression of a place, not involving other dimensions that are less visible such as those involving historical, mythological, or ecological context. As Ong implies, the visual sense, for all its immediacy, is not one that *connects* us to place in a more than superficial sense, nor is it one that demands that we open ourselves and expand our own boundaries to include what we are seeing. Vision in the Western world has developed in a way that tends to distance the "observer" from what is observed, and this style is particularly well suited to a life-world that assumes a separation between subject and object, and in which appearance is prioritized over other sources of meaning.

Similarly, as Merleau-Ponty and, following him, Lacan have emphasized, our identification with our own visual appearance has the effect of denying our own felt sense of who we are. In effect, we become objects to ourselves, to be controlled, shaped, decorated, and observed, distancing ourselves a stage further from our bodily involvement with the world and facilitating a mechanistic view of our relation to it. Again, we see that a particular sort of experience of the world also implies a particular, complementary, experience of the self.

Perversion of Experience

Although vision and its various metaphorical extensions have played a central role in industrialism's colonization of the world, it still retains the potential for referring to and evoking a poignant anticipation or remembrance of a world that incorporates the integration lacking in modern industrial society, as David Levin has convincingly demonstrated.[39] But this integration must also occur simultaneously within the self, so that vision becomes part of a whole that also includes hearing, smell, and somatic feeling. It matters little whether it is a "past" world that is recalled or a future one that is fantasized: in both cases the images begin to articulate needs that call for something absent from our current situation. These felt images are whispers of possibilities beyond those offered by industrialism, visions of alternatives that point to a world beyond that of freeways, supermarkets, and television. They are also yearnings for something absent, and can be regarded as symptoms of a significant deprivation that can guide us toward reintegrating those dualistically separated opposites that structure the modern world. But there are traps here too for the unwary theorist. The romanticization of past epochs may, for example, cause us to deny their less attractive characteristics, binding us into false solutions. And images of alterity are often assimilated for commercial purposes, using a nostalgia for the past in order to sell commodities that are thoroughly modern. Terms such as "home made" or "farm fresh" seduce us into associating factory-produced foods with nostalgic images of community-based agriculture and home baking—an association that manages to suggest that these traditional practices are somehow alive and well within the system of production and consumption that has in actuality all but destroyed them.

Commercialism becomes ever more sophisticated at recognizing and harnessing those human needs that it has itself generated through its destruction of the natural order, slipping commercially advantageous experiences into places previously occupied by a taken-for-granted relationality with the world outside ourselves. For example, it is becoming increasingly difficult for most of us to escape from the hectic pace and noise of the modern world into places of quiet and relaxation; but there are a growing number of commercial alternatives such as stress-management classes, self-hypnosis tapes, exercise machines, and so on that offer to help us to relax—at a price. These enlarge the industrial sphere by incorporating *within* it experiences such as relaxation that previously were associated with escape *from* it. Increasingly, there are few aspects of our lives that are genuinely outside this industrialist sphere, and our awareness of the natural order atrophies accordingly as the world defined by industrialism becomes, simply, *the* world. But commercially generated experiences can only temporarily fill the gnawing neediness produced by the demolition of natural structure, and we will need to repeat and repeat again these pseudosolutions in a sort of consumeristic addiction. As Philip Cushman puts it,

> The paraphernalia of a commercial model are . . . a poor substitute for the tools traditional cultures use. . . . Because advertising cannot cure by invoking a workable web of meaning, I believe ads substitute the concept of life-style. . . . It is a kind of mimicry of traditional culture for a society that has lost its own."[40]

Thus the shadowy subjective awareness of need for a context that could integrate us into time and place, instead of encouraging us to nurture such a context, is used in the service of commercial interests that further subvert its possible realization.

And in a further ironic twist, our very need to escape is itself commercially exploited, as is illustrated by the advertisement shown in figure 1. Here, in an unusually frank representation of industrialism, we look out of a prisonlike cell toward the natural world beyond, the word "escape" engraved above the bars. In the second image, the bars have gone, and we are driving up a deserted track. The implication is clear: buy the vehicle, and escape the stresses of industrialism. And this is a message that is not entirely false; for we can indeed, at least for a time, escape the traffic jams in this way. But what is concealed is that our escape will itself further the spread of industrialism, making escape more difficult for others and squeezing the natural world a little closer to extinction. Like medieval carriers of the plague, our flight from disease ensures its diffusion into hitherto healthy areas. Such solutions then, in the long run exacerbate the problems they attempt to solve, using our neediness to bolster the system that caused it in the first place. And here the car is a potent symbol of other vehicles of ideological infection; for the word and the thought are, in their own way, equally powerful instruments of industrialism as any mechanical contrivance.

Such images suggest that our industrial society thrives on, and can only function in the presence of, a narrowing of awareness that prioritizes certain, often crude, forms of relation to the world while repressing

Escaping the plague, spreading the plague... © WWAV Rapp Collins

others. The anthropologist and psychiatrist Arthur Kleinman has remarked on the absence within the Western world of states of consciousness that are almost universally recognized and accepted in other cultures, and that can be roughly categorized as those involving trance. The Western self, says Kleinman, is a construction of modern culture that has

> deepened discursive layers of experience . . . while paradoxically making it more difficult to grasp and communicate poetic, moral, and spiritual layers of the felt flow of living. Trance and possession are not, as some psychologists and psychiatrists aver, "primitive" forms of pathology. Rather, they are ways of experiencing and articulating the body/self in nondualistic, archaic tropes.[41]

Kleinman is here discussing the subjective side of a cultural predicament that also, and more tangibly, involves the physical transformation of the world from a diverse wilderness into a rationally ordered monoculture. The senses ensure that the body is multiply connected to the rest of the world; and so to the extent that we lose touch with our bodies, substituting a "psychological metalanguage," so we also lose touch—literally and metaphorically—with our rootedness in the world. The separation of the "human" realm from the "natural," then, is also one that shatters the integrity of the self, allowing our disembodied identification with a constructed virtual world that is related to nature only in the crudest ways. This is therefore both a psychological and an environmental loss, although the dissociation between disciplines has usually prevented us from addressing the root causes of either. This dissociated stance forms the basis of a relation to the earth that is instrumentally powerful but partial, a combination of qualities that is particularly damaging to the natural world. The problem is compounded by the fact that the partiality of our understanding has been forgotten, so that we exist within a universe that, for most of us and for much of the time, *appears* to be whole, but that is in actuality a reduced, fragmented universe generated by our forgetfulness of other subjective possibilities.

Human adaptability, and in particular our reconstitution as the thinking, disembodied beings of Descartes, ensures that our demolition of the natural order is superficially less toxic to us than to most other species. Nevertheless, as each other form is stilled, so something within us also dies, and the resulting loss of resonance between self and world increasingly defines a form of selfhood that contracts into the confines of the self-contained individual. This is another form of extinction, but a slow one in which those human characteristics incompatible with technique are lost over the course of the centuries, shriveling inward to condense in the uncommunicative, childlike "libidinal ego" of the object relations theorists, where they lie dormant as unexpressed potentials. Somehow, a process of cultural evolution that once seemed to hold human interest at its core has mutated toward one whose lethality we sense but struggle to explain, to the extent even that some have described it in terms of the disappearance of those essential characteristics that make us human. This transformation, suggests Arthur Kleinman, "can be of a kind to cancel, nullify, or evacuate the defining human element in individuals—their moral, aesthetic, and religious experience."[42] And if we are no longer "human" in this defining sense, then what *are* we; and how does our emergent character interact with the industrialist landscape that we have evolved to complement, and constrain our ability to envision a natural world that is not determined by industrialism?

Some have argued that these changes are inevitable and not necessarily unhealthy. Donna Haraway's "cyborg vision,"[43] for example, advocates the abolition of the boundaries separating the human from the machine, the human from the animal, and the technological from the natural as a necessary transcendence of the dualisms that she and other feminists have seen as a prime cause of environmental and other problems. A dualism, however, is a distinction that has become reified, a frozen fragment of a prior symbolic diversity; and to collapse distinctions such as that between the human and the machine is to undermine such important facets of our self-identity as animality, gender, and, in Kleinman's sense, humanity. Such distinctions, then, should not be identified with the dualistic forms that result from their reification: they are also differences essential to the *natural* order. Biological sex, for example, has been used as the basis of dualistic and oppressive gender roles; but it is also basic to our identities, whether as children, parents, or sexual partners, and any sustainable and profound cultural order must be built on these natural foundations. The frequent perversion of the relation between the cultural and the natural spheres of human life does nothing to contradict the necessity and potential of this relation. Differences between man and woman, like those between predator and prey, parent and child, or day and night cannot be collapsed into an amorphous epistemological soup; and the attempt to abolish those distinctions on which industrialism is selectively based is a blunt instrument that sweeps away essential structure as well as its perverted forms. Haraway's approach is partly ironic; but it is an uncertain irony that teeters on the brink of nihilism, and such approaches run the risk of becoming self-fulfilling prophecies. As such, its allegiance is ultimately, if unwittingly, to industrialism, since the abolition of structure, whether epistemological or natural, is always the first step in the imposition of an industrialist order.

The already industrialized world in which the convergence between the human and the cyborg is becoming an established reality often appears harmonious and lacking in violence. Its violence is the violence of the cemetery or the museum rather than the battlefield: it is based on a violence already accomplished, a wildness already exterminated, a "nature" already replaced by a simulacrum; and our easy habitation of this world says a good deal about the extent to which our character has been reshaped to fit it. In this industrialized world, an ostensibly self-sufficient, humane, and democratic system feeds off the wildness that still exists at its distant interfaces with the natural world. This replacement of existing natural structure by industrialist forms is the fundamental ecological change that underlies the destruction of the natural world, and it is a change that reconfigures selfhood as much as any other aspect of the world. Even our awareness of the violence that characterizes the assimilation of the wild, for example, is frequently assimilated to and used for commercial purposes. The images of starving children, of land-desperate peasants, and sweatshop laborers producing trainers for First World markets are assimilated as material to be sold, or are themselves used to sell, as in the notorious Benetton advertisements. The experience of loss, and the representation of suffering and violence by the media, are themselves grist to the economic mill, so that, as Kleinman puts it, "experience is being used as a commodity, and through this cultural representation of suffering, experience is being remade, thinned out, and distorted."[44] Even at the moment that we become aware of suffering, our experience is already being commodified to suit commercial interests, and the industrialization of our personalities is continually reinforced. Media images of nature also accord with this "reconstituted" selfhood: wildlife documentaries often reflect the pressure to show more and more spectacular shots of charismatic fauna, sometimes resorting to techniques involving faking, tethered prey, sequences filmed in zoos and game parks, or mini-helicopters to produce the desired images. As Martin Colbeck, a leading wildlife cameraman complains: "Some of the highly constructed films seem to be determined more by the imagination of the film-maker . . . than the wildlife."[45] Sir David Attenborough, justifying the filming of the birth of a polar bear cub, ostensibly in the Arctic but actually in a Belgian zoo, defends the artifice, arguing that "the reality of the situation . . . was the birth of the polar bear, not where it was taking place."[46] Reality has become a collection of individual biological events each of which is divorced from ecological or cultural context, a framework constructed by the editor from expendable biological fragments. As one ex-BBC producer complains: "Are we to go on endlessly producing more and more living Edens when they are disappearing faster than we can edit the film?"[47] Our acceptance of wildlife documentaries as authentic portrayals of nature may therefore reflect our assimilation into the industrialist order rather than our appreciation of the natural order.

The natural world that is the subject of these media images is not simply being destroyed in ways that are obvious; it is also being more subtly reconfigured to fit into the industrialist landscape, sometimes by those who claim to defend it. For example, organizations such as the Nature Conservancy appeal to our wish to preserve nature; but they do so through means that are firmly anchored within the world of commerce and investment. In bartering those areas that are "truly ecologically significant" for those that are merely "trade lands," the latter, as Tim Luke has pointed out, become "denaturalised zones whose main value is that they can be sold, like old horses for glue or worn-out cattle for dog-food . . . in seeking to preserve nature, [the Nature Conservancy] oversees its final transformation into pure real estate."[48] Increasingly, too, nature is packaged and sold to us as a form of substitute satisfaction for the real nature we are losing. "Make contact with another world" suggests the advertisement for Sea World, offering forms of communication not ordinarily possible. Having become alienated from nature, we can now—for a price—be reconnected to it. The advertisement, which includes pictures of happy children and adults hugging each other as they watch the whales and dolphins, also implies the satisfaction of more general longings. "The other world of nature," suggests Susan Davis, "is also an interior world, one of emotions and feelings . . . the theme park offers customers access not only to nature and exotic animals but to themselves. Asking us to 'remember the feeling of wonder' and 'bring back the smile' suggests that we need to return to authentic feelings" of relation;[49] and such needs, by being denied satisfaction and harnessed for commercial ends, are made to serve the industrial system that keeps us needy, in a process loosely analogous to the way water's natural tendency to cascade downstream is harnessed to generate electricity. Even the wilderness preservation movement, as we saw earlier, cannot entirely escape the influence of industrialist taxonomies. The age of innocent environmentalism, in which a demonised technology could be contrasted with more "natural" bases for living, has long since given way to an era in which our theories of and attempted solutions to environmental destruction have themselves been colonized by a destructive agent that is elusive to consciousness.

Under these circumstances, in which our attempts to relate to, understand, and protect nature are implicated in its destruction, our reactions may be more complex than the outrage and mourning that are the more expectable reactions to avoidable loss. Over years of discussing environmental issues with groups of undergraduate and postgraduate students, the feeling that I have most often encountered is helplessness: there's no point in talking about these issues, the students suggest, because there's nothing we can do anyway, and to discuss them is to open oneself to feelings of depression that are best not stirred up. The "numbing feeling, almost one of boredom," which Anthony Giddens identifies as a common reaction to the threat of environmental disasters may not only derive from the belief that environmental changes are on such a scale that the only possible reaction is a sort of fatalistic acceptance, but also from a suspicion of rationality itself as we sense the infection of our thought-processes by industrialist ideology. "A sense of fate," says Giddens, ". . . relieves the individual of the burden of engagement with an existential situation which might otherwise be chronically disturbing. Fate, a feeling that things will take their own course anyway, thus reappears at the core of a world which is supposedly taking rational control of its own affairs. Moreover, this surely exacts a price at the level of the unconscious, since it essentially presumes repression of anxiety."[50] An article in the *London Sunday Times* recognizes both this anxiety and the attempt to conceal it:

> Some educationists say the green lobby's zeal is poisoning young peoples' attitude to the world, leaving them frightened and confused. "A mixture of half-truth and propaganda fed to children at too early an age could distort their development," said Martin Turner, senior educational psychologist with the London borough of Croydon.
>
> "This is a form of institutionalised child abuse. Children are learning that the world is a threatening place and that is paralysing their confidence" he said.
>
> The new emphasis on the environment is the modern-day cousin of anti-racism and anti-sexism dressed up with a new respectability, said Dr. Dennis O'Keefe, an educational psychologist at the Polytechnic of North London. "We live in a world of cheap scares, and children are the most exploited by them. Environmentalism is just political radicalism wearing another hat. It is becoming much stronger than previous fads because it touches on a deep-rooted anxiety about the future of the world," he said.[51]

The tacit recognition of our ensnarement within events that are outside our understanding and control is itself difficult to articulate, so exaggerating the gulf between conscious and unconscious, and generating confusion and depression. All too often ecogenic depression is experienced—and treated—as a psychological problem requiring individual treatment; but antidepressants, unfortunately, cannot prevent the destruction of wilderness, nor can they address the unrecognized causes of depression with which this destruction may sometimes be associated. But although depression is usually understood in a narrow psychological or biochemical sense, the *experience* of depression is often still one of bodily weightedness, as suggested by the derivation of the word depression from the Latin *deprimere*, to press down; and so the depression that originates in our awareness of the destruction of the natural world is an emotion that potentially reconnects us to the earth, in which we share the suffering of the earth, recognizing that the world we inhabit is "a world of wounds."

And while environmental realities may indeed frighten children (as well as adults), and while it is important not to oppose widespread fatalism with exaggerated proclamations of environmental doom, we cannot *begin* from the repressive dissociation of human interests from environmental interests. To "protect" children from an awareness of environmental degradation is to deny them the opportunity to develop the means through which this degradation could be rectified. In assuming the prior separation of the child from the environment, claims such as those in the *Sunday Times* article above imply that children can flourish while the natural world disintegrates, and that their destiny or well-being is somehow independent of the health of nature as a whole. A superficially humane impulse is therefore yoked to a reductive agenda that abandons those essential elements of humanity that are not socially constructed. This is an inherently repressive approach that denies children's own felt experience of the earth and the threats to it which, in my experience, they are quite capable of sensing. Our children will have difficulties enough relating to the physically mutilated world they will inherit. Let us not magnify their problems by allowing our own fears and dishonesties to distort their growing awareness, assuming too quickly that they will be unable to recognize and resolve the predicament that we bequeath to them.

More generally, if we are colonized beings whose rationality is deeply suspect, then our efforts to transcend industrialism cannot be located only in the intellect that is the primary seat of that colonization. This is not to reject the intellect permanently or completely, but rather to make it more accountable to those other faculties through which our relation to the natural order is sensed and expressed, so that it regains its consistency with these other faculties and so become integrated within a rediscovered whole. Recognizing that "environmental" and "psychological" problems generally emerge out of the same traditions and ideologies as the discourses that are intended to offer analyses of them, we become aware that finding solutions demands that we transcend these separate discourses as well as necessarily working within them. For example, while we may usefully look at depression in conventional psychological terms, or at conservation of the natural world in terms of the value of its separate parts, we *also* need to recognize the common roots of both these issues, implying that our detachment from the world has both psychological and environmental consequences. In the absence of such a radically critical perspective, personal, social, and environmental problems become isolated facets of a discordant and ultimately meaningless collage. Experiencing the world as a set of largely separate domains, we remain ignorant of the meaning of the symptoms that tell of its disintegration. From a conventional Western perspective, there *is* no environmental crisis, although there may be gas shortages, polluted beaches, or vanishing topsoil, in addition to "unrelated" problems such as rising crime rates and increasing evidence of personal distress. It is only when we can recognize the common roots of these problems that we begin to glimpse the ghostly, shattered forms that lie in the shadows of industrialist "rationality."

In the chapters that follow, I explore these roots that underlie both our experiences in the modern world and the industrialist transformation of nature. But this sort of analysis, in itself, would be merely another addition to the growing mounds of writing on environmental topics that are proliferating within the comfortable sphere of academic discussion, safely insulated from the violent edges of the industrialist empire inhabited by the Brazilian sugarcane workers, the Bolivian tin-miners, or the relocated Navaho to whom I refer later. Academic writing is only of value if it reaches out beyond academia, confirming rather than ignoring the existence of a world that is real and embodied, and that is being destroyed not just by bulldozers and chainsaws but also through our disingenuous collusion in the widespread denial of its reality. And simply acknowledging that this is happening will not do, either. Recognizing the potential integrity of the world implies a moral imperative to *defend* this integrity, which is the ultimate basis of any moral authority. A theory that has a merely intellectual resonance may be academically successful but still morally bankrupt; and our aim must be to reach out toward those more distant resonances that are the echoes of the natural order. If industrialism has picked apart the fabric of the world—a fabric that is *simultaneously* physical and subjective—then academic writing must extend beyond a gloomy commentary on this dismemberment, actively promoting a vision that would reweave the warp and weft of materiality and subjectivity into the wholeness of the world. If we succeed in this aim, then theory can be fully embodied, articulating the pain and protest of the natural world as it is destroyed by industrialism, and obstinately holding out a vision of integration between psyche and materiality that can catalyze the realization of this integration. The objective of this book is to outline just such a vision. But before we attempt this, we need to consider precisely why existing theory, and particularly existing psychological theory, fails in these respects—a task to which we turn in the next chapter.

Chapter 2

PSYCHOLOGY'S BETRAYAL OF THE NATURAL WORLD

"I should really *like* to think there's something wrong with me
Because, if there isn't, then there's something wrong,
Or at least very different from what it seemed to be,
With the world itself—and that's much more frightening!"

—Celia, in T. S. Eliot's "The Cocktail Party"

The Colonialist Attitude

When European adventurers began to cross the oceans in search of the New World at the end of the fifteenth century, the maps and stories they brought home with them were often constituted in roughly equal proportions by observation and fantasy. But fantasy is not created merely out of the individual imagination: it is generated ideologically, and the European navigators were the unwitting agents of an ideology that has since infected most of the globe. The reports of strange creatures—either human or nonhuman—that were alleged to populate the new domains, were symptoms of the defensive reaction of this ideology to the radical otherness of the New World, indicative of the inability of the European psyche to recognize and accept anything beyond its own boundaries. What is remarkable about the histories that were produced during this period is how little light they throw on the cultures, customs, and characteristics of the native inhabitants themselves; but this is consistent with the ideological character of their enterprise, since ideology is blind to the structure of the raw material it assimilates to itself. Columbus, for example, insisted that the native peoples he encountered should learn Spanish, and made no attempt himself to learn their languages. Indeed, he fails even to acknowledge that native speech represents a valid language, writing on one occasion that: "I shall take from this place six [natives] to Your Highnesses, so that they may learn to speak."[1] In the same vein, Columbus insisted on renaming all newly discovered lands, even though he was well aware that they already had perfectly good Indian names. As Tzvetan Todorov remarks, "nomination is equivalent to possession." And the presumed power of language is bizarrely illustrated, on one occasion, by Columbus' insistence that each member of his crew swear an oath asserting that their landing place (which today we would recognize as part of Cuba) was "the mainland and not an island, and that before many leagues, in navigating along the said coast, would be found a country of civilised people. . . . A fine of ten thousand maravedis . . . is imposed on anyone who subsequently says the contrary of what he now said, and on each occasion at whatever time this occurred; a punishment also of having the tongue cut off, and for the ship's boys and such people, that in such cases they would be given a hundred lashes of the cat-o'-nine-tails, and their tongues be cut off."[2] One might suppose that such extreme measures would be rendered redundant by a bit of empirical observation; but, as Todorov remarks, "ideological certainties can always overcome individual contingencies."[3]

Moreover: "having learned the Indian word *cacique*, he is less concerned to know what it signifies in the Indians' conventional and relative hierarchy than to see to just which Spanish word it corresponds."[4] He is concerned immediately to relate each new discovery to his own world of meaning rather than in seeing native understanding as a valid system in its own right, articulating the physical and spiritual world it related to. Columbus' activities are less a discovery of something new than a reproduction, in a novel realm, of the old. Each thing that he finds is immediately assimilated to familiar structures rather than allowed to be part of the new. The new is *deconstructed*, its systemic character ignored and therefore demolished. Todorov summarizes: "The interpretation of nature's signs as practised by Columbus is determined by the result that must be arrived at. His very exploit, the discovery of America, proceeds from the same behaviour: he does not discover it, he finds it where he 'knows' it would be."[5]

Columbus' approach seems outrageous to us; and yet his basic methodology is still replicated in so many ways, both within academia and in the world outside, that we might suspect it to reflect fundamental principles of the project of modernity. Todorov posits two major types of communication: that between "man and man," and the other between "man and the world," the former reflecting the language of the colonialist explorers, and the latter that of the Indians. The victory of the Spaniards over the Indians, he suggests, was also a victory of the former style of communication over the latter; but it was one that carried with it ominous and quite unrecognized implications for the world that the conquistadors were bringing into being, for "man has just as much need to communicate with the world as with men . . . this victory from which we all derive . . . delivers . . . a terrible blow to our capacity to feel in harmony with the world, to belong to a pre-established order."[6] We are accustomed, says Todorov, "to conceiving of communication as only interhuman, for since the 'world' is not a subject, our dialogue with it is quite asymmetrical (if there is any such dialogue at all)."[7] The world of the conquistadors is a more or less autonomous realm, protected from its inconsistency with the world outside by its aggressive refusal to recognize any structures alien to its own. The native universe of meaning, in contrast, refers directly to the outside world, finding its own logic and structure within that world, and maintaining a dialogue with it.

Within the European frame, understanding therefore becomes an act of hostile assimilation, requiring the liquidation of any structure incompatible with Eurocentric ideology and the simultaneous imposition of a conceptual monopoly. This "understanding-that-kills"[8] is, as we will see, a perennial if hidden motif of the method by which the assimilation of the world to technique is accomplished, in spite of its claim to objectivity; and it is one that we can discern throughout the spectrum of industrialist practices. For present purposes, however, we will restrict ourselves to one of these practices: the "science of mental life," psychology. We will see that in the name of objective science, psychological methodology actually reproduces many of the distortions perpetrated by Columbus and those who came after him; and these distortions include the refusal to recognize and acknowledge other forms of language, the related assertion that reality is a by-product of language, the lack of openness to alternative forms of structure, and the readiness to assimilate the new to existing ideology. As James Hillman remarks: "While other Nineteenth Century investigators were polluting the archaic, natural, and mythic in the outer world, psychology was doing much the same to the archaic, natural, and mythic within."[9] And this methodological convergence between psychology and geography is no mere coincidence; for as Hillman suggests elsewhere, "the gradual extension and civilisation of outlying barbarous hinterlands is nothing else than ego-development."[10] Equally telling is Freud's own metaphor for ego-development as the draining of the Zuider Zee. The colonialist process, then, is defined not simply by the use of physical force, but, more accurately, by the repression of any structure that differs from that of the colonizing ideology; and by this definition, colonialism continues and flourishes in many unrecognized forms.

The realisation that the objectivity to which psychology often aspires in its claim to scientific status may be a facet of this lethal understanding is only possible within a historically and culturally informed perspective.[11] As we will see, the roots of the schism between the person and the natural world are deeply entwined within the historical evolution of technological society, and these same roots gave rise both to the flowering of the industrial society for which environmental problems are a potential nemesis, and to the discipline of psychology itself. There is thus an ideological convergence between psychology and industrialism that makes the former peculiarly blind to the problems generated by the latter, and binds it within the ideological orbit of modernism. This chapter explores the character of psychology partly in order to assess its usefulness (or otherwise) for an environmental agenda; but also because the character of psychology's inadequacies are instructive in understanding the form that might be taken by a more fertile comprehension of the natural dimension of human life, a possibility that we will explore in the second part of this book.

Destruction of the natural environment is due to human behavior; so one might, on the face of it, expect that psychology, which has defined itself as the science of human behavior, would be able to offer a powerful and far-reaching analysis of our relation to the natural world. If so, one would be sorely disappointed. A review of the psychological literature reveals a remarkable absence of concern about or comment on our environmental problems. Searching the journal abstracts listed on the CD-ROM version of *Psychological Abstracts*, for example, shows the popularity of customary areas of psychological research such as memory (12,428 entries) and problem solving (3,771 entries). In comparison, psychological interest in the natural world is barely perceptible: "natural world" occurs twenty-one times, and "environmental crisis" five times. True, there is a fast-expanding field known as "environmental psychology"; but this field is almost completely concerned with the effects of variation in environmental conditions on human behavior and performance, and offers no analysis of the origins of environmental degradation in human praxis. And here we can recognize a parallel with the conquistadors' attitude towards the world. They were accustomed, as we saw above, "to conceiving of communication as only interhuman, for since the 'world' is not a subject, our dialogue with it is quite asymmetrical (if there is any such dialogue at all)";[12] and this nicely sums up psychology's attitude toward the natural world. For example, in one influential and in many respects excellent psychology text, the only discussion of our interaction with the natural world is in a section headed "Recognising Natural Objects," these being "animals, plants, people, furniture, and clothing"[13]; and what follows is a discussion of the problems of perceiving complex shapes. If the possibility of communication with something or someone outside oneself is methodologically denied, then the possibility of a relationship does not exist; and the world according to psychology is thus a world defined according to the perspective of a detached observer rather than that of a participant. To the extent that we deviate from this perspective, our functioning becomes "deluded," "inefficient," "subjective," and so on; and psychology's purpose is the policing and diagnosis of such deviation. Just as the geographical mapping of the world made possible the idea of navigational error, so, as Bernard McCrane perceptively points out, "the task of Newtonian anthropology was . . . the Enlightenment formulation of a *psychology of error*."[14] During the Enlightenment, McCrane notes, there was a switch from emphasizing the contrast between Christian and non-Christian to that between ignorant and nonignorant, and from viewing myths as living or dead to an understanding of them as true or false. Henceforth, a particular form of selfhood, and a particular style of relation to a world defined complementarily, became the grain of sand around which crystallized the European sciences, whether their subject matter was the innermost recesses of the psyche or the most inaccessible lands and peoples; and it is this same crystalization that today objectifies what is "intelligent," "reality oriented," or "valid."

Psychology as Colonialism

While the singular term "psychology" suggests a unified, integrated discipline, there are, in fact, a bewildering diversity of psychologies, many of which are barely on speaking terms with each other. In British and American universities "experimental psychology" is dominant, an approach that prides itself on using scientific methodology in the investigation of human behavior. Although "experimental psychology" encompasses considerable methodological diversity, experimentalists are united in their rejection of "unscientific" alternatives such as transpersonal, humanistic, or psychodynamic psychologies; and perhaps this has something to do with the way the human mind has always sat uncomfortably astride one of the major fault lines that defines the philosophical terrain of our times—that which divides Descartes' categories of *res extensa* and *res cogitans,* or their various ideological successors. The dangerous no-man's-land between categories is inevitably taboo to those who police the boundaries of the permissible, in scientific realms as well as in the religious spheres that were in earlier times patrolled by institutions such as the Inquisition. In the long term, the blindness that ideological structures defensively maintain with respect to such taboo segments of the world bodes ill for their capacity realistically to represent reality, as the dwindling ability of institutionalized religion to articulate human spirituality forewarns us. The further any ideological structure is developed, the more strain is placed on its assumptive foundations, and the greater the potential consequences of any fault in these foundations. That scientism generally is haunted by the ghosts of what it has omitted is demonstrated with increasingly urgency by global environmental problems such as climatic change and pollution; but while the physical and natural sciences have made some progress toward recognizing the necessity for reductionism to be complemented by radically holistic and systemic notions, many types of psychology have shown a robust immunity to such changes. Experimental psychologists have been particularly fervent in their defense of traditional "scientific" principles against "mystical," "unscientific," or otherwise blasphemous ideas originating within fields such as psychoanalysis or existential psychology, bringing to mind Frederick Turner's comment on the psychodynamics of the Christian colonizers of the New World: "It is the classic reaction of those who have lost true belief . . . that they must insist with mounting strenuousness that they do believe—and that all others must as well."[15]

Ironically, at a deeper level experimental psychology has a good deal in common with the approaches it rejects, as we will see; and a little historical digging reveals that psychoanalysis and existential psychology have their roots in the same bedrock, often defending their particular techniques with an almost cultlike intolerance. The rivalries between these various psychological factions, however, has had the effect of focusing attention on issues related to their *differences*, and in doing so often has blinded psychologists to those important assumptions that are shared by most or all of them. This internecine squabbling therefore colludes in covering up the deeper taboos that hold psychology within a constricted world defined by its covert denial of what exists outside its boundaries. This is particularly the case with experimental psychology, given its survivalist tendency to withdraw into a well-defended coterie that is preoccupied with distancing itself from threatening pretenders. This inwardly directed gaze has had a crippling effect on experimental psychology's imagination for change and development, as Don Bannister suggested:

> Had Christopher Columbus . . . possessed the mind of many modern psychologists . . . he would never have dis-

> covered America. To begin with, he would never have sailed because there was nothing in the literature to indicate that anything awaited him except the edge of the world. Even had he sailed, he would have set forth bearing with him the hypothesis that he was travelling to India. On having this hypothesis disconfirmed when America loomed on the horizon he would have declared the whole experiment null and void and gone back home in disgust.[16]

But while Bannister was right about experimental psychology, he was wrong about Columbus, who, as we have already seen, can be understood as inventing America as much as discovering it. The myth of discovery, serving to hide the latent function of assimilation, operates similarly in intellectual as well as geographical realms. Just as Columbus and his men interacted with the inhabitants of the New World only in the most predetermined and limited ways, so the world of the experimental laboratory has only the most restricted forms of relation with the alien lands beyond its boundaries. This restriction is, of course, necessary in order to control the many influences and interactions that would occur under the normal conditions of our everyday life, and that we couldn't hope to operationalize and take account of. Only in the most tightly controlled conditions do people behave in the ways expected of them in the experimental laboratory.

Now, such restrictions are not necessarily fatal to the experimental approach so long as they are recognized as such. *Any* model of reality, simply by virtue of being a model, will accurately represent certain features of what it is modeling while being necessarily inaccurate in other respects; and the particular drawbacks of the experimental approach are the unavoidable price we pay for limiting our analysis to a few variables and associated forms of statistical analysis such as the analysis of variance. The reduction in meaning that is an inescapable aspect of positivist methodology is quite acceptable *as long as* we recognize the model of the world that results from it as the poor thing it is, and as long as we bear in mind Korzybski's warning about the dangers of confusing the map with the territory.[17] Unfortunately, the divergence of experimental conditions from anything approaching "reality" is frequently forgotten by experimentalists, who have often, on the basis of their findings in the laboratory, drawn wide-ranging conclusions about life in the world outside. This forgets those characteristics of the world that are least sensible to the experimental method; and it is no coincidence that the sophisticated interactions that underlie the systemic character of nature and that potentially relate us to it are among the factors that are most often forgotten.[18] This forgetfulness, according to Sigmund Koch,

> Has had . . . crippling entailments for the character of the psychological enterprise. If empirical decidability . . . is the criterion for bounding the meaningful, one then has a perfect rationale for selecting for study only domains that seem to give access to the generation of stable research findings. If any domain seems refractory to conquest by the narrow range of methods (usually borrowed from the natural sciences and mutilated in the process) held to be sacred by the workforce, then, obviously, *meaningful* questions cannot be asked concerning the domain, and that domain is expendable.[19]

Part of what is thrown out with the "mystical" bathwater, according to Koch, are the babies of experience, meaning, and the whole subjective realm. Since any noninstrumental relation that we may potentially have with the natural world is almost certain to be part of this forbidden realm of subjectivity, it is hardly surprising that experimental psychology has on the whole behaved as if such relations are best swept under the carpet.

Replacing the Person by the "Subject"

In effect, then, experimental psychology is the attempted fulfilment of Descartes' dream of inventing a more "perfect" world than the one that actually exists—a world in which the environment is the source of "information" that is processed according to more-or-less successful "strategies" by "subjects," watched over by emotionless, perfectly efficient "experimenters" who record a steady stream of quantitative "data" before calculating the "significance" of their results. Kurt Danziger suggests that the tale of the Sleeping Beauty is an apt parable for the experimentalist's approach: relevant characteristics of the "subjects" are simply seen as already "there," fully formed and immaculately conceived, devoid of sociohistorical context, ripe for discovery and awakening by the magic kiss of his methodological manipulations.[20] This invented world is often elegant and, in terms of psychological data, productive; but sadly, it constitutes a mediocre model of the actual world. As Seymour Sarason has put it, "Psychology has for too long sought to measure a world of its own contrivance, and this it has done extremely well—so well that for decades it did not have to face the possibility that ingeniously measuring a world of one's own making is a mammoth waste of time."[21] This world that has been contrived by the experimental psychologist is, of course, uninhabitable by those folks who actually survive outside it, in what is still, in many respects, the "real" world; so it becomes necessary to eliminate those human characteristics that are inconvenient to the experimental method and to construct a new individual more consistent with the world of the laboratory. As Foucault perceptively put it: "Western man could constitute himself in his own eyes as an object of science . . . only in the opening created by his own elimination."[22]

This substitute person created by the psychological experiment, the modern counterpart of the dispassionate Cartesian "knower," is a thinking, deciding creature which (I use the sexless pronoun deliberately) is relatively unemotional and unsocial and is notably detached from the world it relates to as a passive, formless background to its decisions. As this "environment" supposedly lacks any intelligent structure, changes that occur as a result of learning and experience are necessarily *intrapersonal* ones; and the emphasis of the experimental psychologist concerns the strategies adopted by the subject, how they "develop" with experience and age, and how they can be more or less "efficient." Such views emphasize the supposed psychological properties of the person while rejecting the potential existence of any structure larger than that of the individual thinker. The social, cultural, and natural worlds are therefore not seen as intelligent, evolving, reacting, or as in any way constituting the person; and so the way the individual interacts with the world becomes less a "relationship" than a one-way process of manipulation. The potentially communal processes that would unite the individual and the natural world within a shared project of joint creative evolution are replaced by the colonialist enterprise of egoic expansion; and in Vygotsky's words, an "*interpersonal* process is transformed into an *intrapersonal* process."[23]

For example, the notion of the 'soul' is not accepted within experimental psychology as a way of understanding human experience: the soul-less individual is taken for granted prior to any experiment, and is not understood as a matter of possible discussion between a "subject" and an "experimenter" whose views and experience may differ. What could, therefore, be a matter of interpersonal debate is instead foreclosed by the prior designation of the subject as an "information processor," or whatever. This approach is reminiscent of the conquistadors' certainty that the "false attribution of soul life ('animism') to nonhuman species of the phenomenal world . . . prevented primitives from seeing the world as it really is."[25] In its most extreme form, this intrapersonal emphasis can be seen in the search for brain locations that determine particular experiences—an approach that altogether bypasses our relations with the world outside, making subjectivity a function solely of the brain.

Individualism is, of course, not limited to psychology, being a pervasive feature of the modern ideological landscape as the person becomes increasingly identified with internal experiences dissociated from the world itself. Adventure or wilderness, pain or ecstasy, for example, are more often "experienced" televisually than through physical engagement with the world; and the current emphasis on personal autonomy also implies a withdrawal from the social, cultural, and natural worlds outside. Psychology is not, therefore, an aberrant enterprise when seen in cultural context, but rather a particular expression of this context; and this ideological convergence between psychology and industrialist society has two major effects. Firstly; it makes psychology's assumptions about a particular style of individualism seem natural and unremarkable, since we ourselves are constantly making the same assumptions; and secondly, it obliterates psychology's potential capacity to comment on and critique the particular forms of personhood, behavior and experience that are accepted as "normal" within the modern world. And this is the major source of psychology's inability to contribute to any radical environmentalist critique of modernity; for it is precisely these "normal" forms of personhood, behavior, and experience that are implicated in the exploitation and dismemberment of the natural order.

The realization that this individualist stance is far from universal, which emerges clearly from a plethora of geographically diverse ethnographies in the anthropological literature, was brought home to me with particular force when I was teaching at a college in Colorado several years ago. Around 10 percent of the students in my Human Development class were Navahos from the large reservation that covers much of the Four Corners area. I found that when I asked the class a question, the Anglo students were often forthcoming and keen to answer; while the Navahos never answered. This puzzled me until one day after the class, one of them approached me and explained that it was the tribal tradition not to make decisions individually, but rather to discuss an issue as a group before arriving at a collective answer. The emphasis, then, is on social rather than individual process, and the social world is a primary reality to which intrapersonal development owes allegiance. Furthermore, this social reality is one that, as we will see later, is strongly interwoven with the natural world; so the whole emphasis of Navaho ontology diverges markedly from that of industrial society.

In contrast, within those areas of the world that show the historical imprint of the Enlightenment tradition, *individual* characteristics, and their divergence from the norm, are what matter. In Galton's "psychometric laboratory," the "performances . . . defined characteristics of independent, socially isolated individuals and these characteristics were designated as 'abilities'."[25] Context, and its contribution to performance was generally discarded; and this is still a prominent characteristic of psychology's search for "universals" of human behavior. The "individual" who results from this process of subtraction is a sort of "lowest common denominator," what remains after the removal of all those aspects that are co-created through the person's interaction with the world. The historically evolving split between the modern individual and the rest of the world is therefore the initial dissociation in the more general fragmentation of the cultural and natural fabric of the world: and from this initial dissociation emerges both the egoic individual and a world that seems to invite exploitation. Moreover, a world that is detached from us and so fit to be unemotionally exploited is also one that we are likely to experience as hostile and alien. Consequently, we withdraw inward into ourselves, further exacerbating the individualist turn—a vicious dialectical spiral that is given apparently unstoppable momentum by our physical construction of a world that seems to embody our alienated fantasies about it. Increasingly, we *escape from* the worlds of work, rushing traffic, deadlines, and so on into our own self-constructed retreats, be they physical retreats such as our own home or forms of psychological solace constructed through soap operas, fantasies, or meditation, in effect abandoning the world outside as already lost to us.

Models of Nature

It is important to acknowledge that experimental psychology has made significant contributions to the understanding of behavior in a range of tightly defined conditions—mostly involving the study of vigilance, perception, memory, choice reaction time, and other aspects of human performance that can without too much difficulty be operationalized. However, since the experimental approach is based on only the most partial and selective forms of interaction with the natural world, and is in fact zealously involved in the project of attempting to replace this world with an alternative, technical one, it is only to be expected that it has made extraordinarily little contribution to the environmental debate. But to put it like this is still to misunderstand the extent of the problem; for it is not merely that experimental psychology has failed to foster a realistic awareness of our place in the natural world, but rather that it has *actively contributed* to the construction and legitimation of a form of personhood that is inherently hostile to nature. We can most realistically regard experimental psychology, therefore, as part of industrialism's challenge to natural structure in all its forms.

If this embeddedness within a particular ideological tradition is, as I have suggested, a defining feature of the experimental approach, then whatever its claims to the contrary, experimental psychology cannot realistically pretend to generate findings that are in any sense absolute, transhistorical, or valid across cultures. Rather, the findings of psychology are the products of a quite specific historical project. Kenneth Gergen, in a paper that is now a classic,[26] has shown that the concepts produced through psychological experiments, while often presented as part of an "objective" scientific process of discovery that pushes back the boundaries of ignorance, can more accurately be viewed as historically and culturally relative, and furthermore, as subtly prescriptive in their effect. Experimental results, particularly if they accord with previous work in the area, become part of an expanding knowledge structure that comes to seem increasingly real and universally applicable, and that structures experience and legitimates certain forms of behavior. In short, the generation, communication, and acceptance of experimental findings in psychology can most accurately be viewed not so much as the result of open, dispassionate enquiry, but rather as the outcome of a covertly motivated process in which existing ideological forms ensure that knowledge produced is consistent with their own continuation and expansion.

I want to illustrate these characteristics by referring to one particular experimental study of the relation between people and the natural world. It would be easy enough to select one of the many studies that acknowledge the natural world inadequately or not at all; but this would be a fairly tedious exercise given that the decontextualized character of psychological experimentation is so apparent. So I am selecting one of the best of those very few experimental studies that gives serious thought to our relationship with the rest of the natural world: Rachel and Stephen Kaplan's *The Experience of Nature: A Psychological Perspective.*[27] The Kaplans are aware of many of the drawbacks of the experimental paradigm, and chart a middle path between the requirements of this approach and the need to encompass aspects of the natural world that are elusive to categorization. Nevertheless—as I suspect the Kaplans would be the first to acknowledge—their study, although a significant advance in this area, ignores dimensions that are of crucial importance for our purposes. I will pass over those commonly acknowledged weaknesses that have often been pinpointed within experimental studies, focusing instead on those more subtle implications that have usually passed unnoticed.

The Kaplans' approach is largely based on having subjects express their preferences, on a 5-point rating scale, for photographs of scenes that the researchers conceptualized as varying along two major dimensions. The first of these involves the extent to which the photograph allows the subject to *understand* the scene pictured, and the opportunity for *exploration* that it offers. The second involves the extent to which the photograph allows the subject to *infer* three-dimensional characteristics of the scene portrayed. The preferences that various groups expressed for photographs of varying degrees of wildness and domestication were assessed in terms of these two dimensions.

We immediately notice the emphasis that this technique places on the *visual* sense, which "eliminates some of the experimental 'noise' of the actual physical setting."[28] Now vision, as we have briefly noted and will further explore later, is of particular significance to the "perspective" of the world adopted by modern society. It is a perspective, we have noted, which distances us from the world and fragments it into particular dimensions—an essentially reductionist style that is consistent with scientific and industrial requirements. Thus the selection of *visual* stimuli immediately aligns the meaning uncovered within the study with existing forms of understanding, and leaves unnoticed all that other sensory information that tends to be dismissed as "experimental noise." However, it is precisely this "experimental noise"—the subtle scents, the unfamiliar silence and the faint sounds, the texture and feel of the vegetation—that has the potential to challenge us into finding new ways of relating to the world.

In defence of their methodology, the Kaplans argue that "people's responses to the two-dimensional representation are surprisingly similar to what they are in the setting itself," since "much of the information that we consider all the time reaches us by means of the two-dimensional representations of three-dimensional settings. When watching television or seeing pictures in a book or a painting on the wall people are not likely to say that the representation is deceiving."[29] However, these very revealing remarks locate the boundaries of the study firmly within the humanly constructed world; for it doesn't concern our relation to the natural world at all, but rather our relation to artifactual *representations* of this world. To equate a two-dimensional photograph of a natural scene with direct experience of the scene itself, because this is consistent with "daily human experience"[30] is to beg the wider issue of how this "daily human experience" configures our relation to the natural; for modern consciousness's reduction of nature to its two-dimensional representation is part of the fundamental problem we are here attempting to address. As Adorno pointed out, reification can be understood as the suppression of the *differences* between some aspect of the world and its abstract representation; so to begin by accepting such reification anchors us firmly within industrialism. Furthermore, participants in the study were often shown black-and-white photographs rather than color prints. This, we are told, "poses no problems," for participants "often 'read' color into the scene."[31] Such comments make it even clearer that we aren't dealing with the "real" world, but rather with those fantasies and memories of the world which the photographs stimulate. Although this research claims to be informing us about "the experience of nature," it takes place entirely within the humanly constructed world that is attempting to become independent of nature! The study is therefore permeated by that technocentric ideology that we are setting out to challenge.

The same criticisms, however, cannot be made of a second part of the Kaplans' research, which *does* concern the direct experience of nature.

This research evaluates the psychological effects on participants of an "Outdoor Challenge Program" in Michigan, during which they hiked into the wilderness, camped, and spent two days by themselves. And their evaluation of the effects of wilderness experience is strongly positive, and so potentially valuable to the environmentalist bent on arguing for the preservation of wilderness. These results are viewed from the perspective of the "restorative environment," implying that the function of wilderness is to counterbalance the harmful effects on individuals of a pathogenic society. "Difficulties abound," say the Kaplans, "even in an affluent and enlightened society like ours. . . . Current levels of family trouble, child abuse, and homicides are painful clues that much is not as it should be. . . . Restorative environments offer a concrete and available means of reducing suffering and enhancing effectiveness."[32]

These are worthwhile and important points. Nevertheless, there is a danger that the availability of the restorative environment will perpetuate destructive lifestyles in the same way that sleeping pills can enable people to ignore the fundamentally unhealthy way they are living. If wilderness is viewed anthropocentrically as a crutch to prop up an otherwise insufferable lifestyle, then the availability of wilderness in effect maintains practices that are destructive to wildness, both within ourselves and in the world outside. If the fact that wildness exists in particular "reservations" allows us to abandon its necessity in the remainder of the world, including our own lives, then we are reducing it to a sort of health farm that we can visit occasionally to obtain those intangible and indefinable benefits that industrialism cannot provide. Just as sixteenth-century adventurers sought minerals, plants, and slaves to bring back to Europe, so wilderness today has to justify its existence as a source of relaxation, tranquillity, and restoration. Within this framework, wilderness exists merely for anthropocentric reasons, to restore human well-being. While these purposes are not trivial or wrong, if they are allowed to exclude *other* justifications for the existence of wilderness, then wilderness becomes merely another human resource, having no intrinsic value and existing for no other purpose. The fundamental project of modernity—that of assimilating all to the humanly constructed world, and denying or annihilating what cannot be assimilated—remains unchallenged. However, any radical environmentalism must insist that the reasons for the preservation of wilderness, while they may include human needs, extend far beyond these needs. The threats and the attractions of wilderness derive partly from its embodiment of what *is* outside human understanding and human control; and one of the paradoxes thrown up by the limitations of our own rationality is that a nonanthropocentric appreciation of the world is ultimately necessary for our own well-being.

The acknowledgment of a world beyond socially constructed categories, and the intuition that we ourselves have roots in this world, constitute the bedrock of our ontological security.

The Kaplans' work should be acknowledged as an important attempt to extend the experimental method beyond the restricted confines of the laboratory. However, where it comes closest to escaping from these confines—in its study of the Outdoor Challenge Program—it would, ironically, be rejected on methodological grounds by many experimental psychologists. And where it is more faithful to the experimental paradigm, as we have seen, it construes the natural world as devoid of inherent intelligence and systemic character, as merely a collection of objects or dimensions that together accrue to form the "environment."

These characteristics are not accidental. If they were, experimental psychology's simplified conception of the world and its (human and nonhuman) inhabitants would be readily seen to be fallacious, and the discrepancy with the actual world would result in more accurate models being developed. However, while experimental psychology betrays the wholeness of a world in which the person and the rest of the natural world might be reasonably harmoniously integrated, it is largely consistent with the world of Descartes' dream—the world towards which industrialism is striving, the world of egoic rationality; and as this is the world popularized by advertising and by economic practice, it is not one that we can dismiss as simply "unreal." The foundations of psychology lie within Enlightenment philosophy rather than in the day-to-day experience of the person-in-context; so it should not surprise us that the covert aims of psychology are closely aligned with those of the whole industrial project. Only if we naively accept the experimentalist's claim to be "objectively" studying the person would we anticipate that psychology might offer a welcome and necessary corrective to the personally alienating and distorting influences of industrial life. What we find, instead, is that it is one of industrialism's chief accomplices, proposing and legitimating views of the individual that are consistent with the development of a market economy, and denying legitimacy to any cultural and spiritual structures that might challenge the hegemony of the economic determination of our lives. Experimental psychology, then, has a certain ironical truth value, as David Ingleby has pointed out:[33] it posits as "natural" a model of the person that is consistent with a domesticated world increasingly determined by industrial and commercial requirements, and in so doing it helps to bring about our conformity to this model. It follows that it is covertly ideological in the sense that it is hostile to whatever would challenge this industrialist hegemony, even if this hostility often takes the form of ignoring such challenges rather than actively attacking them. Experimental psychology is subtly antagonistic to all those realms that it implicitly denies; and these include much of human subjectivity and the whole complexly integrated world of the natural.

This situation is one that is problematic for environmentalism, which is faced not only with the gradual disappearance of the world that it seeks to defend, but also with the complementary shrinkage of forms of personhood that are best equipped to defend this world. It is not merely "charismatic megafauna" which are endangered; but also the poets, visionaries, and prophets who still sense their alliance with these creatures. Those aspects of individuality that protest at the devastation of nature are the visceral, intuitive, spiritual faculties; and our reaction to the sight of our favourite wood being bulldozed is not simply a calculated recognition of the loss of woodland or biodiversity, but also as a deeply embodied *emotional* recognition of a heartfelt loss. However, these arational faculties are invariably denigrated within psychology as in the rest of the modern world: we are taught to keep "cool," to think rather than feel, to despise intuition as "unscientific," to argue logically rather than emotionally. Just as what we define as the "natural" world is both denied methodologically and destroyed in physical reality, so the forms of personhood that could and, at least within indigenous populations, often have resonated with and defined themselves through this natural world are themselves repressed and obliterated. Today, the psychologist and the industrialist are as closely allied in the transformation of the world as the conquistador and the missionary were in Columbus' era. As mind and nature are intertwined within healthy ecosystems, so the agents of their destruction are covertly related, working for the same basic aims. These fundamental aims are the destruction of the fabric of the world, the division into mind and nature, human and animal, and so on, reinforcing the illusion of "naturally" separate realms through the occlusion of what unites them. The integrity both of the natural world and of the healthy human animal that could be a constituent of it has in part to do with their interrelation and the way they are conjointly defined; so the denial of this interrelation generates impoverished accounts of both. Psychology's active support for a form of individuality that is consistent with industrialism is therefore misleading in two crucial, and related, respects: firstly, in the implication that the person studied as an isolated entity separate from culture or nature is either whole or healthy; and secondly, that alternative forms of personhood are somehow necessarily deficient. The failure of psychology to offer a framework that articulates what we already, at some level, know leaves us with no basis save our own subjective experience to challenge such repressive definitions.

Social Psychology

Social psychology, it is fair to say, has not been one of the most epistemologically sophisticated areas of the discipline. Its methodology has been, and still is in some departments, conventionally experimental and, ironically, thoroughly individualistic in its frequent recourse to hypothetical internal states such as attitudes. As Gergen assessed this approach at the end of the 1970s, "[the experimentalist] is tragically deceived; experiments are largely worthless, except as descriptions of the way people carry on in trying to make sense of the impoverished environment of laboratories."[34] Such criticisms contributed to the growing unease felt by some social psychologists about the way their discipline seemed to be reproducing all the least attractive aspects of positivist methodology, leading to a crisis within the discipline and a subsequent fragmentation into various theoretical sects. Aside from those who chose to remain true believers in a conventionally empirical style of research, some turned increasingly to language for an alternative structure on which to base an understanding of social behavior. Others welcomed postmodern literary theory and the study of gender as nuclei around which alternative theoretical conceptions could crystalize. What unites these otherwise quite disparate approaches is a common rejection of the "objective" pretensions of empiricism in favor of a view of the individual as constituted socially and, sometimes, historically.[35]

The widely recognized difficulties of an empirical approach that attempts to reproduce the physical sciences' ideal of objectivity should, on the face of it, suggest that environmentally concerned social scientists might welcome these alternative approaches with open arms. Experimental social psychology's obvious allegiance to the methodology of the natural sciences makes it distasteful to most environmentalists, even if its subject matter seems of little relevance to their concerns; so any new approach that might loosen the stranglehold of empiricism on psychology should surely be beneficial. Nevertheless, given that academia is riddled by paradigms that are ideologically suspect to the alert environmental theorist, it is important that we carefully examine the assumptive bases of these new variants of social psychology to make sure that we are not escaping from the positivist frying pan only to plunge into the postmodern fire. As we will see, our caution is well founded.

A social psychology acceptable to environmental theory would be one that healed the disjunction between the social and natural worlds, recognizing that while social factors significantly influence our reactions to and interpretation of nature, the natural world is also an important constituent of our being and of the natural phenomena we interact with.

In Merleau-Ponty's words: "Everything is cultural in us . . . and everything is natural in us."[36] We are not *only* social beings: we are also a particular sort of animal, with an evolutionary history which has unfolded in conjunction with the whole natural world. However, modern industrial society is built on the widespread denial of the second component of this dialectic, pretending that technology allows us freedom from natural forms and constraints—a pretense that environmentalists are in a good position to debunk.

It might be argued, however, that social psychology is justified in ignoring the natural order, since this order now has so little apparent bearing on the way we live. If we were to accept this argument, we would be promoting a psychology that has abrogated all critical potential, accepting its role as a merely descriptive discipline that colludes in the anthropocentric pretensions of industrial society. Such a chameleonlike discipline, which takes on the ideological coloring of its surroundings, is clearly unacceptable to those who challenge the overall direction of industrialism in consuming the natural world. An adequate social psychology, rather, will be one that insists on the potential importance of the natural in human social life, even while present conditions nullify this importance. It will be able to encompass not only what *is*, but also what *could be*, and will be able to recognize and critique the processes whereby the latter is reduced to the former. Anything less will be a betrayal both of human potential and that of the whole natural world.

Social Constructionism

It is disappointing to find, then, that even within those new and comparatively radical critiques of mainstream social psychology that see the individual as existing within and constituted through a social context, there is no suggestion that we are also located within and constituted by a *natural* context; and this limitation is especially marked in the social constructionist approaches that have proliferated in recent years. While such approaches can be seen as reflecting a welcome swing of the pendulum away from the biologism of the first half of this century, they maintain the dissociation between the social and natural worlds as resolutely as before, simply insisting on the dominance of the former rather than the latter, and suppressing any possible dialectic between them. To deny that social behavior may reflect, mediate, and articulate natural forms is a claim of profound significance, since it implies the existence of a realm of human life independent of nature, and so reproduces the centuries-old dissociation between the natural and the social. The emergence of social constructionism may in certain respects amount to a revolution in social psychology; but like many revolutions, it perpetuates some of the crucial features of the approach it claims to supersede.

The particular aspect of social life that constructionists often focus on is language, so that meaning, allegedly, is not *expressed in* language, but is *generated by* it. As Terry Eagleton, a leading proponent of this approach, explains, experience can have no other source but language, and "our experience as individuals is social to its roots."[37] Now, this movement away from an assumed objectivity might seem to offer a desirable liberation from the usual positivistic assumptions; but unfortunately it introduces problems that by now will have an all too familiar ring to them. Not the least of these for our purposes is the denial of the natural world as a potential source of experience, understanding, or morality, however these might be mediated by social factors. The social realm, and language in particular, are conceptualized as comprising a "human" realm from which nature is entirely excluded, locating this approach disappointingly firmly within a particularly narrow version of the anthropocentric tradition. The natural is not merely downgraded as *res extensa*, matter in motion, or whatever; its very existence independent of the human realm is denied.

This exclusion of the natural, for example, is quite explicitly advocated by Michael Billig. He approvingly quotes Karl Popper's opinion that the origin of the social sciences can be traced to Protagoras' distinction between the natural and the social, arguing that "questions about the existence or nonexistence of unchanging realities can be left to one side,"[38] and that "we must concentrate on the one power which separates humans from all those other organisms: the power of language."[39] But why should the social sciences be based on the "one power which separates humans from . . . other organisms"? Why not, instead, base it on the *many* powers that *relate* us to other organisms? The implication of this curious choice seems to be that psychology can take place within a *purely* social realm dissociated from the natural, and that we are therefore justified in ignoring our relation to the natural world together with any potential implications that nature might hold for social behaviour. According to Billig, we should focus on the dynamics of the opposing rhetorical views; and it is the argument and counterargument between such opposed views that is the real stuff of life rather than any natural reality which we might argue *about*. Thought is a by-product of this rhetorical universe, since "[h]umans do not converse because they have inner thoughts to express but they have thoughts because they are able to converse."[40] This view is powerfully repressive in effect, since it denies the existence of experience and thought that cannot be expressed in language, sealing us within this constricted linguistic world. As André Gorz points out,

> Language is a filter which always forces me to say more or less than I feel. Learning one's language is a form of original violence done to lived experience; that process forces these experiences for which there are no words to remain silent, while I am forced to express meanings which do not correspond to my experience. . . . It forces me to substitute a discourse which is not my own for the one it forbids me. It is a form of discipline and censorship and induces us into inauthenticity, pretence, and play-acting.[41]

Gorz here implies that the source of "lived experience" is not language, but is a real world that extends beyond the human; and that the proper role of language is to attempt to convey the character of this world and the forms of life that it makes possible. But the constructionist reverses these relations, arguing that reality is a product of an exclusively human realm. Vivien Burr, for example, suggests that "what we regard as truth . . . is a product not of objective observation of the world, but of . . . social processes and interactions";[42] and Peter Mason asserts that " 'reality' itself is a product of the activity of our imagination."[43] Similarly, constructionists "question the assumption that science is about nature as it exists outside us." Rather, "scientific paradigms are socio-historical constructs—not given by the character of nature, but created out of social experience, cultural values, and political-economic structures."[44] Nature, according to this view, has no inherent structures or patterns of its own—a notion often criticised by constructionists as "essentialism"[45]—but is structured *discursively*. The "dubious" logic of nature, suggest William Chaloupka and R. McGreggor Cawley, must therefore be replaced by "rhetoric."[46] "Nature," then, becomes merely an artefact of this linguistic reality, a flimsy by-product that owes its very existence to our ability to speak; and as Jane Bennett and Chaloupka argue, "nature, like everything else we talk about, is first and foremost an artefact of language."[47] In these terms, language is seen not as *representing* nature more or less adequately, but rather as *constituting* it, so that "any attempt to invoke the name of nature . . . must now be either naive or ironic."[48] It follows that the diversity of languages and cultures will generate a corresponding diversity of "natures," so that as William Cronon puts it, it "hardly needs saying that nothing in physical nature can help us adjudicate amongst these different visions [of nature], for in all cases nature merely serves as the mirror onto which societies project the ideal reflections they wish to see."[49] Nature, then, is simply another artifact, so that wilderness, for example,

> Is not a primitive sanctuary where the last remnants of an untouched, endangered, but still transcendent nature can for at least a little while longer be encountered without the contaminating taint of civilisation. Instead, it is a product of that civilisation, and could hardly be contaminated by the very stuff of which it is made. Wilderness hides its unnaturalness behind a mask that is all the more beguiling because it seems so natural.[50]

Discourse analysis in particular and social constructionism in general, therefore, construe all structure as emanating from cultural forms such as language, so that reality is a product of social life rather than empirical assessment, and the entities and forms that science identifies are "constituted through the artful creativity of scientists."[51] However, as Holmes Rolston argues,

> The sporophyte generation of mosses is haploid. Malaria is carried by *Plasmodium* in mosquitoes. Neither of those facts is likely to change with a new cultural filter. Golgi apparatus and mitochondria are here to stay. There is no feasible theory by which life on earth is not carbon-based and energised by photosynthesis, nor by which water is not composed of hydrogen and oxygen, whose properties depend on its being a polar molecule.[52]

Science, then, may be a partial understanding that we often fatefully misconstrue as being a *complete* description of nature; but it is nevertheless firmly anchored in realities that are beyond the influence of language.

Furthermore, if constructionism denies that social life is framed within a natural context, it also ignores its *historical* context; for language originally grew out of our interaction with the world, and it is comparatively recently that European languages have developed a more nominal relation to reality. This development can be seen as part of a more general process wherein natural structures are replaced by economic, conceptual, and linguistic structures, so defining an allegedly autonomous "human" realm. The constructionist prioritization of language embodies this historical myopia, ignoring language's role during most of history as representing, imitating, and growing out of a preexisting natural order. It is only since mediaeval times that, as Timothy Reiss argues at length in his formidable study of the evolution of modern language, the "discourse that names and enumerates becomes, replaces, the order of the world that it is taken as representing."[53] Social constructionists, therefore, take as a *fait accompli* the outcome of a particular historical project—that of "industrial man"—and make this outcome the foundation of a general theory of human existence. As a result, they unwittingly *embody* this trend and are themselves shaped by it, losing any critical perspective on it as a particular historical development, and making any alternatives appear inconceivable. Ironically, this taken-for-granted primacy of language and the corresponding abolition of natural structure is celebrated by postmodernists as a liberation and by social constructionists as a revolutionary theory. Such theories are blind to long-term historical trends, allowing industrialism to colonize the future by limiting possibilities to those that are consistent with an ideologically defined present. Under these circumstances, the extinction of the natural order becomes a present inevitability rather than a future possibility.

This failure to relate theory to long-term cultural trends—a failure that is replicated fairly generally within psychology—is exacerbated by an almost exclusive reliance on "recent" research, supplemented by scattered references to ancient Greek philosophy. There is therefore a sort of U-shaped curve that governs the acceptability of psychological research: only the most recent or the most antique is acceptable, and almost no reference is made to research carried out in, say, the first seven decades of the last century. In this psychology faithfully replicates the fashions of consumer culture: only the very new and the very old are valuable, and anything that is merely "out of date" is consigned to the rubbish tip. Consequently, there is little awareness of the relations between cultural and psychological fashion, and social psychology, like schools of fishes that suddenly but uniformly change direction, tends to oscillate between poles such as biologism and constructionism. It is as if Gergen's paper on social psychology as history[54] had never been written.

Constructionists' lack of historical awareness is particularly ironic when the natural origins of language are indicated in their own writing by the metaphors that creep in unnoticed like weeds through a paved driveway. Alternative approaches, says Billig, "find themselves *bogged down* in other issues";[55] and the use of a particular phrase will "indicate the seed, if not the flower, of an argumentative position."[56] But the significance of such metaphors remains unrecognized, even as they imply the rootedness of language in a natural world whose existence is simultaneously denied. We ignore this rootedness at our peril, for as Barry Lopez puts it, "[t]he landscape is not inert; and it is precisely because it is alive that it eventually contradicts the imposition of a reality that does not derive from it."[57]

Few environmental writers would claim that direct, unmediated contact with nature is possible, or deny that our understandings of nature are affected by our cultural background, training, language, and so on. By analogy, most of us would accept that our perception of an animal will be affected by the type of binoculars we use. However, many would be less willing to accept the claims that the creature is *constructed* by the act of looking through the binoculars, and that it has no independent existence outside this act—claims that reflect what Roy Bhaskar has referred to as the "epistemic fallacy," or the view that "statements about being can be reduced to or analysed in terms of statements about knowledge."[58] The confusion between these two quite different types of statement has sometimes spread to environmental writing. For example, it is difficult to disagree with Philippe Descola's recent suggestion that "many anthropologists and historians now agree that conceptions of nature are socially constructed"; but two pages later, this statement has mutated into the much more debatable assertion that "nature is socially constructed."[59] As Andrew Collier ruefully notes, "the kind of idealism which treats the world as dependent on our cognitive choices . . . has really come into its own"[60] in recent decades.

Unsurprisingly, social constructionists frequently retreat from these extreme claims toward a view that is more consistent with Bhaskar's "critical realism." For example, Ulrich Beck's influential *Risk Society* vacillates between the claim that environmental problems are social constructions that have no reality independent of our understanding of them, and the quite different view that objectively measurable environmental problems are making living increasingly risky.[61] Similarly, Phil Macnaghten and John Urry, as we saw earlier, oscillate between a dispassionate, relativist social constructionist discussion of "the notion of nature as threatened," emphasizing that "what is viewed and criticised as unnatural . . . in one era or one society is not necessarily viewed as such in another," and more embodied, realist references to "the alarming rate of natural destruction caused by urban growth."[62] John Hannigan also has his epistemological cake and eats it, claiming that "environmental problems and solutions are end products of a dynamic social process of definition, negotiation and legitimation" while also cautioning us on the same page not to "deny the seriousness of the threats faced by our planet."[63]

Furthermore, as the bizarre implications of some of their claims sink in, some constructionists have withdrawn toward so-called "weak" versions of constructionism.[64] In some forms, these claim simply that our construals of nature are influenced by culture and language, and so offer no insights that would supplement a critical realist perspective. Other versions perpetuate in diluted form the problems of "strong" constructionism: Sergio Sismondo, for example, argues that "a distinction can be drawn roughly along the line of meaningfulness: social objects must be meaningful, whereas material objects are only meaningful when they are incorporated into the social."[65] Many environmental writers, however, would object that the meaningfulness of the natural world is not simply dependent on its *social* significance. What is more, even "weak" constructionists continue to dismiss any reference to "nature" as a determining factor in our lives as "essentialism," supposedly suggesting an unchanging core unmodifiable by social factors. This misrepresentation of any model of the human that includes a natural component as "essentialistic" parodies the influence of the natural world, implying that "nature" is a static, isolated factor that has no dialectical relation to the social. Far from emphasizing a more realistically vital, dynamic interrelation between the social and the natural, the influence suggested by the term "essentialism" is one of rigid determinism. Thus the wonderful diversity of nature in all its manifestations is constricted to a single dimension defined as external to social life; and this meager parody of the effects of nature is then accused of ignoring the influence of the social! If there were indeed "nothing beyond the text," then the natural world indicated by this use of language would be a poor thing indeed. One wonders whether social constructionists have ever got lost in a real wilderness, been bitten by a non-discursive mosquito, or taken shelter from a genuine thunderstorm.

Thus the interplay between social life and humankind's place in the natural world remains untheorized and unacknowledged, the implication being that beyond providing a "site for a repertoire of definitional and contestatory activities,"[66] the natural world plays no part in determining the course and nature of our lives. In place of experimental psychology's decontextualized individual, we therefore have the decontextualized society. While there may be a certain ironical truth value to this view, in that our immersion within a largely urban, manufactured environment disguises and violates our rootedness in the natural world, to accept this environment as our starting point for psychological explanation is like viewing the behavior of a caged cheetah as fully expressing its inherent potentialities. While the importance of *social* context is becoming more generally recognized in the understanding of behavior—for example, in the application of systems theory to such areas as family therapy,[67] and in recent attempts to include anthropological insights within psychology and psychiatry,[68] this awareness has yet to generalize to a recognition of the problems produced by separating the "individual" from the "natural environment."

One of the advantages claimed for a social constructionist viewpoint is that in its emphasis on "competing voices" in the "negotiation of reality," it avoids the imposition of any single, monolithic vision, and so "celebrates the diversity" of postmodern life. And within a restricted arena, this is undoubtedly correct—a characteristic that has made it attractive to a few environmental philosophers. Unfortunately, as I have argued above, the terrain that an adequate environmental philosophy must cover needs to be much broader than the specific case of modern life that it critiques. Modernity is in many respects the problem to be addressed, not the context within which we seek a solution. Our imaginative reach must be one that can extend not merely to the diversity of currently existing opinions, but one that is also capable of comprehending the way these opinions are formed by a particular cultural and historical situation. This is a necessary initial step toward envisioning not only those discourses and conceptualizations that are inconceivable today, but also the ecocultural conditions within which these might become possible. Given the widespread destruction both of the natural world and of the cultural structures that are consistent with nature, the aims of an effective environmental philosophy should extend not merely to defending the ecological and subjective diversity that *already* exists within the modern world, but also to nurturing human awareness in order to maximize the ecological and conceptual diversity that can *potentially* be realized in the natural world when social conditions allow. In this respect the maintenance of conceptual diversity is closely linked with the vital work of environmental activists and others in maintaining genetic diversity and preserving endangered species; for if the *psychocultural* conditions that correspond to and resonate with a healthy *ecological* realm are absent, then those natural species that remain will one day be seen merely as relics of an bygone era rather than as vital ingredients of a healthy natural world. It is a commonplace insight that we need to preserve habitats as well as species; but nature is mindful as well as physical, and it is just as essential to preserve the psychical component of the natural world as its material manifestations. Not only must species be preserved, but also their meaning within a natural world that is psychologically as well as ecologically alive.

It is these aspects of a potentially healthy ecocultural world that approaches based on language cannot incorporate. We have to look *beyond* current forms of language and their implications if our imagination is not to be held within the orbit of industrialism. Furthermore, and not coincidentally, since post-Renaissance language has become so markedly separated from the natural order,[69] it is increasingly difficult to articulate natural form through conventional language; and we need to incorporate and develop other ways of communicating such form. As usual, Gregory Bateson was ahead of his time in recognizing this, asking whether we can change our "understanding of something by *dancing* it."[70]

In addition, the deconstructive bent of discursive approaches limits their capacity to challenge the structure of modern industrialism. Just as science has been reluctant to recognize the holistic qualities of nature, so we have been slow to appreciate that the power of industrialism and its resultant near-hegemony in the modern world is largely the result of its ability to integrate science, politics, and everyday social life within a structure that appears complete and self-sufficient. This structure cannot be challenged without reference to alternative structures. To celebrate choice and free play without also celebrating the frames of meaning within which they take place is simply to guarantee our assimilation to and absorption within industrialism, and so represents a philosophy of surrender. For example, "freedom" has little meaning in the absence of a framework of democratic laws which protect the vulnerable against the "freedom" of the powerful to exploit, intimidate, and mislead. Similarly, my freedom to explore an area of wilderness is negated if energy companies and off-road vehicle clubs also have the freedom to use the area as they see fit. Freedom is all too often interpreted as the absence of structure; and structure gives meaning and implies responsibilities and limitations. One of the most insidious aspects of the colonization of the world is industrialism's silent but lethal elimination of structures that could challenge it. The widespread lack of appreciation within academia of the way in which postmodern approaches involving deconstruction promote this insidious *conceptual* assimilation to industrialism is an index of the urgent need to develop a psychocultural dimension to our environmental understanding.

Finally, we should not ignore the possibility that an emphasis on language serves particular defensive functions for the social scientist. Noam Chomsky has noted that if "it's too hard to deal with real problems," some academics tend to "go off on wild goose chases that don't matter . . . [or] get involved in academic cults that are very divorced from any reality and that provide a defense against dealing with the world as it actually is."[71] An emphasis on language can serve this sort of defensive function; for the study of discourse enables one to stand aside from issues and avoid any commitment to a cause or ideal, simply presenting all sides of a debate and pointing out the discursive strategies involved. As the physical world appears to fade into mere discourse, so it comes to seem less real than the language used to describe it; and environmental issues lose the dimensions of urgency and tragedy and become instead the proving grounds for ideas and attitudes. Rather than walking in what Aldo Leopold described as a "world of wounds," the discursive theorist can study this world dispassionately, safely insulated from the emotional and ecological havoc that is taking place elsewhere. Like experimentalism, this is a schizoid stance that exemplifies rather than challenges the characteristic social pathology of our time; and it is one that supports Melanie Klein's thesis that the internal object world can serve as a psychotic substitute for an external "real" world that is either absent or unsatisfying.[72] Ian Craib's description of social constructionism as a "social psychosis"[73] therefore seems entirely apt. But what object relations theorists such as Klein fail to point out is the other side of this dialectic: that withdrawing from the external world and substituting an internal world of words or fantasies, because of the actions that follow from this state of affairs, makes the former even less satisfying and more psychologically distant, so contributing to the vicious spiral that severs the "human" from the "natural" and abandons nature to industrialism.

Although I do not have the space here to offer a thorough critique of constructionism,[74] I have attempted to show how this approach reproduces many assumptions that are crucially implicated in environmental destruction. We see the denial of natural realities and the corresponding allegiance to an implicitly separate human realm. We see a covert consistency with industrialism. And we see how social constructionist methodology enables the practitioner to remain detached from the emotional realities associated with degradation of natural form. We will have to look elsewhere for an understanding of the human psyche that does justice to the interplay between social and natural spheres.

Mental Health and the Natural World

Within the fragmented epistemological landscape of the modern psyche, clinical psychology and psychiatry might at first glance appear to have little relevance to the destruction of nature. Only at first glance though; for the predicament of the natural world is associated, as we noted in chapter 1, with a transformation of "the defining human element in individuals—their moral, aesthetic, and religious experience";[75] and this transformation has clear implications for mental health. As we will see, both these forms of destructive change can be traced to the colonization of the life-world by the exotic ideology of technologism. But this is to get ahead of our story. Let us begin by exploring the psychological effects of the ideology that separates the "individual" from "nature."

As we saw earlier, the view that there is no alternative to the individualistic style of selfhood is, given the current social context, not without a grain of ironic truthfulness. If the structure of selfhood is dialectically co-defined with the structures of other realms such as the political, the social, the economic, and so on, then the whole forms a hegemonic system that is necessarily difficult to challenge or escape from. But the awareness that we inhabit an ideological prison should not be allowed to limit our imagination to what exists within that prison. It may indeed, in certain respects, be "unhealthy" to disagree with the authorities who maintain the prison, but this does not imply that health should be defined by the extent of our acquiescence to them; and a more authentic health may flow from an entirely different kind of existence within an entirely different kind of environment.

For example, Sampson has pointed out that modern society highly values the capacity for a cool, abstract detachment from some desired situation or entity, and that psychological research such as Mischel's work on delay of gratification legitimates and reinforces this social value. Thus as children grow up, says Sampson, they learn "how to transform an object of interest into an object of disinterest. Cool ideation substitutes thinking for having." Thus "Mischel's hero [is] the person who denies reality in the name of a cognitive dream,"[76] so that the refusal of gratification by some outside authority such as the parent is gradually replaced by the *internalized* denial of gratification that we regard as an index of maturity. Sampson accepts that such "cool" ideation is in various respects better adapted to existing social conditions; but he also points out that the cognitive abandonment of what is desired aligns individual psychology with these particular social conditions, which hereafter are simply regarded as "reality." "What has occurred is a psychological reification," argues Sampson. "A social and historical process has been translated into a fundamental psychological process."[77] What is abandoned during this learning process, then, is not merely a particular form of gratification, but more generally, the impulse to strive toward a world that is more attuned to one's needs and desires. Thus mental "health" is defined simply in terms of our allegiance to a specific social order, and fails to recognize that social reality, in a healthy world, should *also* express and articulate biologically given needs, structures, or preferences. Clinical psychology views "normality" and "psychopathology," then, as derived from a particular, industrialist order; and so the discipline is oblivious to forms of health that could become possible within alternative, imagined, contexts.

But in this respect we need to proceed carefully. I have already pointed to the schizoid dissociation between the intellectual realm and the remainder of the world that the intellect takes as its object; so in envisioning "alternate, imagined contexts," are we not aligning theory precisely along the schizoid lines that I have criticized? The answer, I think, depends on how we go about the process of "envisioning." If this process takes place largely through autonomous intellectual processes rather than through a searching dialectic with the natural world, then the criticism is justified. We would, in effect, be perpetuating Descartes' search for a "more perfect world" that displaces the existing one. But if, on the other hand, our 'envisioning' derives from a dialectic with nature—nature both in the natural world beyond our bodies as well as within our own physicality—then this becomes a project very different from the schizoid fragmentation of "Descartes' dream." It is, rather, a journey of reintegration, of recollection, of reincarnation, based in an embodied experiencing that anchors the intellect to natural structure. In chapter 7, I will explore ways in which just such an embodied experiencing can reawaken our participation in the natural world.

For clinical psychology, however, the mentally "healthy" individual is simply one who functions effectively within the modern industrial world, and whose life is therefore based on alienation from and exploitation of nature. Any possible connection between mental health and one's relation to the natural world is obscured within clinical psychology by *defining* psychopathology as essentially individual in nature. Thus DSM IV identifies "mental disorder" as "a manifestation of a behavioral, psychological, or biological dysfunction *within the individual*",[78] so severing our understanding of problems such as depression from their natural and cultural contexts, despite research that clearly demonstrates the primacy of context in determining the onset and character of such disorders.[79] But if the normality of our day-to-day lives depends on exploiting and degrading the natural order, then *psychological* "health" will embody an intrinsic *ecological* pathology; and human life is defined as a form of parasitism. Recognizing the destructiveness of this inverse relation between individual "health" and the welfare of the natural world, then, it becomes clear that we need to redefine individual health to include a constructive resonance with the world "outside" us.

This is not, of course, to suggest that *any* divergence from current definitions of individual health will necessarily be ecologically desirable; and the issue of precisely *what* sort of divergence is healthy is a complex one. What *is* clear, however, is that clinical psychology's portrayal of the statistically normal present-day configuration of selfhood as "natural" and healthy rests on an ahistorical perspective that is blind to the way current styles of subjectivity have been caught up in the industrialist transformation of the world. Thus a form of self that may be seen as a less-than-happy compromise between the biologically given predispositions of the person, on the one hand, and the demands of an economic system that is increasingly unconcerned with these predispositions, on the other, is presented to us as the only possible configuration. Deviations from "normality" are seen as due either to the intrusion of specific pathogenic influences from outside or to innate predisposing factors, so concealing the possibility that psychological distress may often be largely due to the basic incapacity of a social and economic environment run for the benefit of capital to meet our fundamental psychological needs. If, for example, my need to pay the mortgage demands that I work in a telesales center, cut off from natural rhythms or light, my behavior controlled by rules set by supervisors and by incoming calls, then my tendency to feel depressed may have less to do with any latent psychopathology or specific "stressor" than it has to do with the impossibility of fulfilling my deeply intuited need to be part of a world that makes sense in a bodily, spiritual, and ecological way. Complementarily, the "ability" to tolerate such inhuman conditions without becoming depressed may also be understood, when seen within a broader ecological context, as a sort of numbed acquiescence to a slow spiritual and ecological death.

While the *immediate* cause of psychological distress, therefore, may without too much inaccuracy be identified as an event such as redundancy or death of a loved one, and while a "genetic predisposition" to a specific type of disorder may influence the particular form of somatic protest that occurs, this whole drama is played out against the backdrop of a society in which individuals are *already* dissociated from sources of meaning and support such as a stable natural environment and reliable communal links, and in which the meaning structures that might diffuse the stress of traumatic events are dilapidated or moribund.[80] Within such an individualistic society, we will inevitably be vulnerable to ontological insecurity; but the sources of this insecurity, instead of being clearly articulated, will be mapped onto landscapes of consumerism and psychopathology in such a way that "feeling good" becomes identified with possessing the right consumer goods, moving in the right social circles, or seeing the right psychotherapist. None of these, however, is in the least bit likely to enable us either to understand the origins of our psychological distress[81] or to begin to act in ways that might transform our situation.

A basic problem here is that the reach of consciousness is usually too feeble for us to comprehend the role of those indirect factors that David Smail has referred to as "distal."[82] The idea that my unhappiness will be relieved by "proximal" changes such as a new car, a new lover, or a course of psychotherapy is much more consistent with "common sense" than the notion that it has something to do with my alienation from the natural order or the absence of cultural frameworks that could give my life a more profound meaning. Similarly, practices that are promoted as "green" but that are nevertheless ecologically questionable, such as buying lead-free petrol, are simple enough to grasp consciously; but understanding that our transport system and the way of life it supports are inherently damaging is a more elusive idea that is less amenable to rational analysis. Interactions between the psyche, the social order, and the natural world are inherently systemic, involving an enormous complexity of causation that makes them difficult to grasp consciously. The commercial world's recognition of these limitations of consciousness makes us easy prey for advertisers who sell goods and services that seem to offer solutions to personal and environmental problems, but in fact don't deal with the real causes of these problems. If, for example, I am feeling guilty, worthless, or anxious, a course of cognitive therapy may help me to understand these feelings as "irrational"; but it is unlikely to enable me to recognize the source of my disquiet over the way my lifestyle lays waste the world my children will inherit, or my awareness that the designer clothes I wear are made by people living in poverty, or because my deeply perceived disharmony with the natural world makes me feel terrifyingly, if apparently inexplicably, isolated.

Even when the significance of social and occupational factors in the causation of "mental illness" has been recognized,[83] the roles of "genetic" and "environmental" factors are often mapped out in a way that assumes that their prior separation is a fact of life rather than an ideological construction. For example, the aetiology of schizophrenia is conventionally viewed in terms of an interaction between genetic predisposition, on the one hand, and "psychosocial stressors," often located within the nuclear family, on the other. But what is concealed here is the relationship between these two classes of influence. A particular "genetic predisposition" may incorporate an incapacity to adapt to the specific, and arguably less than ideal conditions offered by urban society, insisting on the need for a way of life that is currently unavailable. This genetic predisposition therefore carries a concealed *social* component. It is not a genetic predisposition to, say, schizophrenia as such, but rather a genetic preference for one sort of environment rather than another. Genetic predispositions are therefore relative to particular ways of living: genotypes that are adaptive in one environment may be disastrously unfitted for another. For example, one of the "risk factors" in schizophrenia has been found to be "emotional responsiveness."[84] In an urban world saturated with fast-moving traffic, huge amounts of information, ever-changing occupational demands, and a good deal of sheer physical noise, it is not surprising that an emotionally responsive individual would often feel bewildered and overwhelmed by the number and intensity of stimuli demanding their attention. Under these circumstances, the ability to ignore a large proportion of this input and to remain detached will be essential for emotional survival. On the other hand, if the individual inhabited the sort of environment within which our nervous systems have mostly evolved—that is, one that is closer to a wilderness situation where attentiveness to every sound might prolong one's chances of survival—then the "risk factors" identified by researchers such as Mednick might instead become predictors of *survival*. Such examples make clear what the abstract discussion of "genetic predispositions," "risk factors," and so on conceal: that risk is relative to the lived character of specific environmental conditions, and that to separate "genetic" factors from "environmental" ones is radically to misunderstand the character of both.

A further aspect of Mednick's research leads to related conclusions. Those at risk for schizophrenia tend to have flatter "associational hierarchies" than those who are not at risk: that is, they have a greater tendency to follow associational rather than strictly logical patterns of thought. Now, rational thought requires that we exclude "unwanted" associations, since it involves holding in mind particular words or concepts while excluding other ideas that are not "rationally" related. For example, if we are assessing the quality of timber that we could extract from an area of forest, it is not helpful to allow our thoughts to wander to the dream we had last night, or the effects of extracting the timber on the local wildlife, or to the noise of the chainsaws. Nevertheless, in other cultural contexts, such associations might be anything but irrelevant. Among the Lele of the Congo, for example, behavior in the forest is guided by a range of factors such one's state of physical health, or whether one has been recently bereaved or experienced a nightmare, or the day of the week, or by the taboo on noise.[85] Such arational associations may be more than the mere superstitions we often regard them as. For example, a taboo on feeding the remains of animals to herbivores—which would not until recently have been regarded as having any rational basis—would very likely have prevented the BSE crisis in Britain during the 1990s. In some situations, then, the attentiveness to associations exhibited by Mednick's "high risk" group might be a positive *advantage*, allowing insights and awarenesses that are more than rational, and that may integrate the individual into social and ecological context. Indeed, Mednick himself recognizes the advantages of associational awareness, proposing that creativity is based on associational fluency.[86] Thus what is a genetic *disadvantage* in situations demanding conventional "rational" thought can become an *advantage* in those other situations that require us to recognize and articulate patterns that go beyond the simplified order of technological rationality; and adding a dimension of historical and cultural variation to the static dichotomy of environmental and genetic factors allows us to perceive that both are engaged in the same dance.

In a similar way, "environmental" factors are conventionally limited to deviations from a norm that is taken for granted within modern society. Thus job loss, the absence of nurturing relationships, or our constant exposure to a relentless deluge of trivial information may rightly be recognized as pathogenic; but the factors that underlie them, such as the fragmentation of cultural structures, the absence of relation to a particular place one can call "home," or our subjection to a reality that is increasingly economically defined, are not. It is as if we recognize the ripples on the surface of the ocean, but are blind to the oceanic currents that ultimately have a much greater influence on the course of our lives. And it is within such "oceanic" currents that nature and the cultural world are most intimately entangled. For example, seldom recognized factors such as loss of contact with the rhythms of the natural world—or, for that matter, recognised ones such as the lack of adequate mothering—cannot simply be pigeonholed as "environmental" factors, to be rectified by an occasional walk in the country or more frequent physical contact between parent and child. They are, rather, pointers to the much more general integration that we unconsciously demand with the natural order, and so imply a necessary consistency between the social and natural matrices within which we live.

The separation of "environmental" from "genetic" factors suspends the eons-old dialectic between genotype and context, so that the self's relation to its environment becomes unhistorical, suspended in a sort of statistical aspic. Thus "environment" becomes a static entity; and "genetic" factors "interact" with this "environment" only in the present, and in a way that says nothing about their prior history. The dance of genes and environment is frozen, their coupling reduced to trivial atemporal effects; and the person becomes a lonely actor within a world that is passive and motionless. The modern self, together with modern society, are taken to be free-standing entities without prior history or context; and the dynamic processes of their co-evolution, contemporary dissociation, and potential reintegration are stilled within a snapshot image that seems to invite our conscious intervention. Within this frame of reference, *current* forms of selfhood become the only possible ones, and so any criticism of these forms can be attacked as "misanthropic," as a rejection of humanity itself. The accusation of misanthropy is therefore blind to the enormous variation in personhood—both existing variation and potential variation. Human adaptability and variety are simultaneously our greatest strength and our Achilles' heel: our strength, because we can ingeniously adapt to many different habitats and living conditions, and a weakness, because we can survive in conditions that are both inhuman and unnatural. Those conditions under which we can, in a purely numerical sense, flourish may also be those under which aspects of life—both human and nonhuman—that are most precious will become extinct. No longer stabilized and located by an unconscious rootedness within natural structures which surround us, we are instead marooned within the anthropocentrically constructed island of economic rationality with which consciousness has become aligned. And in destroying the natural world that is co-extensive with what is "unconscious" within us, we are burning the boats that might enable us to escape from this island.

However, since we are largely identified with this rationality, we try to explain our feelings of insecurity in terms of the conscious concepts and language it offers. Hence the spectrum of psychological disorders is understood as originating only within the familiar anthropocentric world of neurotransmitters, family relationships, stress reactions, "endogenous" (i.e., inexplicable) depression, genetic predispositions, Oedipal conflicts, and so on. Such notions *are* sometimes useful and valid in understanding psychological distress; but the understanding they convey is a partial one that ignores important dimensions of our existence. Even when the world speaks to us through unconscious processes such as dreams, it is usually assumed within the psycho-professions that these are symbolic representations of intrapersonal or interpersonal conflicts. As psychoanalytic orthodoxy would have it, for example,

> The great dinosaurs become the devouring mother, and the 5-year-old's fascination with them reflective not of his awe at these magnificent primitive beings, but of salient Oedipal conflicts. Bears, alligators, and snakes as they appear in our dreams and fantasies are reduced to symbols telling us of our preoccupation with human affairs—the only preoccupations deemed valid or even possible.[87]

The meaning of such symbols, then, becomes a merely human meaning; and such anthropocentric explanations struggle to explain the disorders characteristic of our materially abundant civilization: the widespread depression, the narcissism, the alienation. All these can be understood in terms of our loss of contact with the natural structures and processes within which we are unconsciously grounded, and that are themselves fading with the destruction of the natural world. We are finding less and less resonance between ourselves and what is beyond ourselves as natural structures are denigrated or ignored within most academic writing and physically destroyed in the world beyond academia. We project the consequent feelings of loss onto any situation that seems to embody the lost qualities—the extinction of a species, the absence of a lover, or the death of a princess. Each of these, in their way, embodies something of what has been lost; but the strength of feeling generated indicates a deeper loss than we are aware of, a loss that extends beyond the boundaries of conscious awareness.

And where the world does still speak to us, its voice is likely to be pathologized by the definers of normality and abnormality. Hearing voices is one of the classic symptoms of psychosis; and yet perhaps madness is as appropriately viewed as our failure to listen to or to make sense of what we hear. The commonplace view that the human intellect is the only source of structure has made us deaf to what the earth is saying; and "environmental" problems are regarded simply as practical problems to be solved by the application of science. But they are also communications that have other dimensions: spiritual, emotional, intuitive. Viewing a clearcut, we don't merely calculate the loss of wildlife habitat.

We also feel within us a sense of guilt or mourning. While the practicalities can be articulated and debated through existing forms of language and science, other dimensions are much more difficult to express, and so seek expression through symptoms that conceal their origin: depression, anxiety, eating disorders, antisocial behavior, and so on. If the dominant symptoms of Freud's age expressed sexual repression, those which characterize our era speak of loss of meaning, of disconnection, of nihilism—symptoms that reflect our deprivation of those basic structures and values which cannot be consciously generated, but which can be rediscovered within the deep structure of nature.

Given the assumptions that pervade conventional social science, it is all too easy to misinterpret this situation in a way that is both regressive and dangerous, and that is expressed most crudely in terms of applying "natural" laws and principles to human life. Thus competitive economic processes are seen as somehow "natural" in that they reflect the "survival of the fittest"; or a connection with the earth is seen in terms of a "manifest destiny," justifying aggression. The basic error of such misinterpretations is their uncritical acceptance of a dualistic split between nature and culture, so that the term "natural" reflects a simplistic understanding of nature as *already* alienated from the social. To repeat: the natural and the social realms can only be healthy within a context in which they are fully integrated. The dissociation of either from the other results in a stunted understanding—and practice—that is a travesty of what it could be. For example, a disenchantment with modern industrialism may give rise to misguided suggestions that we should retreat to earlier forms of living—a viewpoint that limits the possibilities for change to those that lie somewhere in the past along an unilinear pathway of industrial development wherein an original "natural" state is historically transformed into one that, instead, is supposed to be "cultured."

The challenge, however, is not to reject the notion of a developed form of "culture" and to return to allegedly more "natural" forms, but to rejuvenate the dialectic between nature and culture that is forgotten by both these alternatives. Anything less than this will introduce new repressions to replace the old ones. Nature and culture are not alternatives, nor are they rivals; they are, rather, mutually essential components of a healthy ecosphere. Nature's influence is seldom as straightforwardly predictable as concepts such as "instinct" would have us believe. For example, to refer to the highly sophisticated and often altruistic ways in which a wolf may defend the den or feed the young as "instinctive" is to deny the complex *social* dimensions of this sort of behavior.[88] Nature's dialectic with culture, when it is allowed to exist, expresses itself in ways that are fully integrated with intelligence and self-identity. And, as the reader will be aware by now, problems on one side of the environment-individual divide are inevitably accompanied by problems on the other side; in the schizoid separation of conscience from feeling, in the eating disorders that are symptomatic of the poverty of structures that constellate "food," or in the illusory choice between "self-control" and an indulgent, narcissistic, self-expression.

Let us explore the latter example a little further in order to illustrate how the nature-culture dialectic—or, rather, its absence—can both predispose us to specific problems, and make the character of such problems invisible to conventional psychological explanation. Freudian psychoanalysis takes for granted as normal the personality structure that results from the collapse of this dialectic, polarizing what should be a sophisticated interplay between social reality and biological patterning into the opposition of controlling ("conscience," "superego") and controlled ("id") parts of the personality. The idea of "self-control" therefore assumes an ontological split within the self, setting a controlling, largely social part of the self in opposition to a controlled part that is more directly connected to biological functioning. This ontological split is more clearly recognized as a problem, however, in Gregory Bateson's rather neglected theory of alcoholism. According to Bateson, alcoholism can be understood as involving a kind of epistemological error: the notion that the "self" (as in "self-discipline") can control "the bottle," so that the alcoholic can engage in "controlled drinking." This epistemological error has also been termed "alcoholic pride": "If I am really in control of my life, then I can drink in a controlled way"—a belief that, in the case of the alcoholic, is repeatedly proved wrong. According to Bateson, this is an "unusually disastrous variant of the Cartesian dualism,"[89] in which "mind" can supposedly control "matter," or the "self" seems to control the "body." However, this style of sobriety contains the epistemological error for which drinking is "an appropriate subjective correction": in other words, while drinking is obviously problematic in certain respects, it also obliterates the dissociation, and the distinction, between the mind and the body, restoring a psychologically desirable unity by relaxing the grip of "self-control." In the absence of cultural forms through which we might achieve more constructive forms of integration, it is unsurprising that so many resort to such alcohol-induced reintegrations.

For better or worse, however, the dualistic error is likely to reassert itself the following morning. The alcoholic vows never to touch the bottle again, and the principle of self-control works for a while. However, as long as this principle operates, it does so by fragmenting the experiential world of the person into (controlling) mind and (controlled) body. According to Bateson, following the approach of Alcoholics Anonymous, the only way out of these episodic cycles of drinking followed by temporary sobriety is to change the epistemology that underlies them. This change requires the abandonment of alcoholic pride, the rejection of self-control, and the acceptance that one is helpless before the bottle, and so cannot drink at all. The idea that there is a part of the self that can control the alcoholic part is replaced by a realisation that *the whole* of the self is alcoholic; and so the only way to avoid being drunk is to avoid being in an alcohol-laden environment.

An epistemology that splits the world into two parts will inevitably lead to problems as the repressed connections between the parts return to haunt us; although the form in which the problem expresses itself will vary according to the location of the dissociations. In the case of the alcoholic, relocating the ontological discontinuity that exists within the self to the space between the self and an alcohol-laden environment may be the only easily available way out of the destructive cycle of sobriety and drunkennness, at least within a society that fails to mediate our relation to what is outside us in any more sophisticated way. Such solutions, however, exist only within the individual life span, since they do not address the *original, cultural* dissociation between individual and environment that caused the problem in the first place. The basic difficulty, therefore, is not an individual weakness, but a *cultural* deficiency.

The alternatives of sobriety and drunkenness in some ways resonate with the equally problematic alternatives of alienation from nature and fusion with it. I will explore the shortcomings of these alternatives in chapter 6; but for the moment, let us note the similarity between the "solutions" of the alcoholic and the approach advocated by those types of environmentalism that suggest that the arrogant dissociation from nature that is the epistemological Titanic of the modern era can be rectified by an uncritical, regressive fusion with it. Both these solutions short-circuit the need for sophisticated structures that could reintegrate us with what is outside us; and the claim that we can "have a relation" with nature in the absence of these structures is a regressive wish-fulfilling fantasy. The dilapidation of cultural structures has led, in the social sphere, to the entirely understandable fear of acknowledging sexual or racial differences, since differentiation without structure tends toward prejudice; and in environmental matters, too, we have tended to shy away from acknowledging the differences that exist between ourselves and many other natural forms, recognizing on some level that within an industrialist context, such differences easily become alienations. One of the central arguments of this book, however, is that the repression of difference is not an adequate answer, whether it is achieved by drinking or by attempting an unmediated relation to nature. Rather, as I will argue in later chapters, we need the cultural structures that could realize the possibility of a world that is both highly differentiated and fully integrated.

We might summarize the argument of the past two pages in the realization that both "psychological" and "environmental" problems stem from epistemological discontinuities or from misguided attempts at solving them; and since these discontinuities move according to ideological fashions, at a more basic level these two classes of problem are the same problem. Thus my anxiety may be understood as an "internal" conflict to be treated by psychotherapy or drugs; or it could be seen as an understandable reaction to the degradation of the landscape I inhabit. Only by reintegrating the psychological and ecological fragments of the life-world within a common ecopsychological frame can we address the dissociation that underlies both these apparently distinct problems, and so avoid "solving" the problem by exporting it to another realm.

It is true that there are a few approaches such as family systems theory[90] that successfully challenge the individualistic stance of most psychology. However, family systems theory redirects our focus away from the individual and toward the family, and this throws up two problems: firstly, we lose sight of *individual* structure,[91] and secondly, larger structures such as culture and nature tend to be ignored as much as before. In effect, then, family systems theory refocuses our attention from one level of structure to another; but we are still no nearer to articulating the dialectic *between* levels of structure, such as those between culture and nature or between the family and culture. There remains little awareness that our lives can be constituted by structures outside the social sphere, or that family and individual problems could be related to this lack of awareness. In structural family therapy,[92] for example, the lack of structure found in "enmeshed" families is resolved by the therapist imposing structure of a fairly conventional sort; and the broader cultural reasons for the structural vacuum that brought the family into therapy is never explored. A cultural absence, then, is defined as a *family* pathology.

The price of ignoring natural structure in the external world will be apparent to most readers of this book. But nature exists *within* the person as well as outside us; and the tacit assumption that we can ignore this "internal" nature as we adapt to our present social situation and, over historical periods, to a developing industrialist context is clearly erroneous. Unfortunately, Freud's hope that civilization should conquer nature has come close to realization in academia; for, as we have already seen, the very reality of the natural world is being extensively denied. Nevertheless, our ability to suppress these natural structures must be limited, and must imply a great psychological price, as Freud recognized, particularly in relation to sexual repression.[93] Most of his more "liberal" followers, such as Erich Fromm or Erik Erikson, however, play down the importance non-ego structures such as the id, viewing the process of development as essentially one of adaptation to a preexisting and taken for granted social world. In other words, they maintain the principle of nature-culture dualism, denying the significance of natural structure and, in some cases, its very existence.

If neo-Freudians such as Erikson and Fromm are correct, then modern industrial society can develop according to the economic and humanistic principles that we accept as the basic framework of our lives without obvious damage to human potential or fulfilment, just as the argument that *external* nature is socially constructed implicitly denies the existence of violence toward the natural order. Human adaptability becomes essentially unlimited, and we can open and close factories, require people to move from one part of the country to another, live in crowded housing estates, or have them alternate between day and night shifts with no ill effects. Emotional reactions such as the "fight or flight" syndrome, which developed in earlier and quite different circumstances, may be regarded as anachronistic psychological relics to be dealt with by effective "impulse control"; and emotional attachment to place can be disregarded in the face of the necessity to move freely to other areas in search of jobs. If, on the other hand, such emotional realities are significant aspects of our presently existing makeup, then this denial of natural structure would be expected to lead to the kind of psychological and social problems that are commonplace today, even though the mechanisms involved are difficult to specify.

It is only quite recently, within the modern era at least, that we have begun to recognize the importance for mental health of factors that are subtle and often hard to grasp within psychology's individualistic focus. It was as recently as the late 1940s, for example, that social scientists began to describe a syndrome in which young children who were adequately clothed, fed, and otherwise materially looked after, but received little in the way of physical contact or affection from parental figures, became withdrawn, emotionally disturbed, and frequently died from no very apparent cause. Today, "maternal deprivation," as it has become known, is a recognized cause of psychological and physical ill-health.

However, simply to ensure that a child is cuddled and hugged enough may be to miss the full significance of these findings. Psychologically, we know that the term "mother" evokes many associations, as psychoanalysts such as Ferenczi point out:

> Individual observations of the symbolism of dreams and neuroses reveal a fundamental symbolic identification of the mother's body with the waters of the sea and the sea itself, on the one hand, and on the other, with "mother earth," provider of nourishment.[94]

If we take seriously the symbolic reality implied by such observations, then the scope and significance of the psychoanalytic concept of "oral neediness" becomes clear. For the "orally needy" individual, as Fromm notes, "the world is one great object for our appetite: a big apple, a big bottle, a big breast; we are the sucklers, the eternally expectant ones . . . and the eternally disappointed ones."[95] Within such a symbolic reality, then, the common roots of the ostensibly distinct realms of "individual" psychopathology and "environmental" problems are revealed.

Jung, similarly, viewed the "mother archetype" as constellating a variety of areas of life, such as a forest or valley where one lives, one's place of work, one's community or tribe, the house one lives in, and so on, as well as one's biological mother.[96] Interpreted in a way that respects the symbolic wholeness of life, we might view "maternal deprivation" as a deprivation not merely of one person, but also of contact with the meaningful and nurturing world she represents. Conventionally, however, the referent of the term "mother" is a biological mother only; hence "mothering" refers to a very specific form of individual contact, rather than to a psychological element that knits together the wholeness of one's environment. Jung felt that this conventional interpretation rested on a nihilistic literalization that led to a loss of coherence, community, and groundedness. This "biologization" of mothering is also a rejection of the cultural realm that could articulate the process and experience of mothering and interwine it with other aspects of life, giving it a fullness of meaning often missing today. Conventional understandings of the term "maternal deprivation," then, imply the acceptance of a reductionist and literal paradigm that denies reality to symbolic dimensions of experience. A drastic narrowing has occurred in the concept of "mothering" that secludes meaning within a single relationship and places the rest of the world at a distance from the child, rendering it largely meaningless and understandable only in only instrumental terms. In this situation, the mother may become the *only* source of psychological nourishment, and so any inability on her part to provide these will be disastrous for the child. To view this situation simply as a deficiency in the mother's behavior, however, is to ignore the *prior* deprivation that has stripped the world of its nurturant qualities. In a symbolically healthy world, therefore, the notion of "motherness" might be better expressed as an adjective that can be applied to many of its characteristics, or as a verb that communicates the process of nurturance, support, and "fit" that occurs between a life form and those "others" that it is ecologically related to. Only in a world that has already been made cold and lacking in such qualities will the biological mother be viewed as necessarily the sole source of nurturance.

This being the case, the schizoid personality structure that almost inevitably results from the mother's inability to satisfy the child's need for meaningful relation is invariably blamed on "inadequate parenting," to be rectified by better "parent training" or "awareness." Such explanations, however, fail to recognize that the situation in which the mother has become the child's only source of meaning *is itself* pathological. Thus the problem and its perceived "solution" all remain within the ideological sphere of a decontextualized rational analysis that is incapable of perceiving its own limitations and implications. The schizoid individual, who can neither relate empathically to the natural world nor transmit any meaning-laden view of that world to his or her own children, is both the inevitable product and the unwitting agent of a self-replicating schizoid ideological system. Through such invisible psychological processes is the web of ideology that assures the destruction of the earth transmitted to future generations.

"Intelligence" and Anthropocentrism

The concept of "intelligence," which is often portrayed as an "objective," widely applicable construct reflecting universal aspects of human functioning, exemplifies psychology's colonialist tendencies in a particularly pure form. Indeed, many psychologists regard it as applicable, through the use of so-called "culture fair" instruments, to *any* culture. Such a definition of intelligence seems both too broad and too narrow: too broad because of its unawareness that other societies often make quite different assumptions about the character and location of intelligent behavior; and too narrow in that it excludes much of what allows the natural world to function as a highly differentiated and sophisticated system of ecological interactions. Even if we limit ourselves to *human* behaviour, the concept of "intelligence" denies that large segment of human potentiality that is not consistent with industrialist processes and demands. In this book, the term "intelligence" will be used to indicate a specifically psychological understanding based around correlational statistics; whereas intelligence without the inverted commas will refer to a more open and undefined notion of human abilities.

To take the first point first, there are fundamental problems in applying the concept to nonindustrialized societies in which human life may be more highly integrated into the natural order—which is hardly surprising when one considers the origins of intelligence testing as an instrument of social stratification within Europe and the US.[97] Jacqueline Goodnow has pointed to various value orientations, such as speed, generalization, abstraction, and a "no hands" approach, which are taken to indicate "intelligence" within the industrialized world, but which may not be highly valued within nonindustrialized cultures.[98] A crucial difference between these two contexts lies in the extent to which we in industrialized societies assume that the characteristics of individuals can meaningfully be isolated from those of the world they inhabit, and that there is no dialectic between these dissociated fragments of reality. An ecological conception is necessarily one that regards intelligence as a property of natural systems; and there are therefore three possibilities with respect to humanity. Firstly, there may be some types of intelligence that are possessed exclusively by humans or by groups of humans. Secondly, there may be other forms of intelligence that exist only in the nonhuman world. And thirdly, there may be forms of intelligence that characterize ecological and cultural systems that include humans in interaction with other aspects of the natural world. In viewing intelligence as a property of individuals that is qualitatively independent of their social and natural environments, psychology recognizes only the first of these possibilities, and assumes that the only conceivable form of relation between individual and world is one in which the former manipulates the latter. Consistently with this assumption, "intelligence" exists in measurable degree in the individual who does the manipulating, but is absent from the entities that are being manipulated, which are seen in terms of matter, or at best, mechanism, in a manner reminiscent of Descartes' *res extensa*. These assumptions are usually simply taken for granted, although occasionally they are explicitly stated.[99]

While the technological worldview generally portrays a cosmos shorn of qualities that would indicate its holistic nature, spiritual significance, or the interconnectedness of its parts, those characteristics that are necessary to the functioning of the technological/economic system—such as quantity, physical properties, or chemical composition—are emphasized. Take, for example, an item from the Wechsler Adult Intelligence Scale: "Eight men can finish a job in six days. How many men will be needed to finish it in a half day?" Here, we are expected to convert the situation into a purely numerical one—that is, $6 \times 2 \times 8 = ?$ The physical aspects of the situation (sweat; grime; the texture of the rock; the heat of the sun); the social aspects (what do the men say to each other? Are they volunteers? Convicts? How do they share the work?); their relation to the work (Why are they digging this ditch? How do they feel about the project?);—this whole world is lost. All that is left is the equation $6 \times 2 \times 8$. The multidimensional nature of the situation has disappeared; and any reference to "nonessential" aspects would be regarded as indicating a lack of intelligence. The ideological preference that this situation unwittingly embodies can be traced to the Cartesian divorce of rationality from other human faculties; for the form of subjectivity that can focus on a few physical characteristics while repressing others is entirely consistent with that suggested by Descartes' description of the earth "as if it were merely a machine in which there was nothing at all to consider except the figures and motions of its parts."[100] The ontological reduction that is implied by this is the basis of a material reduction that results from its enactment: the reduction, for example, of complex fossil deposits to "fuel," or of a forest ecosystem to "grazing land." Technological power cannot exist without such reduction, and so rests on a simplification of the world's structure from that which we can barely sense to that which is consistent with rational consciousness.

Unsurprisingly, there is no psychological test that measures the ability to locate oneself within a cultural or natural context, in contrast to the numerous tests that assess the ability to isolate "essential" elements of a situation from those that are "nonessential." The mentality that can regard a forest simply in terms of board-feet of lumber is legitimated by a psychology in which, for example, performance may be assessed in terms of the "ability" to disembed items from context, as in the Embedded Figures Test,[101] and in which "intelligence" is largely viewed in terms of the manipulation of symbols independent of specific content. The sort of performance that reflects an unwillingness or inability to view a problem in purely abstract or quantitative terms, divorced from the everyday world, is perceived as "concrete" and therefore inferior. However, as Shweder and Bourne have argued, "concrete" thinking, rather than reflecting a "cognitive deficit," may be a by-product of the commitment to a worldview in which one's felt embodiment within the world is accorded as much significance as the intellectual elegance or instrumental power of an abstract model. From this perspective, it is abstraction rather than concreteness that is problematic, since the former style implies the splitting of item from context, of action from morality, of intellect from feeling that underlies environmental destruction.[102] The psychological concept of "intelligence" is one that assumes a fragmented world, and it is inherently opposed to a holistic, ecological conception of the person as part of nature. As Kummer and Goodall complain: "We almost completely lack an ecology of intelligence. No other dimension of behaviour has so systematically *not* been studied."[103]

Some indigenous traditions, however, view intelligence as involving a recognition of and a responsiveness to the preexisting structures of the natural world rather than the imposition of a human order on it, so that intelligence constitutes a genuinely ecological "second nature." As Edward Casey points out:

> Since the Navaho conceive their land as an ancestral dwelling place and since all significant learning proceeds ultimately from ancestors, culture is almost literally *in the land.* It follows that to learn something is not to learn something entirely new, much less entirely mental; it is

> to learn how to connect, or more exactly how to reconnect with one's place. At the same time, to reconnect with that place is to engage in a form of collective memory of one's ancestors; to commemorate them.[104]

In such cultures, then, intelligence has something to do with the capacity to integrate oneself within a larger ecological and spiritual scheme; and the individual is both recognizably separate from natural context and intertwined within it. This sounds paradoxical from an industrialized perspectives; for unless one is securely located by a living culture within the overall fabric of the world, the only safe form of self-definition is one that distances self from the world. This is usually achieved by viewing the world from a perspective that is intellectual, detached, and abstracted.

Those who inhabit less industrialized realms, however, are often reluctant to abstract. Aleksandr Luria, who interviewed the inhabitants of remote farming villages in Uzbekistan and Kirghizia in the early 1930s, reported that these "subjects used concrete, 'situational' thinking," and constantly "slip back into arguments based on experience." For example, Kamid, a peasant aged thirty-seven from a remote collective farm, is asked:

Luria It is four hours on foot to Vuadil, and eleven hours to Fergana. How much more of a trip is it to Fergana?

Kamid "Vuadil is halfway there. It's three hours from here to Vuadil, and another three from Vuadil to Fergana."
Change of conditions in conformity with actual experience.

Luria But what is it according to the problem? *(The conditions of the problem are repeated.)*

Kamid "Three hours farther."

Luria How did you know?

Kamid "I tell you, Vuadil is halfway, and then the road from Vuadil to Shakhimardan is poor, and beyond that it's good."
Justification of solution by concrete conditions.

Luria	And what was the problem?
Kamid	*Subject repeats the conditions of the problem correctly.*
Luria	How much farther is it to Fergana?
Kamid	"Three hours farther!"
Luria	How did you figure it out?
Kamid	"It's a bad road from here to Vuadil!"[105]

. . . and so on. In example after example, we find Luria's informants insisting on the validity of experience over abstraction:

Luria	The following syllogism is presented: In the Far North, where there is snow, all bears are white. Novya Zemlya is in the Far North and there is always snow there. What colour are the bears there?
Kamid	"There are different sorts of bears." *Failure to infer from syllogism.*
Luria	The syllogism is repeated.
Kamid	"I don't know; I've seen a black bear, I've never seen any others. . . . Each locality has its own animals: if it's white, they will be white, if it's yellow, they will be yellow." *Appeals only to personal, graphic experience.*
Luria	But what kind of bears are there in Novya Zemlya?
Kamid	"We speak only of what we see; we don't talk about what we haven't seen." *The same.*
Luria	But what do my words imply? The syllogism is repeated.
Kamid	"Well, it's like this: our tsar isn't like yours, and yours isn't like ours. Your words can be

answered only by someone who was there, and if a person wasn't there he can't say anything on the basis of your words."
The same.

Luria But on the basis of my words—in the North, where there is always snow, the bears are white, can you gather what kind of bears there are in Novya Zemlya?

Kamid "If a man was sixty or eighty and had seen a white bear and had told about it, he could be believed, but I've never seen one and hence I can't say. That's my last word. Those who saw can tell, and those who didn't can't say anything!" *(At this point a young Uzbek volunteered, "From your words it means that bears there are white.")*

Luria Well, which of you is right?

Kamid "What the cock knows how to do, he does. What I know, I say, and nothing beyond that!"[106]

It is clear that Luria and his respondent live in two different worlds, as the latter eloquently suggests! Luria inhabits a universe of elegantly logical abstraction, and so can discuss the coloring of bears without it being necessary ever to have seen one. His informant, in contrast, lives in a more experientially based world within which one is unwilling to draw conclusions from abstract principles. Each seems quite sure that their viewpoint is the correct one. Contrary to Luria's assumptions, however, it is not clear that his informants' style of thought is any less adequately adapted to the demands of their day-to-day lives, or has less survival value, than a more abstract one would be.

Goodnow offers another, equally telling, example in discussing the work of Cole and his colleagues in Liberia:

> The investigators had carefully gathered a set of 20 objects, 5 each from 4 categories: food, clothing, tools, and cooking utensils. Both the objects and the names for the categories were familiar, i.e., all the proper spadework had been done. Despite the precautions, however, many of the Kpelle produced, when asked to put together the objects

> that belonged together, not 4 groups of 5 but 10 groups of 2. Moreover, the type of grouping and the type of reason given were frequently of the type we regard as extremely concrete, e.g., "the knife goes with the orange because it cuts it." Glick (1975) notes, however, that subjects at times volunteered " 'that a wise man would do things in the way this was done.' When an exasperated experimenter asked finally, 'How would a fool do it?' he was given back [groupings] of the type . . . initially expected—four neat piles with foods in one, tools in another."[107]

Such examples suggest several lines of thought. One is that even in cultures that do not encourage abstraction and generalization, it is not so much the case that people are *unable* to abstract, but rather that they *prefer* not to. Secondly, it may be that a relatively concrete mode of thought is well adapted to agrarian or nomadic styles of existence, whereas abstraction is likely to be less useful or even counterproductive. To ascribe this sort of preference for a concrete relation to the world to a *deficit* is therefore misleading and ethnocentric.

Furthermore, it appears that even where abstraction does occur within non-Western styles of thinking, it tends to be directly related to concrete aspects of the life-world, and so "abstraction" in such cases, in contrast to the situation in the developed world, reflects an *elaboration of the world that already exists* rather than an attempt to *replace* this world. For example, in Thomas Gladwin's discussion of Puluwat navigators—who successfully sail between tiny islands separated by stretches of empty ocean extensive enough to frighten even those experienced in yachts—the abstract concepts that the islanders use are clearly and directly related to aspects of the physical world such as waves, currents, the positions of stars, and the behavior of wildlife.[108] In contrast, the style of abstraction typical of the developed world is one that is *distanced* from the physical realities of the earth. For example, modern navigational equipment, such as radar, inertial guidance systems, and weather information transmitted by radio do not even require the modern captain to venture on deck, relying instead on an imposed system of coordinates and satellite location equipment. A conclusion that we can derive from anthropological research such as Gladwin's is that intelligent and adaptive behavior may be intimately entwined with detailed physical realities as well as abstracted from such realities: while Puluwatan navigational abilities derive from an intimate knowledge and experience of the earth, "Western" navigation depends less on such knowledge than on one's relation to a sort of electronic scaffolding erected over the surface of the earth. Our approach is, in its own way, undoubtedly "intelligent"; but to equate intelligence *generally* with a style of abstraction that distances us from the earth is to foist our own values onto cultures and contexts in which they may well be inappropriate, quite apart from the serious doubts about whether our particular style of abstraction is necessarily healthy even within the culture that spawned it. While the psychological conception of "intelligence" reflects an instrumentally powerful understanding of the world, its appropriateness lies *within* the context of modern industrial society, and it is powerless to prescribe a healthy relation *between* modern industrial society and the natural world that is ostracized by this society.

On the other hand, if we see the world as containing its own forms of natural intelligence, then a fully adaptive human intelligence must partly reside in our ability and willingness to learn about and embody this broader intelligence. This implies a quite different attitude to the world than that required by technological power, and one closer to Puluwatan or Navaho understandings than those of the industrialized world, involving an openness to structures and processes beyond the self, and a recognition that wisdom resides partly in our ability to live consistently with these structures and processes. An example of such an attitude is given by Edmund Carpenter in his discussion of the Aivilik word "sila," which

> . . . Means both thought and outside .. . . In one sense, it refers to the world outside man, especially weather, elements, the natural order. . . . But sila also refers to the state of the inner mind; "silatunerk," has intelligence, shrewdness; "silaturpok," prudent, thinks ahead. . . .
>
> Thought, to the Eskimo, isn't a product of mind, but the forces outside of man. . . . Sila, goddess of the natural order, is also the goddess of thought. The successful hunter is her conscious self: he who obeys her laws, prospers. He who ignores her, suffers and dies.[109]

Carpenter is describing a world in which intelligence, rather than being located within the minds of individuals, is a property of the world that the individual can learn to share in. If we are attentive to the structure of the world, then we can share in its intelligence, like Heidegger's cabinetmaker who "makes himself answer and respond above all to the different kinds of wood and to the shapes slumbering within the wood."[110] This state of attunedness to the world is becoming increasingly rare as industrialization spreads over the globe; but it is well expressed by the behavior of the Aivilik carver:

> [Holding] the unworked ivory lightly in his hand, turning it this way and that, . . . [he] whispers: "Who are you? Who hides there?" And then: "Ah, seal!" He rarely sets out to carve, say, a seal, but picks up the ivory, examines it to find its hidden form and . . . carves aimlessly until he sees it, humming and chanting as he works. Then he brings it out: Seal, hidden, emerges. It was always there: he did not create it, he released it; he helped it step forth. . . . The Eskimo language has no real equivalent to our words "create" or "make," which presuppose imposition of the self.[111]

This openness to the character of the world as it chooses to manifest itself is also a distinctive aspect of traditional Eskimo styles of perception. Carpenter notes that

> [W]ith multiple perspective, the moving eye of the observer himself is drawn unconsciously into the scene. Similarly, Eskimo narrators shun a single perspective, preferring to describe an object from many angles.[112]

Thus the images and ideas that are generated are not so much the products of individual "intelligence," but rather emerge as a result of the joint interaction of the individual and those natural and cultural structures that in part constitute individuality:

> When the task of artistic inspiration and creation is assigned to the unconscious, the images that result are corporate ones: they do not come from the depths of any private, individual unconscious; they come from individual dreams, but from dreams that also belong to the whole tribe. Nothing about them can be called private or individualistic. The dreamer looks inward, but his trip takes him directly to the collective unconscious, that storage system for the collective experience of the tribe. When he returns, he is often better able to handle functions of the mind too obscure for deliberate, conscious activity, and to do so lucidly, communicating easily with those who share these complex perceptions and ancient memories.[113]

We are dealing here with a form of personhood, and a relation to the natural world, very different from that assumed within industrial society:

one in which *inwardness* is also relation to what is *outside*. In contrast, describing someone as "intelligent" on the basis of their having a high IQ score is to assess their success in *separating* themselves from their context and functioning as self-contained manipulators, a stance that clearly embodies the technological-commercial ideology of the "developed" world.

Ethnocentrism and Anthropocentrism

If psychological conceptions of "intelligence" are based on a largely Eurocentric set of values and criteria that have often been imposed unrealistically on other societies, then these same criteria, as I will argue below, have also been used to legitimate the ruthless domination of nonhuman species. Most current measures of intelligence are today, as in Spearman's day, based on the statistical correlations between subtests reflecting apparently different skills, although throughout the history of intelligence testing there has been a good deal of debate as to whether these correlations define a single factor of "general intelligence," or several smaller factors such as those reflecting "spatial" or "verbal" intelligence, or even "multiple intelligences."[114] Whatever the factorial structure of "intelligence," however, the correlations that underlie it are seldom interpreted as suggesting intracultural convergences, but are instead assumed to reflect basic universals of human ability. Despite the evidence that such correlations are often absent among nonindustrialized peoples, and despite the fact that the social ordering of individuals by race and class[115] is no longer socially acceptable, the concept of "intelligence" and the techniques used to assess it are fundamentally the same as they were in the days of Goddard and Terman.

Political correctness today ensures that the club of intelligent beings has expanded—not without some difficulty—to include all psychologically healthy humans from any society. This largesse, however, is not the result of the concept of "intelligence" becoming any more cross-culturally valid, but rather reflects the chameleonlike manner in which mainstream psychology fits into whatever political agenda happens to be in vogue at any particular time. Given the origins of intelligence testing within a white, middle-class, largely male, Euro-American establishment, it would be truly remarkable if the scores achieved by individuals outside this establishment matched those of individuals who are members of it. To acknowledge this is not to denigrate non-Western cultures, but rather to recognize that the types of behavior and thought that we value as intelligent are those selected during a specific process of cultural evolution during which human behavior became aligned with technological and economic agendas.

The claim that the pattern of abilities demanded by this specific avenue of cultural evolution is cross-culturally universal is as absurd as the suggestion that all competent species should have two arms, two legs, and be able to walk upright. Nevertheless, while psychology has, at least for the most part, moved away from classifying races as more or less "advanced,"[116] the methodology and assumptions that underlie such attributions remain essentially in place. In this psychology has been helped by the globalization of industrialist assumptions, so that the cultural diversity that would make obvious the ethnocentricity that characterizes measures of ability has been drastically reduced. Furthermore, poor scores on conventional measures of intellectual capacity are easily ascribed to inadequate educational experiences—providing an opening for Western "aid" designed to bring the Third World into the "technological era." In this way, a diversity of cultural systems, each varying according to its particular historical and natural context, is being rapidly replaced by one hegemonic cultural form. Ironically, then, psychometricians are today in a better position to claim cultural universality for their instruments of assessment than they were in the 1920s, even though these instruments have changed little in the intervening eighty or so years. In other words, the discrepancy produced by psychology's mischaracterization of personhood is being reduced not by psychology becoming more accurate and generally applicable, but by the people across the world increasingly conforming to this mischaracterization. This echoes the process whereby the world is reduced to its technological image by the elimination of wildness, allowing the increasingly accurate claim that nature is "socially constructed." However, even were the hegemony of industrialism complete, to claim that psychometric measures are therefore valid across all cultures would be to collude in the repressive limitation of human potential to the particular capacities and faculties required within a technological context. An adequate psychology is one that can recognize not only the patterns of ability that may exist within one particular currently existing setting, but also those latent potentialities that could flourish within currently unimaginable contexts—unimaginable because they may involve not merely other forms of manipulation of nature by humans, but also the *interaction* of humans with currently unrecognized forms of intelligence in the outside world.

Animal "Intelligence"

If past and present psychological conceptions of mental ability are rooted firmly within the ideological realities of modern industrial society, and so are inherently prejudiced against forms of culture that diverge from industrialism, then they are even more strongly prejudiced against the nonhuman members of the natural world. "Intelligence" is seen as being a quintessentially *human* property, possessed by animals only to the extent that they approximate certain aspects of *our* behavior. Rats, for example, have often been tested in Skinner boxes according to an operant conditioning paradigm in which they receive "reinforcements," usually in the form of a food pellet, for learning simple behaviors which have been pre-selected by the human experimenter. This impoverished environment, not surprisingly, produces an impoverished version of rat behaviour—although this is an animal with remarkable sensory powers that can survive in the most inhospitable conditions, and can even self-select a diet containing adequate thiamine and sodium.[117] These latter types of behavior have received only passing attention from psychologists, however, who have preferred to study simple, isolated actions such as bar-pressing that have little if any relevance to "real world" rat activity. It is interesting to speculate about whether a human could or would behave any differently under such conditions; after all, there's not much one can do in a Skinner box except either push the bar or not push the bar. But the inadequacy of behaviorist psychology is now widely recognized, and so to devote space to criticising it is to risk accusations of cruelty to dead horses. Nevertheless, there are underlying characteristics of the behaviorist paradigm that have been inherited by apparently more enlightened approaches: in particular, a separation of the organism from its natural context, and a recognition only of a drastically reduced version of behaviour. Probably the only time a laboratory rat could tell us much about its abilities would be if it escaped from the laboratory—a situation that, unfortunately, is unlikely to be recognized as methodologically valid.

If the treatment accorded to animals such as apes, which are genetically closer to humans, has been more imaginative, it has been even more transparently anthropocentric than has been the case with rats. For example, a number of experiments have attempted to teach apes to communicate using some variant of *human* language, as if this was a suitable measure of their abilities. The success or otherwise of these odd ventures is not the point here; what *is* notable is the refusal to acknowledge that the behavior of the animal within a *natural* context might be intelligent, and that intelligence is effectively a property of a whole, well-functioning *system.* In effect, what is denied is the possibility of any form of intelligent structure other than that of Eurocentric rationality. It is also noteworthy that the vast majority of experiments with apes and other "higher" animals embody the assumption that while it makes sense to have animals learn skills that are considered significant in human life, there is nothing that such animals could teach *us*—except, of course, about their ability, or lack of it, in relation to us. This is an example of what John Rodman has termed the "differential imperative"—that is, the need to identify differences between ourselves and other creatures that confirm our assumed superiority. This is a general principle of colonialism, whether it is applied to other life forms or—as we saw at the beginning of this chapter—to other cultures. Whatever the domain in which this imperative operates, the other is seen as a source of information; but not of a type that could change or enrich the character of our own cultural tradition, which is tacitly assumed not to be in need of any modification. Rather, the other is assimilated, digested, and used to feed the assumptions of existing ideology. For example, both cultural artifacts and spiritual practices tend to become commodified within the industrialized world and thus assimilated to the market economy. Similarly, research assimilates other structures as grist for its own particular mill, becoming a *celebration* and an *expansion* of existing ideology rather than an opening up of it to alternatives.

Abilities, then, are recognized only to the extent that they fall within the specific world that we have constructed. Seen this way, the anthropocentrism implied by most animal experiments is just another facet of a deeply entrenched chauvinism which insists that one's own experience of and behavior toward the world represents the only valid viewpoint—a chauvinism that tries to hide under the label "objectivity." Now objectivity, if one adopts a moderately realist understanding of the world, is not a characteristic that can be simply rejected; but when used to camouflage our withdrawal into a prejudiced and unfeeling relation to the nonhuman realm, it amounts to a failure of integrity that Theodore Roszak has aptly captured in his description of it as "the academic uniform of moral cowardice." In this case, it legitimates an extension into the nonhuman world of the type of prejudice that assumes that the Western conception of intelligence is in principle applicable to all cultures. Overall, with very few exceptions, the psychology of animal behavior embodies a deeply anthropocentric attitude to the whole natural world, and in this respect animal psychology is consistent with farming, hunting, genetic programing, and the use of animals as entertainment. By means of this attitude, we deny what Locke was prepared to admit: the existence and validity of otherness, the inexplicable. Psychology, along with much of present-day science, often has difficulty in admitting this; for with the crumbling of traditional sources of certainty, we have little else to cling to other than *scientific* certainty; and so every aspect of existence must be portrayed as at least potentially explicable by science. Thus the "abilities" of other forms of life are assessed according to the extent that they may be reduced, molded, and rewarded into becoming unwilling imitations of ourselves, and research institutions become the academic equivalents of the kind of theme parks that offer demonstrations of dancing by porpoises and killer whales.

A classic example of this is the well-known case of Clever Hans, the horse that appeared to be able to solve simple arithmetic problems, the answers to which it would communicate by tapping a hoof the appropriate number of times on the ground. It was eventually found that Clever Hans' abilities lay not in arithmetic, but in a highly developed sensitivity to subtle cues that were unintentionally given by its owner. In this case, the talents of Clever Hans were considered to have been disconfirmed, and the remarkable skills that the horse *did* unwittingly demonstrate were scarcely noticed; for the intent to bring the animal's behavior into the realm of human rationality had failed. Such anthropocentric judgments implicitly define intelligence as an abstract ability that exists entirely *within* the individual, denying the validity of any intelligent structure that is larger than the individual. The world is thus made to appear structureless, lacking in intelligent form or purpose, and so fit only to be manipulated for human needs. This is entirely consistent with Descartes' invalidation of animal intelligence:

> [While i]t is . . . a very remarkable fact that although there are many animals which exhibit more skill than we do in some of their actions, we at the same time observe that they do not manifest any at all in many others. Hence the fact that they do better than we do, does not prove that they are endowed with mind, for in this case they would have more than any of us, and would do better in all other things. It rather shows that they have none at all, and that it is nature which acts in them according to the disposition of their organs.[118]

This tortuous line of reasoning can only be seen as a transparent ploy to maintain the distinction between the human "mind" and a "nature" that is *defined* as mindless—a distinction that is faithfully maintained by most contemporary animal researchers. As Tim Ingold has recently pointed out, for example, while the hunting and foraging behavior of humans is usually considered to be the result of consciously formulated cognitive strategies, the often comparable and "eminently rational" strategies of nonhuman animals are generally considered to have been "worked out for them in advance, by the evolutionary force of natural selection."[119] This taken-for-granted discontinuity between the "animal" and "human" realms is beginning to come under fire, however. Beatrix Gardner has suggested that the "uses and misuses to which we put animals have to do with lines that we draw, differentiating ourselves from them. I'm certain that even within human populations, when we behave in a way that is not humanitarian, it is because we draw a distinction—'If these people are not like me, then they don't have the same rights'."[120]

In a similar vein, the developmental psychologist Patricia Greenfield has pointed out how an anthropocentric perspective on the world is consistent with the currently fashionable view of language as primarily *internally* connected rather than as a means of articulating our relation to something beyond the human realm. Noticing how her daughter combined words with *things* rather than with other words, she realized that this challenged the current orthodoxy that language, and by extension the whole of human social life, constitutes an autonomous realm separate from the rest of the world—a challenge that was not welcome to many in academia:

> [My findings] were very unpopular, and [were] very heavily criticized—I think, to a large extent because of the bias that words are more real than non-verbal elements. That is, if someone expresses something in a word, you know it was really there. . . .
>
> Children can do something, and it's called language. . . . A chimpanzee does the same thing, and it's not language. And I think the reason is, there's a double standard, and where the double standard comes from is the fact that we all know that children will grow up and speak full-blown human language. . . . And so there's a bias in the interpretation of the data.[121]

Just as Columbus insisted that the speech of his native informants, because it differs from European language, did not constitute a language at all,[122] so animal researchers often make similar claims about nonhuman animals. In both cases, we can detect the need to exclude and invalidate the other, which, in turn, is a precursor to the assimilation of the other into a privileged reality. The intent of bringing the behavior of animals within the purview of the scientific vision that we have constructed is thus not to bring us closer to them: on the contrary, it is a way of distancing them from us—assigning them a place within an anthropocentrically-constructed world so that we can use them as we wish. In short, they will no longer be wild—for wildness is anathema to technique, and risks powerful passions, identifications, feelings of relation, and the recognition of structures and processes beyond the narrow confines of egoic rationality. Understanding is safe; empathy is risky, since it threatens the boundaries we have set between us and the wild world—which is why anthropomorphism is the great sin in animal psychology.[123] Feeling dimly the dangers and the promise of our own wildness, we cannot bear to see wildness in the world around us; so like Odysseus' crew, we stop our ears and refuse to recognize the call. And most significantly, our fear of wildness has effects more practical than the distortion of awareness; for in our efforts to assimilate wildness to our domesticated world, we obliterate it. In so doing, we destroy forms of intelligence that in offering alternatives to our own rationality, threaten its perceived monopoly.

The idea that intelligence, rather than being located within the brains of humans, may be understood as a property of well-functioning natural systems, represents a fundamental change. Within an ecologically aware subjectivity, there may simply be little place for a concept that assesses one's prowess at manipulating other parts of the world, because understanding the world as a natural *system* is to suggest that it embodies an intelligence that encompasses the whole rather than any one fragment. The notion of intelligence as primarily individual rather than collective implies a fragmented, competitive society consisting of individuals striving against each other, and is hostile to the possibility that systems, be they social or natural, can embody a harmonious integration that results in tendencies toward constructive evolution. The exquisite balance between the behaviors and characteristics of the members of a natural community, rather than being seen as intelligent, tends to be viewed only as the outcome of a vicious process of natural selection, so maintaining the view that nature "out there" is not intelligent. This image of the natural world as a fight for survival between members of competing species, while it is obviously not totally erroneous, is a partial view that disguises the cooperative, purposive qualities of the whole. In recent years it has become more widely recognized that evolution itself possesses characteristics that are difficult not to acknowledge as intelligent, as Jonathan Schull suggests in arguing that, "plant and animal species are information-processing entities of such complexity, integration, and adaptive competence that it may be scientifically fruitful to consider them intelligent."[124]

Psychodynamic Approaches

Within experimental psychology, the separation of the person from the natural context is so complete that no recent theorist has argued for the necessity of this separation; this issue is simply not addressed. In order to find an explicit rejection of the natural world as a constituent of individuality, one needs to turn to psychoanalytic theory. As we will see later, some recent variants of the psychodynamic approach have a good deal to offer with respect to our understanding of the human relation to the natural world; and we will explore these in chapter 5. In classical Freudian psychoanalysis, however, the ego is regarded as striving to outgrow its original connection with nature, suppressing and distorting those arational elements that remain within the psyche. In this respect psychoanalysis, like experimental psychology, is consistent with the individualistic ethic discussed earlier. Freud, for example, argued that

> Originally the ego includes everything, later it detaches itself from the natural world. The ego feeling we are aware of now is thus only a shrunken vestige of a far more extensive feeling—a feeling which embraced the universe and expressed an inseparable connection of the ego with the external world.[125]

This "extensive feeling," which he described as "oceanic" and which he associated with religion, Freud saw as an atavistic remnant of an earlier period, both phylogenetically and ontogenetically. He regarded the separation of the ego from the external world as an essential part of both individual development and the progress of civilization, arguing that the way forward required "combining with the rest of the human community and taking up the attack on nature, thus forcing it to obey human will, under the guidance of science."[126] The "nature" to which Freud refers here is both "human nature"—that is, the "uncivilized" impulses that he regarded as existing within the id—and also nature in the external world. The connection between the technological domination of nature, on the one hand, and the distortion of consciousness, on the other, is clear: in the case of both the individual and the landscape, an alleged lack of natural structure justifies the imposition of an industrialist structure, the myth of the chaotic unconscious playing a similar role within the psyche to that played by the myth of the savage wilderness in the conquest of "outer" nature.

Freud's descendants have, for the most part, followed his lead in this respect. Erich Fromm, for example, although (even in 1941) expressing ambivalence about the character of modern individuality, argues for the necessity of the separation of the individual from the rest of the world:

> The emergence of man from nature is a long drawn out process; to a large extent he remains tied to the world from which he emerged; he remains part of nature—the soil he lives on, the sun and moon and stars, the trees and flowers, the animals and the group of people with whom he is connected by the ties of blood.

> These primary ties block his full human development; they stand in the way of the development of his reason and his spiritual capacities; they let him recognise himself and others only through the medium of his, or their, participation in a clan, a social or religious community, and not as human beings; in other words, they block his development as a free, self-determining, productive individual.[127]

Given this legitimation of individuality as the most important, or even the only, structure to which all else is to be reduced, it is not surprising that psychoanalysis denies the existence of larger structures such as those of culture or the natural world. Even in considering peoples whose relation to the natural world is clearly expressed through ritual and mythology within a coherent cultural system, the majority of psychoanalysts have invariably chosen to depict behavior associated with such belief systems in terms of individual, or at best social, processes wherein aspects of the natural world are merely incidental content. For example, Erik Erikson discusses the Lakota Sun Dance in terms of atonement for guilt feelings associated with "rage at the mother's breast during a biting stage which interferes with the long sucking licence,"[128] so reducing the Lakota's greatest religious ceremonial to a matter of mother-infant relations and locating the meaning of the ritual within the minds of individuals. This viewpoint also embodies the reductionist doctrine that whatever is "lower"—that is, smaller, more reduced, more apparently devoid of feeling or sentience—is somehow more "real." As Denis de Rougemont puts it:

> The superstition of our time expresses itself in a mania for equating the sublime with the trivial and quaintly mistaking a merely necessary condition for a sufficient cause. The mania usurps the name of "scientific integrity," and is defended on the ground that it emancipates the mind from delusions about "spirit." Yet it is difficult to see how there can be any emancipation in "explaining" Dostoievsky by epilepsy or Nietzsche by syphilis.[129]

Such reductionist doctrines are common within social science, although what it is that the world is reduced *to* varies a good deal, biology and language being the two most popular candidates. Consciousness has always found the systemic, dialectical character of nature difficult to deal with, so it is perhaps not surprising that many theorists are tempted to retreat into a single causal perspective that fails to recognize that structures can interact on many levels, and, moreover, can alternate dynamically between alternatives.

Fortunately, some variants of psychoanalytic theory have in recent decades broadened their approach toward a more relational view of the self. One such variant is that suggested by Joel Kovel, who argues that an ecological awareness, and a corresponding framing of modernistic science, are necessarily spiritual in nature.[130] Instead of the division of life into religious and nonreligious domains that is characteristic of the developed world, Kovel suggests that for tribal peoples, spiritual power such as the "mana" of Melanesia pervades the natural world, and is itself experienced as "natural" rather than "supernatural." Such psychological defences as splitting, compartmentalisation, and intellectualization that separate the spiritual dimension of life from our everyday behavior are strongly associated with the "developed" world, Kovel argues, and can be seen as the mature culmination of Cartesian dualism. A second promising psychoanalytic development is object relations theory, which sees the self as possessing a basic need to be *in relation*, implying a more decentered view of the person than that held by Freud.[131] Unfortunately, as is the case within social psychology, this relationality is usually understood as limited to a *social* universe, and the natural world is not seen as a significant constituent of individuality. Nevertheless, object relations theory, as I will argue in later chapters, can be extended to provide the basis for an understanding of personhood that is ecologically sound.

More often, however, calls for the reinstatement of the relational and spiritual dimensions of human experience have come from outside the field of psychology. David Levin, for example, has argued that:

> Freud is unable to conceptualise a development beyond ego structure which would take the form of a hermeneutical movement: a regressive return to retrieve a dimension of experience left behind not only in the developmental transition of the Western individual from infancy to adulthood, but also in the cultural transition from pre-modern forms of life to forms which are distinctively modern. What I have in mind, then, is a movement which is not (so to speak) "completed" until the "oceanic" experience, the wisdom of interconnectedness and wholeness, has been brought back, brought into the present, and appropriately integrated into present living.[132]

The recovery of this "oceanic" feeling, which Freud described as "a feeling of indissoluble connection, of belonging inseparably to the external world," is thus essential, Levin argues, if we are to reinstate our "interconnectedness with all other beings and our grounding in the wholeness of Being." This is easier said than done, however; and in later chapters I will argue that if we are to avoid an entirely regressive conclusion that advocates a reversal of our evolution toward individual autonomy, then we will need to envision cultural structures intermediate between those of individuality and the natural order.

In many respects, then, psychoanalysis has betrayed the unconscious that it made its focus. Freud is the colonialist explorer, stumbling upon a part of the world previously undiscovered, and using the established methods of science to dissect and analyze the species that crawl and flutter around within it. His purpose is not to empathise with and respect these species in their adaptations to their ecosystemic context, but to separate them from this context in a way that often echoes better-established disciplines such as biology. Freud, of course, was in an excellent position to appreciate the threat to ego organization that the unconscious represented, as his many writings on the subject testify; but what has been less often recognized is that psychoanalysis can itself be seen as a response to this threat, involving the domestication of wildness and its assimilation to human intellectual structures. While more recent theory has begun to challenge this fundamental stance of psychoanalytic orthodoxy, this challenge seldom extends beyond the boundaries of the human world; so that the dissociation between humanity and the rest of nature remains safely in place.

Humanistic/Existential Psychology

The psychoanalytic and experimental approaches have been strongly criticised by advocates of the "third force" in psychology—the "humanistic," or sometimes "existential" approach. Associated with names such as Maslow, Rogers, May, and Perls, humanistic psychology draws on the liberatory discourses of the 1960s and 1970s as well as continental existential philosophy in its attempts to offer a model of behavior and experience that does justice to the breadth and depth of human potential. In several of its manifestations—and there are many—the emancipatory intent of the approach, together with its preference for the data of immediate human experience over the abstractions of complex theorizing, make it a potentially important part of the project of reestablishing the relations between the human and natural worlds, and one that demands a more sympathetic critique than is the case with the approaches so far considered. Nevertheless, it embodies particular problems that need to be seriously addressed. Both the difficulties and the strengths of humanistic psychology will be considered here by reference to the work of one of its foremost exponents, Carl Rogers.

Rogers' account of the character of the "fully functioning" person represents a determined subversion of the repression and intellectualization that ties us to a technological worldview, and the organic metaphor of growth that Rogers was so fond of has been a positive and fertile one that implicitly relates humanity to the natural world. Furthermore, his emphasis on empathy, which he defined as the ability to "perceive the internal frame of reference of another with accuracy and with the emotional components and meanings which pertain thereto as if one were the person, but without ever losing the 'as if' condition," offers us a potential mode of relation to the natural world that differs from both the scientist's "objectivity" and the deep ecologist's "identification." Many humanistic psychologists use a similar concept as a basis for their thought and therapy; Laing, for example, referred to "co-presence" as a necessary part of a healing relationship. Such attitudes toward relationship need not be limited to the realm of human interaction.

There is much in Rogers' approach that is left untheorized, which is both a strength and a source of problems. On the positive side—and this is very important for our later argument—it admits the existence of processes and components of experience that we cannot rationally understand, as Hugh Gunnison has pointed out in drawing attention to the parallels between Rogers' approach to therapy and the methods of that other shaman of the therapeutic world, Milton Erickson.[133] Both Rogers and Erickson were able to admit into awareness types of experience that were not amenable to rational conceptualization, and could work comfortably and effectively with these types of experience. Both therapists also resisted the attempts of acolytes to pigeonhole these types of experience in conventional terms, defending their "otherness" as irreducible to everyday awareness. There is a parallel here with the dilemmas surrounding 'wilderness preservation': both "wilderness" and "trance" are important aspects of the "wild" world beyond rationality; but there is a danger that their categorization within an anthropocentric framework is a sort of colonialist appropriation rather than an authentic recognition, an allowing-to-be. That Rogers recognized this danger and was prepared to defend subjectivity against it is clear from his own writings:

> When I am closest to my inner, intuitive self, when I am somehow in touch with the unknown in me, when I am perhaps in a slightly altered state of consciousness . . . then whatever I do seems to be full of healing. . . . I know much more than my conscious mind is aware of. I do not form my responses consciously, they simply arise in me, from my nonconscious sensing of the world of the other.[134]

This willingness to admit into experience elements of subjectivity that have difficulty finding expression within the industrialized world is an essential first step in the project outlined in this book. However, it *is* only a first step; for unless such forms of experience can be nurtured and supported by a sympathetic cultural framework, they are in danger of remaining at the level of mute, fleeting insights and feelings with little capacity for survival once outside the consulting room. Insights and feelings, *unless* they coalesce to form a stable articulatory structure, have negligible power to challenge an ideological system that has proved overwhelmingly capable of digesting and destroying whatever has challenged it. In this respect, Rogers' theory is less helpful; for the cultural and political implications of his approach are left undeveloped, leaving it vulnerable to accusations of individualism.

What is necessary, then, is a further step beyond the recovery of individual experience: the articulation of such experience into a world that is simultaneously social, cultural, and natural, so that the world is reintegrated. If my experiences are percolated through a cultural framework of stories, customs, literature, rituals, and so on, their meaning becomes more-than-personal, extending out into the world so that they become intertwined with the meanings of the world. For example, the Lakota vision quest is not a matter of *using* the landscape to promote an individual vision, but rather an opening of the self to a vision that comes from the landscape, thus integrating self and landscape within a common frame.[135] This is an integration that is more-than-intellectual, drawing out of us feelings and intuitions that cannot be articulated in a rational way. If we lack such cultural means of expressing these feelings and intuitions, they necessarily remain part of that innermost, private, personal realm that therapists have sometimes referred to as the "inner child" or the "libidinal ego." As James Hillman has pointed out, however,

> [If] you go back to your childhood . . . you're not looking around. This trip backward constellates what Jung called the "child archetype." Now, the "child archetype" is by nature apolitical and disempowered—it has no connection with the political world . . . so when the adult says . . . "All I can do is go into myself, work on my growth, my development . . . ," this is a disaster for our political world, for our democracy.[136]

To summarize: lack of cultural structure represses aspects of self that potentially relate us to what is outside self. The "environmental" problems that result from this therefore appear as unconnected with our own mental health; and the "personal" problems that also stem from this repression appear to have no social or environmental implications. While this is, as Hillman suggests, a disaster for our democracy, it is also a disaster for the natural world; for it isn't just nature within us that finds difficulty in finding its voice within the anthropocentric order; but also nature in the world outside the self. However, the previous sentence itself illustrates this difficulty in articulation, since it reflects the mapping of an integrated nature onto a system of conceptualization within which it is dissociated into "nature within" and "nature in the outside world." This dissociation mutilates both fragments; "nature within" becoming reduced to such fragments as "instinct" or "genetic factors," and "nature outside" to an accidental collection of flora and fauna, robbed of their intelligence and communicative ability by our claimed monopoly of those qualities. And here the metaphor of "inner" and "outer" breaks down; for the "innermost" layers of our being are also those that are potentially most capable of communicating with the world *outside*. Our "inner," subjective experience often reflects a potentially *relational* mode of being, overflowing with feelings and passions that embroil us in the flow of life in a way that allows ego boundaries to become less impregnable. One of the unfortunate consequences of the inconsistency between natural structure within us and the largely economic "reality principle" that prevails in the world outside is that subjectivity splits into a technically literate and articulate part that operates competently, if mechanistically, in the industrialized world, and a privatistic awareness, confined to the inner self, that is largely forgotten and mute. This dissociation is also recognized in the object relations theorists' distinction between the "central ego" and the largely repressed "libidinal ego," the latter embodying our unrealized relational potential. These dissociations, however, are not an inevitable part of "human nature"; rather, they reflect the imposition of rationalism onto a subjectivity that can be much more than merely rational. These are losses that could be reversed as we learn to recover the resonances between ourselves and other fragmented parts of the world, and this recovery of wholeness unites the superficially disparate projects of the emancipation of the self and the ecological recovery of the natural world. Opening ourselves to emotions and intuitions that have previously been "denied symbolisation," as Rogers put it, is a potentially relational as well as an individual process: we recover not simply repressed feelings, but repressed relation. As we do so, the world comes alive again, resonating with an exuberant subjectivity. Thus "nature within us" and "nature in the world outside" may become, simply, "nature"; and our own identity and destiny become—or, better, are recognized and felt as—intertwined with those of the natural world.

This, however, is a potential within Rogers' work that he left latent and therefore vulnerable to assimilation by the "personal growth" industry, and more generally, by industrialist realities. Any theory of personal being will be interpreted according to the social realities in which it is embedded; so that in a society which fosters a radically narcissistic, survivalist type of mentality, any theory of personal growth that does not explicitly challenge these qualities is likely to be seen as consistent with them. Ultimately, then, Rogers' work and that of other humanistic psychologists, lacking an adequate way of articulating experience, is in danger of collapsing back into individualism. We will explore its unrealized potential in a later chapter.

Conclusions

Psychology, as we have seen, is made up of a number of approaches that at first glance seem to embody very different assumptions about the person and the world we inhabit. In this chapter, we have looked at several of these approaches, and what this brief review suggests is that in spite of their obvious differences, these various psychological perspectives are in many respects united in their anthropocentrism and in their tacit acceptance of the isolation of the person from the world. These assumptions are probably the two most important bases of our exploitative attitudes and behavior toward the natural world, and in these respects, psychology is entirely consistent with other aspects of the industrialist enterprise, including technology, commerce, and the ideology of "development." As Sigmund Koch ruefully summarizes the situation:

> The mass dehumanisation which characterises our time—the simplification of sensibility, homogenisation of experience, attenuation of the capacity for research—continues apace. Of all the fields in the community of scholarship, it should be psychology which combats this trend. Instead, we have played no small role in augmenting and supporting it.[137]

Thus psychology is, in many of its manifestations, part of the problem that we need to address, rather than a part of any potential solution. Deeply compromised by its rootedness in those same colonialist ideologies and practices that have soured our relation to the world, psychology is capable of suggesting only of the most superficial and ineffective "solutions" to environmental problems. How, then, do we escape from the intertangled "reality" of industrialism in order to recover our resonance with the world? To begin to answer this question, we will need to loosen our identification with the ego, exploring the emergence of the modern self and perceiving the way in which psyche and technology have evolved together to create our present situation.

Chapter 3

THE COLONIZATION OF THE PSYCHE

The snow is falling *out of control* in Vermont!

—TV Weatherman

Why the Self Became a Problem

Any form of selfhood, if the individual is not to degenerate into a chaos of conflicting impulses, must embody some overriding principle of integration to which all behaviors and tendencies must be subsumed, at least most of the time; and in this respect the Western self is no different to any other. Whether we live in Borneo or Bournemouth, mental health has a good deal to do with the extent to which all aspects of our mental and physical functioning are coherently and consistently organized in the service of goals that, while they may evolve, are not haphazard; and in any society, fragmentation of the self is recognized as threatening the very basis of selfhood. The individual, then, will always be recognizable as an entity in his or her own right: one that is to some extent and sometimes and in some ways autonomous, with recognizable characteristics that differentiate us from other such individuals. Although we are not entirely independent beings whose behavior is totally predictable from hypothetical internal mechanisms such as "personality," neither are we puppets who dance in unison to the tunes played by external forces. Furthermore, we are not constructed from varying proportions of each of these two unrealistic extremes; for the range of possible interactions between an individual who is partly autonomous and the diversity of structures that lie beyond our physical boundaries allows a multitude of interactional possibilities that together define the vitality of the world.

From this point of view, then, individual autonomy is not necessarily the problem that some environmentalists and social theorists have suggested. While *individualism*—that is, explaining everything by recourse to structures within the individual—*is* a problem, *individuality* need not be. The reason for this is that any natural evolutionary process, whether ecological or cultural, tends toward greater differentiation; but this differentiation, in turn, evolves within ever more sophisticated integrative structures, so that the parts of any healthy system function through a balance of autonomy and cooperation. If this balance breaks down, as it seems to have done within industrial society, it makes more sense to identify the disintegration of structure as responsible rather than the autonomy of the individual parts. Blaming the individual, in other words, itself reproduces and assumes the individualism that is part of the problem, since in doing so we implicitly deny the significance of structures outside the individual, and so perpetuate one of the key features of industrialist ideology. We therefore need to turn our attention to these domains that individualism denies—those natural and cultural structures whose essence and existence are repressed in the modern world. If the modern self can be said to be destructive, then this destructiveness also has a good deal to do with the annihilation of structure that in a healthy culture would extend individuality into the world "outside." In a process of dialectical co-reduction, this loss of integrative structure narrows the self toward what we today understand as the "individual," and simultaneously reduces the world to a pile of "raw materials." This being the case, the modern self projects its own structures onto the world as a precursor to physically assimilating it, in a process rather analogous to the way the New Zealand flatworm reduces its victims to a sort of homogenous jelly that it then ingests. The self in modern industrial society tends toward this individualistic self, which is also the self assumed by mainstream psychology; and at its most extreme, it is the self of the conquistadors, the colonizing self.

It is this easily identifiable individualism that has seduced many environmental writers into suggesting the need for a relaxation of the boundaries between self and world, so that we become more continuous with it. But as I will argue later, a less regressive response to individualism is to rectify the lack of integrative structure, enabling individuals to rediscover their resonance with structures in the external world. The Aivilik term *sila*, which we discussed in the previous chapter, expresses this attitude well. I refer to this form of selfhood as "resonant," because it is capable of resonating with what is outside itself, so defining larger structures that include the self while not completely defining it. In later chapters, I will develop this initially puzzling notion of resonance, and the forms of self and world that it implies.

Given that we are socialized into a form of individuality that mostly denies this possibility of resonance between ourselves and the world, extending ourselves into structure beyond our own boundaries requires a considerable voyage of personal discovery, the prototype of which is Aldo Leopold's conversion from the simplistic calculus of conventional game management in which "fewer wolves meant more deer, [and] no wolves would mean hunters' paradise,"[1] to a more-than-rational awareness that he described as "thinking like a mountain." Such conversions represent the glimmerings of an awareness that selfhood can be defined not only by its *contradistinction to* the world, but also through its *resonance with* the world; and this resonance invites us to perceive and participate in the world in a more-than-rational way, drawing out in us aspects of subjectivity that normally remain unrecognized and dormant in modern society. An ecological subjectivity, then, is both an enhanced awareness of the character of the world and an enhanced self-awareness; and these two enhancements cannot occur in isolation from one another. Rather, they imply each other, revealing a whole in which the accepted configurations of both self and world are simultaneously redefined. If conventional forms of awareness assume a detachment of self from world, therefore, the participative awareness that environmentalism reaches out toward is one that implies a reintegration. But this is to take us a little ahead of our argument, into areas that I will discuss in later chapters; so let us pause to examine the developmental processes—both personal and historical—that have led to the stilling of resonance between the self and the world within the industrialist landscape.

The story of modern consciousness tells of the gradual separation of an increasingly self-conscious individual from its surroundings—a separation that is both the bane and the glory of the modern psyche. If we put aside the discontinuities, regressions, regional variations, and long periods of apparent quiescence in favour of a long-term view, historical studies of European selfhood demonstrate the coalescence, over the past several millennia, of an egoic self that experiences itself not only as separable from its context, but as *permanently* separate. We can trace the origins of this modern self at least to the early Greek era, around one thousand years before the birth of Christ. According to Julian Jaynes, it is in the contrast between the style of the *Iliad* (originating somewhere between

1230 B.C. and 850 B.C.) and that of the later *Odyssey*[2] that the first clear evidence of the emergence of individual consciousness appears. The *Iliad* contains few terms indicating will, mind, or other features of consciousness: human action is typically determined by the gods rather than by individual intelligence. For example, at the beginning of the poem, we are told that Achilles' anger with Agamemnon causes him to reach for his sword. However, at that moment, Athena appears, holding him back and warning him not to indulge his anger. While today we would see this account as a poetic expression of conflict within the individual psyche, in Homer's era such divine action was taken for granted as part of everyday life.[3] Nature, too, was part of this world, in that the gods did not act in opposition to the natural order, but rather through it, so that the early Greek cosmology was in an important sense nonanthropocentric. As Bruno Snell puts it: "Even Hera forcing Helios to plunge quickly into the ocean remains within the limits set by nature since Helios is envisaged as a charioteer who may well lash his steeds on to a greater speed. On no account must she be thought to have sought to disturb the processes of nature by magical means." This contrasts with the God of the Old and New Testaments, who repeatedly demonstrates his power by acting *against* the natural order.[4] The *Odyssey,* however,—which, Jaynes argues, appeared at least a century later—is peopled by conscious beings who seem to have some control over their destinies, who experience pounding hearts and behave deceitfully or honorably, and who are aware of themselves as integrated beings occupying particular locations within space and time. As Snell puts it, whereas in the *Iliad* "each single turn of events is determined by the gods . . . in the *Odyssey* the gods may be said to act as permanent companions."[5] Another indicator of the emergence of a more autonomous self is the concept of "possession"—that is, the idea that consciousness can be taken over by an entity external to the self—which seems to have appeared between the time of the *Odyssey* and about 400 B.C., indicating the presence of a consciousness that *could* be taken over.

During these early centuries of consciousness, what was experienced as within the boundaries of the self and what was outside were defined changeably and fluidly. The emotions were no longer identified with particular gods, as love was with Eros or Aphrodite, and their location was not often clearly specified. Thus grief or joy might be properties of a situation as well as of the individual psyche. As Ruth Padel puts it in relation to early Greek writing:

> When tragic poets wrote about what was inside people, they are also writing about what is outside, as their culture represents it. Outside explains inside, and vice-versa.

> The two-way connection between them is fluid, ambiguous, mercurial, transformative, and divine.[6]

It was not that Greeks living in the fifth century B.C. could not clearly distinguish between inner and outer, but rather that inner and outer did not exist as distinct categories to confuse. Aspects of the life-world that to us would seem quite unrelated or merely figures of speech echoed each other within a fluid, metaphoric world in which the separation between "metaphor" and "reality" was unclear or nonexistent. For fifth-century Athenians, according to Padel, "The image is not a vehicle for explanation. It *is* the explanation. . . . 'Filled with fear'; 'heartsick'; . . . these are metaphors that imply specific images for what is inside us."[7] Language, image, and reality were, therefore, overlapping parts of a whole. They operated within a close-knit, systemic world where no part was separate from any other part, where the structure of language resonated with the structure of the world, and where an image was not so much a phantastic product of an imaginative individual psyche, but was something that grew out of the world itself. Within such a universe of meaning, metaphor becomes more than a poetic device with no material relevance: rather, it directly expresses something about the character of the world. For us modern earth-dwellers, in contrast, metaphors have become

> Unmeaning fossils that do not match what we now believe is inside us. . . . We tolerate extraordinary dissociations between what we think is inside us and what we imply is inside us when we speak of our feelings. We, not they, are the cultural oddity. . . . We have abandoned most of the physiological perceptions of which these images were a part, yet we have kept on their image-system, calcified in our newer languages.[8]

As George Lakoff has demonstrated, many of the words we use, especially those referring to emotion, originate in bodily experience;[9] and our lack of awareness of these roots of our modern languages indicates the extent to which our relation to the world outside ourselves has been denuded of meaning. If we can be said to have separated from the world in a material sense, given the extent to which we conceal our dependence on many of the natural processes that ultimately support our lives, this material separation is encouraged, confirmed, and underpinned by a *conceptual* separation in which language and thought are seen and experienced as products of the mind, having only a nominal relation to our physical embodiment or to that of the external world. Today, and especially since Saussure, the formal structures of thought and speech have declared their independence from the organization of the world; and they often intersect with it in ways that hide rather than illustrate the long-forgotten structures of the world.

If early Greek civilization was mostly systemic and organic in its organization and experience, later Greek society changed in directions that are more consistent with the deep-rooted assumptions of our own era. As Val Plumwood has shown,[10] many of the dualisms that underlie modern rationalism permeated the fourth-century B.C. philosophies of Plato and Aristotle, resulting in prejudices and inequalities analogous to those that have existed during the current century. These dualisms, as Plumwood shows, include the familiar ones that underlie what she refers to as the "hyperseparation" of man and woman, human and nature, and mind and body that has so profoundly influenced the course of the modern world. Nevertheless, it was not until the Enlightenment that Europe once again moved decisively, and—from an environmental viewpoint, fatefully—in the direction spearheaded by later Greek philosophy. During the centuries between, the modern self, with its distinct boundaries and lessening sense of relation to the world "outside," gradually emerged in a process that was neither smoothly continuous nor easily traceable; and the process of change seems to have slumbered or gestated during the centuries between the demise of Greek civilization and the eleventh-century A.D. "rediscovery" of self.[11] By the twelfth century, however, individual consciousness was becoming clearly established; and complementarily, the idea of the external world as a separate and stable entity was accepted. This external world, nevertheless, seems to have been an order of which humankind was still an integrated part. As Marie-Dominique Chenu argues, the relation between humankind and the world contained a hierarchical order that implied "a continuity that is at once dynamic and static in principle. . . . We are far from a discontinuous universe in which each being possesses its own dynamism and intelligibility wholly and only within itself."[12] In other words, the separateness of the individual was not an ossified, assumed separateness, but was a temporary, provisional stance that was framed by a larger integration.

Twelfth-century Europeans, then, were not "individuals" in the modern sense of beings whose experience of self presupposes a high degree of autonomy and independence from what is "outside" ourselves. Rather, they inhabited a world in which meaning, as well as material relations, were jointly defined through the cooperation of many parts that would today be regarded as largely independent. The human mind was an intrinsic part of this ensemble as well as an organ that was capable of sensing its wholeness. In turn, this was a wholeness that was grounded in the Divine: as Lee Patterson notes, "the Middle Ages is a time in which all forms of human activity were understood in relation to an original perfection. There is hardly an area of life—whether it be political, institutional, intellectual, spiritual, or artistic—in which medieval people did not legitimise their activity by reference to transcendent values and first principles. Medieval culture understood its own activity as the effort to ground itself upon a divinely authored originality."[13] Today, in contrast, this world has fragmented, resulting in an enormous increase in technological power and a more covert diminution of meaning: the "laws of nature" are no longer experienced as pregnant with spirituality, and "intelligence," as we saw in the previous chapter, is viewed primarily as an attribute of humans and humanly constructed machines, but not of the world outside ourselves. If there is a wholeness to be sensed, it resides increasingly within economic globalization rather than within any residual integration of mind, spirit, and cosmos. Expressing this contrast between the modern person and the pre-Renaissance European, Owen Barfield argues that

> [I]t is clear that [man] did not feel himself isolated by his skin from the world outside him to quite the same extent as we do. He was integrated or mortised into it, each different part of him being united to a different part of it by some invisible thread. In his relation to the environment, the man of the middle ages was rather less like an island, rather more like an embryo, than we are.[14]

This organically integrated model of the universe continued to structure human experience until it began to give way during the Renaissance to the more mechanistic conceptions discussed later in this chapter. While this change, however, has sometimes been seen in terms of a revolutionary transformation of the organic social paradigm to the mechanistic one, medieval society undoubtedly possessed many of the characteristics of the later period, although these were balanced and contained by the frame of a supposed divine order. What happened in the sixteenth and seventeenth centuries was a bursting of this frame, so that the changes that had gestated within the European cultural scene for centuries came to fruition in an explicit recognition of the emergence of the new order of science and commercialism. This enormous social upheaval was simultaneously psychological and cultural; for the fading of the organic cosmos also relinquished the wholeness of the world, so that all its aspects came to be structured by the huge splits that were emerging, such as those between culture and nature, conscious and unconscious, and masculine and feminine.

The tearing apart of the world that these developments imply was a violent and destructive one. While medieval Europe had embodied tensions between the poles of the dualisms noted above, these tensions were more often than not held in check by their overall framing within the organic wholeness of the world. After the middle centuries of the second millennium, however, social reality itself became structured by these dualisms, and the integrative structures that balanced them faded, so that today we tend to think of conscious and unconscious, self and other, and particularly culture and nature, as *opposites*. Comfortably inhabiting a world where these dualisms are taken for granted, those of us who live in industrialized society tend to forget the brutality of its birth—the witch burnings, the Inquisition, the slaughter of animals—which originated in the need to extinguish those concepts, entities, and cultural forms that integrated the poles that were about to become dualistically separated. Barry Lopez, contrasting the integration of the type of world accepted in many tribal societies with the violent persecutions that characterized emerging modernity, recognizes the connection between this violence and the human (and largely masculine) attempt to achieve a distance between the emerging self and what became not-self:

> In a hunter society, like that of the Cheyenne, traits that were universally admired—courage, hunting skill, endurance—placed the wolf in a pantheon of respected animals; but when man turned to agriculture and husbandry, to cities, the very same wolf was hated as cowardly, stupid, and rapacious. The wolf itself remained unchanged but man now speaks of his hated "animal" nature. By standing around a burning stake, jeering at and cursing an accused werewolf, a person demonstrates his allegiance to his human nature and increased his own sense of well-being. The tragedy, and I think that is the proper word, is that the projection of such self-hatred was never satisfied. No amount of carnage, no pile of wolves in the village square, no number of human beings burned as werewolves, was enough to end it.[15]

Such violence is today distanced from consciousness conceptually, geographically, and temporally. We forget the violence that still occurs in the modern world, at the unseen fringes and beneath the glossy surface of our affluent lifestyles, exported to places and situations that we prefer not to be aware of—a point I will explore in the next chapter; but we also ignore the violence that is sedimented into such aspects of our lifestyles as the "objective" vision of science, or our own predominantly intellectual orientation, or the domestication of the landscape. This is the violence of imprisonment rather than of warfare, expressing itself in the permanent denial of potentialities, in the accepted suspension of vitality, rather than in the crushing of already flourishing life. It is an imprisonment that confines both jailers and jailed, for the drastic simplification and ordering of the landscape is reflected in a complementary psychological and spiritual reduction. And if our detachment from the "nonhuman" world was achieved at such cost, what, we might wonder, is the character of the latent emotionality that might be released in their recombination?

The time scale of the events involved in the birth of modernity is necessarily vague, uneven, and often controversial. But for our purposes, the significance of these developments lies in the overall direction rather than the details; and this direction is fairly clear. With the emergence of consciousness, Europeans moved away from an unreflective immersion in the world, and discovered the world as something external that could be manipulated for their own benefit. In this, they gained power over what they distanced themselves from; but the price they paid was that of estranging themselves from that part of themselves that existed beyond the mind and beyond the physical boundaries of their bodies. Thus the ability to experience self as separate from the world became not merely one stance among the many allowed by the assumed wholeness of the world, but rather one that ossified into an *alternative to* this integration, trapping Europeans into a permanent separation from which there was no way back.

Today, we inherit the legacy of these trends, and we are born into a cultural context in which the separation of the individual psyche from the world is experienced—at least after the initial processes of socialization are complete—as a natural reality. As we saw above, separation is not in itself necessarily pathological, *as long as* there also exist integrative structures that can rejoin these separated elements of the world into some sort of meaningful order. And as we will see in chapter 4, these integrative structures are those of culture, so that in a world where the sense of individual autonomy is strong, there is a need for correspondingly powerful cultural mechanisms that restore the wholeness of the world. The balance, or lack of balance, between these separative and integrative tendencies provides the essential backdrop to all debate about the health of the natural world, and also to all comprehension of the character of individuality: it makes no sense to consider "environmental" issues separately from the organization and experience of selfhood. "Environmental" issues, then—to recall a motif that we will encounter frequently throughout this book—are simultaneously *psychological* and *cultural* issues, and must be addressed conjointly with them.

The Effects of Colonization

The development of modern industrial society has been traditionally understood in terms of such concepts as "progress" and "civilisation," through political mechanisms that are "democratic" and that allow personal "freedom," and through philosophies that are "rational." However, as we enter the twenty-first century, the limitations of modernism become more apparent, and these concepts, and the certainties that underlie them, have come under increasing suspicion as the future that we are "progressing" toward comes to seem less like a technological and social utopia and more like a descent into the sort of debacle illustrated most graphically in films from *Modern Times* to the *Mad Max* trilogy. Under these circumstances, it is easy to get swept up in one of several unthinking directions, ranging all the way from dogged defence of the traditional values of modernity (i.e., we need *more* technology, progress, etc.; or *better* technology, progress, and so on) to the totalistic rejection of modernism or Enlightenment thought. These polarized opposites lie along an industrialist dimension that is defined by such dualisms as "primitive" versus "modern," all other dimensions being excluded. According to this view, we can choose between going "forward" into a world that is more technological, more individualistic, and more rational; and going "back" into a world that is more primitive, less differentiated, and governed by "myth" rather than by "rationality." This view, therefore, excludes alternatives such as that suggested here: moving toward a world that is more than rational, at least as differentiated, and also more structured—an alternative that I will develop in the course of this book. The choices that are available to us are not only those of continuing the historical separation between self and world that we traced above, on the one hand, and returning back down the same historical path toward a less differentiated world, on the other. We can be, and must be, more imaginative than that, recognizing that an understanding of the changes in selfhood that have occurred over the past several millennia may imply a need for other, compensatory changes that are at present beyond the conventional political imagination. I have explicitly pointed to the need for integrative structures as one such change.

Within the academic world, the current fashion involves a sort of deconstructive feeding frenzy in which any values or structures are grist to a mill that is capable of undermining any faith in its zeal to reveal illusions, but that is utterly unable to recognize the need for structures of some sort if our lives are to work at any level. This is a fashion that embodies industrialism's denial of the need for reintegration, and as such it is clearly inconsistent with an environmental awareness. That the nihilism of this stance, and its contribution to our current problems, have yet to be generally acknowledged says much about the extent to which we are still held in thrall by the values of individualism and atomism, even when, as is usually the case today, they are ritualistically denounced. The pervasive influence of these values, in attacking traditional cultural structures and preventing the formation of new ones, ensures that the resulting vacuum will be filled by those powerful and degraded structures, such as those of capitalism, which are simultaneously destructive to human integrity and to the integrity of the whole natural world.

This is not to deny that traditional structures have often become repressive, supporting practices and beliefs that seem indefensible today. However, our failure to recognize that we simply cannot function adequately in the absence of at least minimal cultural frameworks has led us to see structure itself as repressive, so that emancipation, according to this view, simply involves releasing latent individual potential from the oppressive weight of moribund convention, allowing each individual to flourish in their own unique fashion. Such views confuse the vital role that cultural structures can and must play in our lives with the repressive functions of institutionalized religion or elitist class frameworks that serve the affluent and the powerful. The denial of culture is a limitation of 1960s philosophies such as those of R. D. Laing and Carl Rogers which operate in terms of a "real self" that is ready to flower when repressive social forces have been removed; and such philosophies fail to recognize the validity of anthropological insights that selfhood cannot function in a cultural vacuum any more than a plant can grow in the absence of soil. That cultural anthropology is the only academic specialism fully to have acknowledged this is a measure of the extent to which the industrialist severance of the individual from our natural context pervades academia. We will have more to say about the significance of culture in the next chapter; but for the moment, let us simply note that the healthy reintegration of the self and the natural world problematizes the accepted boundaries of each of these, and suggests the need for inclusive structures that permeate both.

Conventional science's narrow focus blinds it to those inclusive structures that exist, and those that could exist, in the biosphere. Its reductionism foregrounds the structures of small parts of the world, and it becomes progressively less able to recognize structures as they become larger and more systemic in character. This is the necessary price we pay for the exactitude and power of its vision. It is not accurate to blame the scientific approach for its omissions, since it is our cultural misapprehension that the scientific view of the world is the way the world *really is* that is the problem. If science was seen realistically in terms of its potential and its limitations, within a wider frame of reference that included those things that science omits, then we would not be misled into blaming science for its inevitable partiality. However, the post-Enlightenment misapprehension that science accurately reflects the structure of the world in all its manifestations is only one aspect of a more general scientistic blindness that pervades our whole ideological milieu and that, consequently, has infected and distorted the psyche. The problem, then, is not so much that science is limited; but rather our blindness to its limits, and our consequent inability to perceive the need for those other forms of structure that are necessary in order to make our relation to the cosmos more complete. The utterance of a recent British prime minister expresses this social and cultural atomism succinctly: "There is no such thing as society. There are individual men and women and there are families."[16]

The history of subjectivity, then, especially since the Enlightenment, reflects the increasing colonization of the human psyche by economic and scientistic structures and the corresponding denial of the natural order. This colonization has enabled us to make enormous and genuine advances in the quality of our lives. However, its very success has allowed us all the more easily to forget that science's partiality should be complemented by an awareness of what it omits, and that its reductionism has to be complemented by a counterbalancing integration. The assimilation of the natural world as "raw materials" for industrial processes is part of a more general assimilation that also encompasses human subjectivity, and that reflects the elevation of what should be a temporary, pragmatic orientation into a philosophy of life. This colonization of subjectivity can be illustrated by reference to various areas; and in the following two sections, I will focus on vision to illustrate the extent to which the ways we perceive have become aligned with the requirements of industrialism.

Vision and Technology

We have already noted in chapter 1 that modern industrial society's emphasis on the visual sense introduces certain problems for the environmental theorist. These problems are compounded by the particular style of vision that became dominant in Europe around the time of the Renaissance, and that has been seen as a prerequisite for the development of technology.

Linear perspective vision—that way of representing the world in which greater distance from an observing individual is suggested by proportionally diminishing size—was familiar to certain ancient Greek and Roman artists and technicians, notably Ptolemy; but it was not until the fifteenth century that it was adopted by artists such as the Italian painter Brunelleschi, and formalized by Alberti in his *De Pictura*.[17] Today, it is only with difficulty that we can empathize with painters of this period who needed the assistance of grids and frames in order to perceive the world in a way that we take for granted as "natural"; and yet the survival even to the present of alternatives to this type of vision reminds us that perception is heavily saturated with cultural ideology. For example, Colin Turnbull's study of the Ituri pygmies of the Congo relates how this forest people's visual style differs from our own. In the following extract, Turnbull has driven Kenge, an Ituri, to the edge of the forest, where for the first time, he looks out over the miles of rolling savannah:

> Then he saw the buffalo, still grazing lazily several miles away, far down below. He turned to me and said: "What insects are those?"
>
> At first I hardly understood; then I realised that in the forest the range of vision is so limited that there is no great need to make an automatic allowance for distance when judging size. Out here in the plains, however, Kenge was looking for the first time over apparently unending miles of unfamiliar grasslands, with not a tree worth the name to give him any basis for comparison. . . . When I told Kenge that the insects were buffalo, he roared with laughter and told me not to tell such stupid lies.[18]

Such perceptual styles as Kenge's, which do not depend on linear perspective, have become rarer as Western viewpoints have become increasingly universal, although even today they survive in a few isolated areas of the globe. Edmund Carpenter relates that "native artists of British Columbia represented a bear, say, in full face and profile, from back, above and below, from within and without, all simultaneously. By an extraordinary mixture of convention and realism, these butcher-draughtsmen skinned and boned . . . to construct a new being . . . that retained every significant element of the whole creature."[19] Such representations suggest a form of consciousness that, rather than being located in a single place from which the world is viewed, roams around the scene in a more intimate fashion. To many inhabitants of the developed world, however, this style of painting simply looks odd: in spite of any intellectual awareness we might have about the possibility of other perceptual styles, the world appears to present itself to us in a given way, and we accept this presentation as natural and accurate. Even when we are aware of the distortions involved, we are still subject to visual illusions such as that produced by Ames' Room. A story that Gregory Bateson told about Picasso makes a similar point. The artist was traveling in a train, when a stranger asked him why he didn't paint things as they actually appeared. Picasso said that he didn't understand what the stranger meant; so his accuser pulled a photo of his wife from his wallet. "There," he said, "That's how she *is*." Picasso replied, rather hesitantly: "But she's very *small*, isn't she? And rather *flat*?"[20]

It is as difficult for us to perceive in a manner that does not involve linear perspective as it was for Brunelleschi's predecessors to experience a modern Western style of perception; and, as Robert Romanyshyn notes, "what originated with Alberti and his times as a way of seeing has become for us a world that is seen."[21] We have come to believe that the form taken by the world when viewed according to linear perspective represents *the way the world actually is*, rather than one mode of vision selected from a number of possible modes. This draws our attention to the difficulties we have in escaping from the ideological "pull" of assumptions that have become sedimented into our cultural context over many centuries. Yet it is just such assumptions that environmentalists need to challenge if our solutions are to be more than cosmetic.

For our purposes, the significance of the development of linear perspective lies in the location and attitude of the perceiver that it implies. For whereas earlier modes of perception typically imply an "immersion" of the perceiver in the context being perceived—as was the case with the Ituri—linear perspective vision implies that there is a definite point from which a view is seen. In other words, the perceiver is *separate from* what is seen, a stance that is consistent with Descartes' identification of a *res cogitans* that is separate from and can manipulate *res extensa.* If we view the world as Descartes suggested—as simply matter in motion—then it is experienced as more psychologically distant from us than would be the case if we saw it as part of the same spiritual, cultural, and natural community within which we ourselves existed. Compare, for example, the Cartesian vision that sees landscape as if viewed through a window, with the Australian aboriginal sense of being immersed in a timeless landscape that one is integrated into through the media of song and myth.[22] Another example is given by Romanyshyn: although Galileo's telescope caused the moon to appear closer to us in a purely visual sense, it also distanced it from us in less obvious ways; for we are in an important sense more intimate with a moon which is part of a cosmology and a mythology that pervade one's universe of meaning than with one which is perceived simply as a chunk of rock hurtling endlessly and meaninglessly around the earth. Phenomenologically, nearness consists in the quality of felt resonance that something has with our own experience. An essentially mechanical cosmos made up of chunks of inert matter to which we are related only by the laws of physics is one in which it is difficult to feel "at home"; and within such a cosmos we experience ourselves as "ontologically insecure," as shrinking toward isolated points of experience within a spiritually empty world. Our experience of the world today is as often mediated by the hours we spend watching TV as it is by direct, participatory experience; and our reliance on such electronic mediation makes us aliens within our own planet, heightening our sense of distance from events and processes.[23]

The distancing of the self from the world implied by linear perspective vision, which today is accepted as "normal" throughout the developed world, is associated with an equally unquestioned *conceptual* orientation. Vision and understanding are mutually dependent parts of a conceptual system that anchors us within a particular, constructed reality; and in the case of a society in which the dominant mode of perception is one that fragments the world and distances us from it, the obvious danger is that this constructed world loses touch with the real, physical world, for we cannot simultaneously be part of the world if our orientation is that of an observer. John Dewey recognized this in his critique of what he called the "spectator" theory of knowledge, arguing that if we are to take part in the life of the world, "then knowledge is a mode of participation . . . it cannot be the idle view of an unconcerned spectator."[24] In the event of this loss of contact with the real world, then likely symptoms might include unpredictable reactions of natural systems to human intervention, the development of social forms bearing little relation to natural structures, and a configuration of human selfhood characterized by ontological insecurity and schizoid withdrawal. Each of these symptoms has been commonly alleged to characterize modern industrial society. Linear perspective vision, because it tends to portray the world as fragmented, is "an analytical vision which decomposes the whole into parts, a vision whose power lies in its ability to isolate [and] decontextualise."[25] Allied to the maintenance of a stance of separation from the world, this fragmentation is an essential prerequisite to a technology that is grounded in what Romanyshyn refers to as "the violence of a reductive vision."[26] It is difficult for us to identify with, or empathize with, or relate with passion to, a world made up only of inert "things," a world that presents itself as merely raw material for human purposes; and in such a world our psyche shrinks into the mind rather than reaching out into the world. Thus the cycle of alienation and exploitation is completed.

Today, this reductive vision is taken for granted to such an extent that any other viewpoint—for example, one that would experience the world as alive with spirit and intelligence—would be seen as fanciful and "animistic." Some have argued, however, that the scientific worldview, as well as implying that human consciousness is the location from which all else is perceived, also had an opposite effect; that is, it displaced humankind from the center stage, as when Copernicus established the heliocentric nature of the solar system; or when Darwin applied evolutionary theory to the whole natural world, including humanity; or when Freud emphasized the power of the unconscious to exert control that is beyond human awareness.[27] Thus "the same vision that places us in the centre of things also displaces us."[28] This insight recognizes that we are merely temporary agents of mechanism, central to the establishment of a mechanistic world only in the early stages of colonization. Indeed, we will survive in a mechanized world only to the extent that we ourselves become mechanisms; although the term "survival" in this case begs the question of what exactly it is that survives. Can we, as mechanism, as cyborg, be said to be still human?

Art and Technique

But surely, one might object, one cannot damn the visual realm in so blanket a fashion? Surely visual art, and especially painting, represents a realm in which one can comment on and critique established assumptions, and so offer a locus of resistance to industrialism. Certainly, as Ernst Bloch has argued, such art can provide a vehicle for the conscious recognition of the "not yet," of potentialities not yet realized, without which technology appears to define reality in a complete and so totalitarian way, leaving us only with a vague, unarticulable feeling that "something's missing."[29] Similarly, visual art which refers to a more-than-visual reality, which uses vision to point symbolically to something beyond normal awareness, is intrinsically integrative; an integration that Picasso alluded to when he said, "If you paint, close your eyes and sing."[30] This ability to point to possibilities that industrialism denies makes visual art a potential ally of the attempt to envision healthy environmental futures which are at present unimaginable. Nevertheless, the visual arts, like most other branches of art, have often been subtly or explicitly drawn into the orbit of industrialism, becoming consistent with it even while appearing to offer alternatives.

The word "culture" was originally closely associated with terms such as "cultivation" and "agriculture," all these deriving from the latin *cultus*, worship, which in turn finds its origins in *colere*, to take care of, till, dwell.[31] Culture was thus originally conceived as reflecting or expressing the natural realm, as indeed it still is in many nonindustrialized societies. However, the fragmentation of the life-world that developed in European life during and after the Renaissance led to the positing of a cultural realm *in opposition to* the natural, so that art became a vehicle for imposing a "humanization" on an otherwise "wild" nature. This opposition itself implies a corresponding dissociation within the self, since "nature" and "culture" are equally parts of the human psyche; and this dissociation is today accepted as part of our everyday life and embodied in theories such as Freud's that posit conflictual components of the psyche such as the id and superego. No longer restrained by or patterned on natural structures, the cultural realm quickly became aligned with an economistic and materialistic ethos that feeds off a nature perceived as something alien to be vanquished, and eventually, a resource to be exploited.

Even if art contains a potential to subvert economistic structures, then, it has nevertheless often been covertly seduced into consistency with the commercial world. As early as the Renaissance, art can be seen as serving class and economic values rather than those of the majority of the population. Often considered to be separate from economic interests, paintings in particular were even then beginning to reflect and embody ideas and priorities that were consistent with the emerging commercial order, even if they did not explicitly support it. As Kenneth Coutts-Smith puts it:

> It would not seem to be coincidental that the Medici and their successors should have chosen and reinforced the medium of the visual arts to express and confirm the justification for their vision of a new, fragmented, and competitive structure of human social relations. . . . A concept of appropriation that is soon to declare itself as colonialist in nature can thus be seen to have initiated its central role in European culture from the very point of the emergence of a continental "European" consciousness.[32]

Furthermore, according to Coutts-Smith, these social developments were associated with psychological changes that, although subtle and gradual, were to lubricate the introduction of a commercial ethos, and so have far-reaching implications for our attitudes toward the natural world. These changes

> Operated inwards, towards a "colonisation," as it were, of subjective mental territory. As the first force [of colonial expansion] can be observed as co-opting the cultures not only of non-European peoples but also of the vanished peoples of the past, so the second force can be seen to launch an attempt to appropriate the whole twilight territory of the mind, the landscapes of dreams and fantasies, the preserves of psychology and psychopathology, the primitivism of childhood, the bizarre territories of superstition, magic, folklore, and the absurd.[33]

The colonization of humanity, therefore, while it most obviously occurred through the aggressive dismissal of other cultures and of any deviation from a religious orthodoxy that itself reflected the growing emphasis on the intellectual and the rejection of the physical,[34] also involved a fundamental mutation of consciousness away from the natural order toward an alleged "rationality." The mode of this colonization involved the creeping reduction of alternatives within the psyche as much as the imposition on it of authoritarian structures from outside, so that an awareness of the mysterious indeterminacy of the world came to be replaced by the "single vision" that Blake railed against. As the visual arts aligned themselves with this vision, the disparities between schools and styles covered up a more subtle convergence within a larger scheme that was not simply artistic. It is in this sense that Edgerton refers to a "renaissance paradigm" that denotes "a cultural constellation of related ideas . . . in which science, art, philosophy, and religion all interact and prejudice one another to the extent that no scientific invention, work of art, or philosophical or religious concept can escape the influence of the paradigm as a whole."[35]

In the artistic sphere, Coutts-Smith argues, this appropriation of consciousness later matured into a subjectivist style of painting that reified the separation of humanity from the natural world, and was thus concerned with representing a mental landscape from the perspective of this reified viewpoint rather than with expressing any more whole-hearted *immersion* of humanity within a natural context. Whereas previously art had been concerned with expressing the glories of a divine natural order through a process of mimesis, artistic creativity increasingly became an expression of individual creativity; and so the primary structure that determined its final form became that of the autonomous human mind rather than the whole natural world. This subtle but profound change

> Reversed the image celebrated by such . . . predecessors as Constable who envisaged a coherent and humanised landscape, and in this way projected an image of an absolute, fragmented, and dehumanised landscape. Immersed in their narrow stylistic concerns, the individual artists, many professing liberal, humane, and even "socialist" affiliations, nevertheless acquiesced in the restructuring of man's relationship with his environment which, ultimately, was profitable to restricted political interests. The capitalist social relations that were consolidating at the peak of the industrial revolution demanded a divorce between man and his natural environment in order that the masses might better accept the artificial environment of the industrial milieu.[36]

The distancing of the painter or viewer from the scene portrayed is therefore consistent with a style of perception that might be familiar to the scientist, and the increasing acceptance of this mode of viewing the world represents a sort of sensory totalitarianism that has become so "natural" to us that we are mostly quite unaware of it. In effect, the world depicted by many post-Renaissance painters, although neither untruthful nor inaccurate, is nevertheless an incomplete world which unwittingly facilitates our acceptance of an economistic social order.[37] Such art takes what it desires from the world, leaving behind those characteristics for which it has no use, in a manner analogous to the extraction of "ore" from "dirt."

That this is sometimes the case even among those who might be expected to be least enthusiastic about such attitudes was brought home to me clearly during a conversation I once had with a well-known nature photographer who organizes photo expeditions in the American West. I was enthusing about the potential of photography to communicate the realities of the natural world; but he insisted that photographic images were simply "raw material" (his term), a starting point from which one could use modern computer graphic techniques to produce an end result possessing commercial potential. Today, the snowballing use of electronic manipulation in landscape and wildlife photography lends itself to photography's assimilation to commercialism, to the production of stimulating images whose relation to the natural world becomes ever more tenuous. While such techniques are not *in themselves* destructive of our relatedness to the natural world, their replacement of more authentic images (and experience) of nature colludes with commercialism in replacing the natural world by ever more sophisticated simulacra.

The psyche that can allow this substitution is one whose sense of integration with the world has already been lost, and it is one that already existed when photography was in its infancy. Oliver Wendell Holmes, as early as 1859, suggested that photography was the most wondrous of our "conquests over matter."[38] Photography introduced an era when the "image would become more important than the object itself, and would in fact make the object disposable."[39] "Form," stated Holmes, "is henceforth divorced from matter. In fact, matter as a visible object is of no great use any longer, except as the mould on which form is shaped. Give us a few negatives of a thing worth seeing . . . and that is all we want of it. Pull it down or burn it up, if you please." In the light of such remarks, Stuart Ewen's comment that "technically reproduced surfaces were beginning to vie with lived experience in the structuring of meaning"[40] seems to understate the rapidity with which the "human" world was parting company with the natural one. The emerging relation of art to the more obviously exploitative aspects of industrialism is stated more bluntly by Holmes. "We have got the fruit of creation now," he suggested, "and need not trouble ourselves with the core. Every conceivable object of Nature and Art will soon scale off its surface for us. Men will hunt all curious, beautiful, grand objects, as they hunt cattle in South America, for their skins and leave the carcasses as of little worth."[41] Seldom has the attempted declaration of autonomy from the natural been stated with such stark clarity.

Similar attitudes, rather more subtly expressed, are dominant within nature photography today, and are revealed by the language used by photographers, who "shoot" a subject, "capture" an image, "freeze" an animal, and so on. As Ansel Adams presciently remarked in 1943, "the common term '*taking* a picture' is more than just an idiom; it is a symbol of exploitation."[42] Furthermore, "wildlife" photographers often use "game parks" in their search for saleable images: in other words, their photographs, while seeming to reveal a world that is still "wild" and elusive, in fact represent a simulacrum of it, a nature that is already imprisoned, a substitute that belongs to the alternative world of virtual reality, even before any further manipulation of the resulting images.

Of course, the replacement of nature by image did not go unopposed by those who were still sensitive to the inarticulate aspects of the natural world. The French poet Baudelaire was fearful that photography, in being taken as capable of an exact rendition of nature, would allow our awareness of "the impalpable and the imaginary" in nature to atrophy, so contributing to "the great industrial madness of our times." "Are we to suppose," asks Baudelaire, "that a people whose eyes are growing used to considering the results of a material science as though they were the products of the beautiful, will not in the course of time have singularly diminished its faculties of judging and feeling what are among the most ethereal and immaterial aspects of creation?"[43] The greatest photographers, too, sensed what was lost by the commercial framing of the natural: as Ansel Adams put it, "the highest purpose of . . . photography [is] to relate the world of nature to the world of man. . . . My approach to photography is based on my belief in the vigor and values of the world of nature."[44] "Group *f* 64,"[45] which included such notables as Adams and Edward Weston, defended the idea that photography should ideally bring to our attention what is latent within nature rather than succumbing to an intoxication with superficialities of technique and equipment. Weston, for example, argued that the restrictions and particular characteristics of equipment in no way diminished the photographer's ability achieve this aim:

> Limitations need not interfere with full creative expression; they may, in fact, by affording a certain resistance, stimulate the artist to fuller expression. The rigid form of

> the sonnet has never circumscribed the poet. . . . The mechanical camera and indiscriminate lens-eye, by restricting *too personal* interpretation, directs the worker's course toward an *impersonal revealment* of the objective world. . . . In the discipline of camera-technique, the artist can become identified with the whole of life and so realise a more complete expression.[46]

Weston here articulates truths that too many social scientists have ignored: that structure is essential for expression; that "freedom" from structure is less a freedom than a form of repression; and that if we enthusiastically ransack those structures through which we might express the "objective world," we will quickly be led toward the conclusion that there is no "objective world" to express. If this is the case, then "nature" can be assigned whatever meanings we choose to give it, and our conversation with the world becomes a monologue devoid of cultural or historical perspective.

What has happened here is that a genuine alterity—nature as mysterious, unpredictable, and sacred, but still entirely real—has been replaced by an alias within the humanly constructed realm, so that the otherness of nature has vanished. The "wild animal" has become a human category; and the beast itself has disappeared. The next step, of course, is the "realization" that "nature is socially constructed," which in effect takes human categories as the real, and denies the existence of what has not been, or cannot be, categorized. The awareness that much of the world is inexplicable and mysterious has been replaced by the illusory self-sufficiency of a manufactured realm that denies the existence of anything outside itself. This invisible *conceptual* assassination of what cannot fit our preestablished human categories is the essential prerequisite to a *material* assassination in which otherness is removed from the world through human action.

An example of the potential problems that await any presentation of nature which ignores such preestablished categories is discussed by Barry Lopez in his *Arctic Dreams*.[47] The painting he refers to is *The Icebergs* by Frederic Edwin Church, completed around 1860. American artists of the period, Lopez suggests, attempted "to locate an actual spiritual presence in the North American landscape," generating an atmosphere that "is silent and contemplative." Several critics "have described a peculiar 'loss of ego' in the paintings. The artist disappears. The authority of the work lies, instead, with the land."[48] Drawn into the landscape, Church seems to have been able to transcend the ingrained conventions of representation, and to have managed to express a more direct, if unconscious, communion with the natural world.

In a society which had extensively lost the cultural structures whereby land could speak directly without the intervention of rational consciousness, it is not surprising that this displacement of the "subject" was mystifying and potentially threatening to a population trained to perceive landscape in ways that had by then become second nature. In the absence of any human dimension which could frame the wildness represented by the painting, audiences were faced with an undomesticated "otherness" for which they were utterly unprepared. Lopez reports that when unveiled to an expectant public in 1861, reaction was less enthusiastic than Church, then at the height of his popularity, had anticipated. So Church took the painting back to his studio and added a part of a shipwreck, a mainmast with a crow's nest. Thus "improved," the painting was successfully exhibited in London, and was eventually sold in 1979 for $2.5 million, the highest price then paid for a painting in the United States. Lopez asserts that

> We can make very little sense at all of nature without resorting to such devices. Whether they are such bald assertions of human presence as Church's cruciform mast or the intangible, metaphorical tools of the mind—contrast, remembrance, analogy—we bring our own worlds to bear in foreign landscapes in order to clarify them for ourselves. The risk we take is of finding our final authority in the metaphors rather than in the land.[49]

Church's added mainmast is a direct pointer or referent to a familiar human world, and it frames the rest of the painting through its relation to the otherwise incomprehensible wilderness portrayed. Church's modification is less an addition than a *reduction*: the reduction of nature as mysterious and greater than the human realm to a "nature" relegated to a comprehensible human category.

Art as Integrator

Even potentially subversive areas such as art, then, are often subtly structured by the so-called "human" realm that attempts to assert its domination over the "natural." Within this realm, the alternatives that are on offer, while appearing to offer us a taste of genuine "otherness," are often generated by the commercial world as *substitutes for* alterity, so reinforcing our distance from and forgetfulness of whatever real alterity has so far succeeded in escaping the grasp of commercialism. The "Jurassic Park" that is presented to us by Steven Spielberg is not so much a "Lost World" as an invented one, reflecting a "past" that is a technological assimilation of the past to the present. From this perspective, most aspects of our lives become suspects within the systemic drama of industrialism. We cannot single out scientism and technologism as being the villains in the destruction of the world; for they are only the most visible manifestations of an ideology that is much more pervasive than we usually suspect. "Separate" areas such as leisure, recreation, and art are more than accidentally related to technology and commerce, and the appearance of choice weakens resistance by fostering the illusion of alterity while maintaining consciousness within the orbit of industrialism.

We cannot, on this basis, regard technological developments by themselves as sufficient to account for destructive attitudes toward the natural world. Rather, it seems more likely that apparently subtle cultural shifts provided a context in which a commercial ideology could displace or assimilate aspects of culture that either conflicted with it or were indigestible to it, so that the technological potential that until then had remained dormant could begin its gestation. We know of other cultures in which a potential for technological development has existed, but in which this potential has been restrained by existing, and counterbalancing, religious and cultural traditions—just as was the case, according to Carolyn Merchant, in medieval Europe.[50] Windmills and water mills, for example, were used by Buddhists in India to turn prayer wheels, but were not developed for industrial purposes.[51] Technology, such as it was, was therefore framed and controlled by a larger context, remaining subservient to the greater resonances that defined this larger context. Only when the cosmos was fragmented and distorted by what was previously merely a part of it—when technological capability, in other words, became technologism—was technology allowed to express itself in a way that was hostile to natural forms and structures. It is therefore apparent that the rapid development of technology within Europe was not due simply to the widespread recognition of the promise of technology to ease the burdens of life, but rather that this recognition was associated with profound cultural and psychological changes. We can discern the emergence of a configuration of selfhood that prepared people to take part in the new economic order, and that, by undermining forms of culture and religion which intrinsically related people to their natural environment, allowed the atrophy of those aspects of subjectivity that stood in the way of an economistic experience of the world.

If the realms of art and commerce, beneath their obvious divergences, have sufficient latent commonality to ensure that the former offers no serious threat to the latter, commercialism also has other ways of minimizing the possibility of subversion. One is the relegation of art to a separate existence, not seriously involved with the "real" business of life. We enter an art gallery as a diversion from this "real" business, not as an integral part of it. Just as the apparently innocuous addition of the mainmast framed Church's painting and located it within a familiar human realm, so the entrance to a gallery marks the boundary between the "real" world and the "artistic" one that provides a temporary refuge from it, rather in the way that the entrance to a national park marks the boundary of "recreation land." The temporary character of the "escape," however, merely confirms the inescapable "reality" of the world outside: "leisure" is defined as a socially approved diversion from the "real" world of work. Psychological life, then, is divided into a series of adjacent domains, each with its own rules and expectations. Connections between these domains—for example, taking "work" on holiday, or attending a job interview in shorts and a T-shirt, or allowing the insights of radical artists such as Blake or Picasso to suffuse one's work—is discouraged or considered frankly bizarre. Psychological dissociations thus converge with cultural, bureaucratic, and eventually environmental ones in the emerging geographical landscape of "separate" urban areas, parks, and industrial land. Potentially subversive aspects of experience are rendered harmless by their location within a frame that detaches them from the "real" business of life, paralleling the apparently benign "saving" of individual members of threatened species by their abstraction from the wild and "captive breeding" in zoos. In each of these cases, something is saved, but at the cost of its isolation from the rest of life, so that it accords with the fragmented ontology of industrialism. What is lost in such situations is not easy for consciousness to grasp, but it has something to do with the wholeness of the world: an integrated life-world becomes transformed into a series of antechambers, each separately labeled, policed by conceptual markers such as "work," "personal life," "art," and so on, reflecting and reinforcing the schizoid splitting of the modern personality. A fundamental restructuring, then, has occurred, both within consciousness and in the outside world; and a natural realm that previously framed and integrated everything else has been replaced by a commercial/technological frame that separates the components of life, replacing the complex and sophisticated interwovenness of life by simpler and more literal relations.

When art becomes split off from nature in this way, it loses the ethical force which derives from its capacity to express the order of nature, leaving a moral vacuum that allows it to be quickly assimilated to economic relations. For example, the pots made by Pueblo Indians of the Taos region of New Mexico are integrated into the life of the tribe in complex ways: the clay is collected on foot from places known only to the potters and their forebears, so relating the temporally distanced lives of different generations; the decoration reflects the history of the tribe and its journey to the mesa that is their home; and the making of the pot echoes the integration of the elements—heat, water, earth—that is fundamental to many other aspects of life. In such cases, art is an integrative activity that resonates with all other aspects of one's life-in-the-world. Among several southwestern tribes, however, artistic activity is in danger of losing this function, becoming a primarily commercial enterprise that is part of the tourist industry; and in these cases, the patterns—cultural, subjective, ontological—that were expressed through these activities have become fragmented. If pots originally carried subjectivity beyond the boundaries of the individual, so that it became continuous with the patterns of the world outside, the pot-as-tourist-artifact exists only within an economic space devoid of subjectivity, isolated from these patterns of meaning, so that subjectivity withdraws into the individual.

One of the more important functions of art, then, is its role as integrator of experiential spheres that are otherwise detached from each other; and an art that fails in this role is trivial and impotent, just as a nature that is fragmented is largely reduced to an artifact of human categorization, a zoo. Whether in external nature or within the human psyche that could be continuous with it, dissociation destroys the integration which distinguishes these realms from their reductionist simulacra. For example, if there is a separation between the aesthetic and the intellectual realms, then the truths of the intellect become an-aesthetic and disembodied; and correspondingly, the aesthetic realm becomes incapable of intelligent engagement with the alienations and omissions of technologism. In David Levin's terms: "just as the modern experience of truth (truth restricted to correctness) is cut off from the richness of a more primordial experience of truth, so our handling and using of things is cut off from the actuality of a more enriching encounter with the depth of their thingly presence."[52] Levin's "truth restricted to correctness" corresponds to the literal, ossified, conscious world of meaning in which technology operates, a world where truth is of an either/or type, where a process of "conjecture and refutation" assures that truth be *singular*. Nature, within this system of categorization, becomes a recreational location, a temporary retreat from the demands of day-to-day life, rather than an all-encompassing ground of our being from which the rest of our life experience derives. Church's mainmast plays a role analogous to the trails and mileage markers within a national park: it reconnects us with areas of experience from which we have become alienated, but in a manner defined by industrialism, by human consciousness. Only when such markers are not available to us—say, when we find ourselves lost in a wilderness area, or faced with an artistic creation that we cannot ideologically pigeonhole—do we run the risk of hearing the echoes of more subtle, more distant, and more ancient resonances that elude rational consciousness, resonances such as those experienced by Jack Turner, as he wandered through the then almost unexplored canyons of the Maze in 1964. Hiking the south fork of Horse Canyon, Turner recalls, he "was startled by a line of dark torsos and a strange hand on a wall just above the canyon floor. I froze, rigid with fear. My usual mental categories of alive and not-alive became permeable. The painted figures stared back at me, transmuted from mere stone as if by magic, and I stared back in terror."[53]

But the hegemony of the style of experiencing promoted by industrialism ensures that "normality" quickly reestablishes itself in such situations; and this occurs at both individual and social levels. Turner reports that after a few seconds, the "torsos became *just* pictures. My mind discovered a comfortable category for the original perception and the confusion passed. . . . Nevertheless, seeing them as representations did not reduce the emotion I felt. I was chilled, shivering, even though the air was warm." Two months later Turner was back in the Maze; but the pictographs had lost their power for him. He had become, he acutely remarks, "a tourist to his own experience—I tried unsuccessfully to recapture the magic of those first moments. I took notes, but they exceeded my power of description. I kept photographing, first in 35 mm, then with my $2^{1}/_{4} \times 3^{1}/_{4}$ Zeiss. But what I sought could not be captured with photography or language." And today, the Maze has become part of a national park, its canyons and mesas mapped and described in numerous guidebooks, its secret parts laid bare for all to stare at. Somehow, and for reasons that are elusive to us, its wildness has been diminished, although Turner hints at these reasons by suggesting that "maps and guides destroy the wildness of a place just as surely as photography and mass tourism destroy the aura of art and nature." The processes that cause us to become "tourists to our own experience," then, are both individual and collective; but each of these processes points to its origins in the common web of industrialist structures that pervade the psychological and physical landscapes of our era.

The world that environmental theory is striving toward re-engages these dissociated realms of life. The emphasis here is on reengagement rather than on the substitution of a "more correct" understanding for a "less correct," perhaps "scientific," one. In other words, scientific understanding should be reassimilated within a larger frame rather than abandoned for an alternative vision—a reintegration which Jay Bernstein points to in arguing that "the discordance between art and truth is misconstrued if regarded as an opposition that simply inverts their relationship: art and aesthetics are true while truth-only cognition [i.e., Levin's 'truth restricted to correctness'], say in its realisation in the rational sciences, is false. The challenge is rather to think through what truth, morality and beauty . . . are when *what is denied is their categorical separation from one another* . . . a separation . . . that is constitutive of modernity."[54] This separation, which implies the devastation of the shadowy middle-ground of culture that is now excluded both from the ecological and psychological realms, allows and normalizes both the physical degradation of the world and the erosion of meaning that makes consciousness consistent with this degradation.

To summarize, then: while art has the potential to challenge the commercial/technological system, it is often subtly assimilated to this system as "art," and as is the case with religion, has become somehow irrelevant to the "real" issues of life. Art that escapes from its frame, like a religion that escapes from the churches or a nature that escapes from its zoo or wilderness area,[55] would be a threatening thing indeed to the egoic self and to the commercial system that maintains it. One might, optimistically, say the same thing about a theory that escapes from the dry pages of academic journals and books.

The Colonizing of Human Intelligence

The form taken by modern subjectivity, then, is heavily influenced by the technological, commercial, and ideological structures that together define industrialism, although consciousness (a less inclusive term) finds it hard to discern this influence. This effect is both ontogenetic and phylogenetic: that is, industrialism shapes an individual's subjectivity during the course of his or her lifetime, and has also evolved jointly with subjectivity over historical time spans. But subjectivity is not *simply* the result of this influence, passively embodying our cultural and technological histories. It is also shaped by its residual resonance with the rhythms, forms, and processes that still exist within the natural world, and in particular, by our embodiment as a particular type of animal. And this is a resonance which is amplified by cultural structures—the rituals, spiritual forms, mythologies, and arts which, in a healthy culture, connect individuality to what individual consciousness cannot express unaided. As Alexander Argyros has put it, "although socio-institutional contexts certainly have a large voice in constituting the world of human beings, theirs is not the only voice. The kind of dualism that postulates an unbreachable gap between human culture and prehuman nature must be replaced by a systemic view of human culture situating it within a larger natural framework."[56]

The recognition that we are not constructed only by existing social conditions rescues us from the impasse of much postmodern theory, because it offers us an opening within which resistance to those conditions can develop. In other words, if existing social realities are out of step with the rhythms and structures of nature that we sense through our bodies, then our awareness of this disjunction becomes the basis of our efforts to change these social realities. My anger at the building of a road through my favorite wood, or my dislike of sitting in crowded trains on the way to work, or oversleeping and so arriving late for work—each of these bodily reactions contains the potential for changing some aspect of life in industrial society. For example, I might decide to join a local group set up to defend the wood, or to complain to the rail company about traveling conditions, or to leave my job so that I can organize my life in a way that is more attuned to my body's natural rhythms. This sounds like a straightforward feedback mechanism that could correct any divergence of the social world from the basic patterns of the natural; but as we have already seen, industrialism contains its own internal "logic" which often enables it to contain and override this sort of feedback. Thus road-builders' lawyers may manage to have objections to the road overruled in court; the rail company may write back telling me that it is "uneconomic" to replace the rolling stock; and I may find life a financial struggle now that my regular paycheck no longer arrives at the end of each month. In short, I may quickly discover that the available ways of aligning my life with my felt sensing of the world are few and inadequate. Not only that, but since I inhabit a discursive universe that clearly articulates the economic rationality which I sense to be oppressive, but that articulates scarcely at all my felt sensings which conflict with it, I may find myself passively conforming to this rationality rather than struggling to express the protest that I feel but have difficulty in giving shape to. I may therefore resign myself to the need for the new road and accept that commuting in crowded dirty trains is the price I have to pay for the security of a healthy bank balance. In other words, I learn to ignore and repress the now-mute awarenesses that previously troubled me, and find it more comfortable simply to live in accordance with an apparently all-embracing economic rationality. My behavior and awareness, then, have been brought into line with industrialism, and to this extent, they are "socially constructed." But our repressed and inarticulate awarenesses do not simply disappear: they are embodied psychosomatically within us, expressed by such concepts as the "libidinal ego" of object relations theory, and the "inner child" beloved of many a New Age therapist.

It is a grave mistake, therefore, to think of such concepts as these as expressing unchangeable and inevitable aspects of the human personality; for they reflect the pragmatic adaptation of the psyche to the current political scene, and the resulting sense of poignancy and powerlessness stems from the residual inconsistency between our archetypal, embodied psychic structure and the ideological structures that surround us. In a limited sense, therefore, this is a configuration of personality which is "socially constructed"; but, like the leopard-in-a-cage, it is a social construction which incorporates within itself the imprisoned subversive potential for living in accordance with the natural world. Carl Rogers' concept of "incongruence" between a conscious "self-concept" and an organically sensed "experience" is perhaps the best conventional expression of this inconsistency. Consequently, what begins in infancy as a tension between the individual and the social is transmuted by socialization into an "inner," "psychological" conflict that appears to exists only within the self. This alienation of the conscious individual from his or her own nature is simultaneously an alienation from nature outside: as consciousness and, increasingly, the world outside assume their industrialized forms, so they become consistent in reflecting industrialist rather than natural forms. In turn, this process concentrates emotion within the individual psyche: the vivid realm of internal object relations, often digitally amplified, becomes the "real" world, leaving the natural world outside drained of feeling and significance. The "rationality" of road-building, rail timetables, deadlines, supermarkets, and financial planning becomes a rationality that I increasingly feel "at home" with; and if I suffer from anxiety attacks or chronic mild depression, then a course of rational-emotive therapy will soon put an end to the "irrational" thoughts associated with these symptoms.

If the colonization of selfhood, then, does not eradicate our emotionality, our intuition, and our spirituality, it achieves its aim by splitting off and repressing these arational characteristics. A fundamental aim of the environmental project must be the recovery and articulation of these characteristics, and through them, the recovery of our resonance with the rest of the natural world. I will discuss these possibilities in the final chapter; but the first step in this direction must be to *recover our awareness of these splits and repressions*. With this in mind, let us return to the concept of "intelligence," and in particular, to its necessity exclusion of feeling, association, or spirituality.

The abilities that the term "intelligence" refers to, such as memory, verbal reasoning, spatial intelligence, and so on, are genuine and important ones, particularly but not exclusively within modern industrial society. Within an industrialist—and particularly a psychological—context, however, these abilities become detached from their emotional or cultural significance, reflecting a fundamentally pathological orientation to the world. For example, psychological studies of memory often use nonsense syllables in order to remove the potentially confusing effects of meaning. However, within a healthy culture, memory and meaning are intrinsically interwoven: among the Aranda of central Australia, for example, memory is experienced as part of physical reality, embedded not only within the songs and mythology of the tribe, but also within the physical features of the landscape itself. It is, perhaps, significant that the role of memory in consolidating the spiritual/physical world of the

Aranda was completely unrecognized by white scholars until the 1970s: as Strehlow points out, one "authority" on the Aranda, who claimed a good working knowledge of the Aranda language, remarked that "the songs of this tribe . . . are merely a collection of sounds and cannot be translated. They have no actual meaning."[57] In contrast to such tribal intelligences, the industrialist conception of intelligence embodies in pure form an attitude in which manipulation of qualities and quantities, abstracted from a mechanical world, replaces the wholeness of relationship that it forgets. As a temporary psychological stance, this withdrawal into abstraction is fairly innocuous; and probably we are all familiar with that psychological state in which we retreat from the sometimes exhausting demands of passionate engagement with the world into the comfortable cool aloofness of pure thought. But the reification of this mode of thought as the only mature and realistic understanding of the world and our relation to it is alienating and destructive; and the "schizoid" personality structure that embodies this reification is profoundly pathological. If this schizoid personality were a rare aberration from a healthier normality, then it would be of little consequence; but the findings of object relations theorists such as Harry Guntrip that the schizoid personality is "a universal phenomenon"[58] suggests that the schizoid splitting of feeling from intellect is a basic cornerstone of modern industrial society. This is a point that we will explore in more detail in chapter 5; but for the moment, we should note that splitting—the pathological mechanism that underlies schizoid personality development—is, in Joel Kovel's words, "the basis of Western civilization's estrangement from nature and attitudes of domination towards nature."[59] The psychological concept of "intelligence," then, and the abstraction of cognitive abilities from any moral or natural framework, are environmentally problematic not because there is anything inherently corrupting about these abilities, but rather because these concepts fragment natural structure into reified components that industrialism can use and assimilate. "Intelligence," therefore, embodies succinctly the mapping of the industrialist order onto the psyche.

Freud, in contrast to later and more radical psychoanalysts, demonstrated his unswerving allegiance to the development of "civilization" in his view that this splitting of intellect from feeling is essential to progress. "Our intellect," he wrote, "can function reliably only when it is removed from the influences of strong emotional impulses."[60] And in any conflict between our scientific and our spiritual aspirations, Freud had no doubt as to which was the ultimate authority: "Whatever may be the value and importance of religion, it has no right in any way to restrict thought—no right, therefore, to exclude itself from having thought applied to it." Freud was explicit in his belief that salvation lies in the hegemony of the intellect: "Our best hope for the future is that the intellect—the scientific spirit, reason—may in the process of time establish a dictatorship in the mental life of man."[61] The concept of "intelligence" assumes exactly this dictatorship of the intellect; and recognizing this raises profound suspicions about the "rational" ways in which we attempt to solve problems such as those we label "environmental."

Few of Freud's followers dared to question this loyalty to the modernistic assumptions of the day; and even in object relations theory, the challenge is more implicit than explicit. One who did dare was the Hungarian psychoanalyst Sandor Ferenczi, who not only resurrected the "seduction theory" that Freud had abandoned in 1897, thus challenging the individualistic foundations of almost all Freud's work after that date, but, consistently with the view suggested here, argued that the development of "intelligence" could be seen as a pathological reaction to a hostile world:

> Intellect is born exclusively of suffering. . . . It develops as a consequence of, or as an attempt at, compensation for complete mental paralysis. . . . The cessation or destruction of conscious mental and physical perceptions, of defensive and protective processes, i.e. a partial dying, seems to be the moment at which . . . there emerges . . . intellectual achievements.[62]

This process, according to Ferenczi, is the result of being overwhelmed by a hostile, mechanistic environment:

> It is to be called intelligent when the individual . . . assessing correctly the proportion of powers, chooses the only way of saving life, that is, giving in completely; it is true at the price of more or less mechanised, permanent change and the partial loss of mental elasticity.

This suggests that the "central ego" of object relations theorists such as Fairbairn is not merely the residue of a potential self—what is left after the arational elements have been shorn off—but is also the result of a more active process, a forced identification with a power so overwhelming that the ego's only hope for survival is to become consistent with this power. Thus the human animal's ability to adapt, in the limit, becomes the source of our demise, and the pathological aspect of "intelligence" is clear: its clarity and power are won at the cost of the integrity of the self and of the life-world. Just as a world made up of "natural resources" presents itself for industrial processes, so does a self fragmented into separate abilities and shorn of its relational tendencies; so that the domination of the world is effected not so much *by* humankind as *through* humankind, in the interests of a purpose that may be beyond our capacity to comprehend. As Ferenczi puts it, "Pure intelligence is thus a product of dying, or at least of becoming mentally insensitive, and is therefore in principle madness, the symptoms of which can be made use of for practical purposes."[63]

Our allegiance to the principles of "intelligence," especially if highly developed, therefore carries a severe cost in terms of our integrity; for only by removing our awareness of the variety, beauty, and spiritual fecundity of the world can we act toward it in a "scientific," "objective" manner. In Lewis Mumford's terms, the scientific vision "was accompanied by a deformation of experience as a whole," since "objectivity" requires that we repress arational knowledge of the world. The instruments of science, writes Mumford, "were helpless in the realm of qualities. The qualitative was reduced to the subjective: the subjective was dismissed as unreal, and the unseen and immeasurable as non-existent."[64]

The psyche, however, does not submit quietly to the elimination of those structures of meaning within which it could live in a fulfilled manner. Much human creativity can be understood as the attempt to fill the vacuum of meaning, to reestablish a relation to the world outside, and so to locate oneself more securely, in ways that are in some respects analogous to the tentative regrowth of forest which occurs after a clearcut. Creativity depends heavily on unconscious, symbolic modes of awareness that are less completely colonized by industrialism; and as such, it can be seen as involving a "regression in the service of the ego," an attempt to return to earlier ways of being in recognition that a wrong turn has been taken. But while the radical potential of individual creativity exists, such regrowth, whether psychological or arboreal, is usually easily accommodated by the commercial system. Just as the regrown forest is allowed to establish itself only until the "timber" it embodies can be harvested once again, so the success or otherwise of human creativity tends to be measured in terms of any resulting *commercial* success. In this respect, the tendencies of psychology[65] and psychotherapy[66] to support and perpetuate existing, individualistic definitions of self parallel the way the "forest products" industry is generally managed so as to maximize timber production. In each case, the tendency toward reestablishing natural structures is hijacked so that it remains within the commercial system rather than being allowed to realize its own tendencies toward a larger reintegration.

It is no surprise that the psychological effects of colonization would be most pronounced among those who are most overtly its agents. The "deformation of experience" to which Mumford refers is often palpable in the lives of scientists, among whom schizoid characteristics are allegedly common.[67] The divorce of thinking from feeling that is a requirement of the objective stance is a defining attribute of the schizoid character, and is often demonstrated by leading physical scientists such as Einstein and Newton.[68] One of the most telling accounts is given in the autobiography of Charles Darwin, who describes how in his youthful years he was led by spiritual feelings

> To the firm conviction of the existence of God, and of the immortality of the soul. In my Journal I wrote that whilst standing in the midst of the grandeur of a Brazilian forest, "it is not possible to give an adequate idea of the higher feelings of wonder, admiration, and devotion which fill and elevate the mind." I well remember my conviction that there is more in man than the mere breath of his body. But now the grandest scenes would not cause any such convictions and feelings to rise in my mind. It may be truly said that I am like a man who has become colour-blind.[69]

Later, Darwin continues:

> My mind seems to have become a kind of machine for grinding general laws out of large collections of facts, but why this should have caused the atrophy of that part of the brain alone, on which the higher tastes depend, I cannot conceive. . . . The loss of these tastes is a loss of happiness, and may possibly be injurious to the intellect, and more probably to the moral character, by enfeebling the emotional part of our nature.[70]

Darwin, with characteristic insight, recognizes the psychological price of a scientific training. While there is a sort of safety in this retreat into rationality, an opposite danger looms: the vertigo that arises out of unconnection, lack of emotional involvement, and isolation. The schizoid individual's vacillation between the two opposite dangers of engulfment by repressed emotionality, on the one hand, and the loss of meaning that arises out of the objective stance, on the other, is extensively discussed by object relations theorists such as Harry Guntrip. For example, the dream of one schizoid patient illustrates the dilemma faced by those of us whose lifestyle simultaneously depends upon distancing ourselves from the earth and deriving our ultimate meanings from it:

> I took off from earth in a space ship. Floating about in empty space I at first thought it was marvellous. I thought;

> "There's not a single person here who can interfere with me." Then suddenly I panicked at the thought "Suppose I can't get back."[71]

As Evelyn Fox Keller has argued, the scientific stance may be understood as attempting to develop a "safe" type of relation to the part of the world being studied, maintaining the scientist's sense of separateness, while allowing an authoritarian type of relatedness.[72] In a similar vein, John Dewey suggested that science has become a sort of "sanctuary" from "the things we experience by way of love, desire, hope, fear, purpose, and the traits characteristic of human individuality."[73] In other words, scientific and technological "objectivity" may support a style of personal functioning in which the relation of humanity to the natural world is one of comfortable domination. As a temporary stance that aims at achieving particular goals mapped out within a larger moral framework, this may be unproblematic; but when it is adopted as a complete account of our relation to the natural world, it becomes a scientism that furthers neither the forms and processes of the natural world nor the integrity of the individual, strengthening only the intellectual "virtual reality" of technological society.

But since science is the taken-for-granted epistemological grounding of the modern world, we are all—scientists or not—subject to the same colonizing logic as the scientist, albeit in popularized and diluted forms. As we hide behind our reflective sunglasses, ingest hypnotic suggestions about our state of calmness or energy from our Walkmans, identify with characters in our favorite soap, or challenge martial arts experts on our computers, so we emotionally invest in this "hyperreality" which is irrelevant or opposed to the natural world. And as the natural world becomes degraded, so we are driven toward dependence on the commercial world. As the water from springs or streams becomes undrinkable, for example, so clean water appears as something we buy from the supermarket or water company rather than as something that occurs naturally as part of our membership in the natural world. Of course, our change of allegiance from natural to industrialist structures is not "all bad": in certain respects, the commercial world *does* offer us possibilities unavailable in less "developed" societies. The problem is that these possibilities come at a cost that we are for the most part blind to—a cost that involves more than the urban sprawl and traffic jams with which we are all familiar, but which also extends geographically, especially to the Third World; temporally, into the lives of our children; and more generally, into those subjective and ecological structures that we all, ultimately, participate in.

This replacement of the natural world, as we have seen, finds its apparently innocuous beginnings many centuries ago; and it is difficult and probably meaningless to attempt to pinpoint any particular moment at which this development became pathological. It is a replacement that may best be considered in terms of a gradual unbalancing and disintegration. But perhaps one distinction that can usefully be made is that between modernizing trends that build on, elaborate, and enhance natural structures and processes, and those that essentially *replace* these processes and structures. As an example of the latter type of trend, Victor Ferkiss relates how our concept of clock-based time originated in the medieval era when "the monks came to believe that prayers said collectively at regular times were more pleasing to God. To some extent the Divine Office could be regulated by nature directly in the form of sunrise and sunset. But this was less than satisfactory, especially in latitudes where the passage of the seasons changed the length of daylight and dark. The medieval monks wanted something that would measure time *independently of ordinary physical nature,* and they found it in the clock."[74] Here, an abstract notion of regularity *replaces* the more complex and variable movements of the cosmos; and God is identified with the former rather than the latter. This is not a change that elaborates natural rhythms, but one that, oblivious to their complex interdependencies and vitalities, ousts them in favor of a simpler regularity. The hidden price of this regularity is a certain psychological numbness as we lose touch with the sensuous aliveness of the world, anchoring ourselves instead to mechanical patterns. The world of nature is not psychologically interchangeable with that of technique; for while the first connects us to the world through our integration into the rhythms and fluxes of the world, the latter *disconnects* us from nature by substituting structures that are *rationally* comprehensible.

In a scene from *Carmen,* a film by the French director Jacques Tourneur, the two leading characters stare across the seemingly endless ocean from their sailing ship. We hear the heroine as she recounts how she gazed at "those great glowing stars, and felt the warm wind on my cheek, and breathed deep, and every bit of me inside myself said 'How beautiful!' " Her male companion reads her thoughts, and replies: "It's not beautiful. . . . Everything seems beautiful because you don't understand. Those flying fish—they're not leaping for joy; they're jumping in terror, bigger fish want to eat them. That luminous water—it takes its gleam from millions of tiny dead bodies that glitter through putrescence. There's no beauty here; only death and decay."

The tensions inherent in this conversation remind us of the gender dimension that permeates the historical process of our colonization by technique. Overwhelmingly, men have been the agents of this colonization, and women have found themselves resisting it, either actively or passively, articulately or inarticulately, as has been noted by a considerable number of feminist and ecofeminist writers.[75] But increasingly, the *current* significance of this gender dimension is fading as both men and women become equally assimilated by dominant ideologies; and, as recent ecofeminists have recognized, pinning one's hopes of an ecologically sound form of consciousness onto existing styles of feminine being may be the sort of wishful thinking that derails the search for genuine alternatives.[76] To reduce the colonization of the world to simple oppressions such as that of women by men, or of nonhuman nature by humanity, is profoundly to misunderstand the extent and depth of our difficulties. All of us—men and women, humans and nonhuman nature, together with the conflicts and relations between them—are redefined and reconfigured by industrialism; so it is this all-pervasive redefinition of the life-world that we need to challenge rather than just the conflicts or alliances that occur as a result of it.

In taking up this challenge, we would do well to bear in mind Pascal's insight that "there are two equally dangerous extremes—to shut reason out, and to let nothing else in."[77] The recovery of repressed modes of being does not imply the *abandonment* of the types of rational thought that underly technology, but rather their framing as one style of thought among many within the "ecosystem" that is the natural order. Overlooking this point would lead us toward an entirely regressive conclusion, implying a simple return to earlier ways of being. However, the destructive power of mechanistic thought and attitudes lies not in their partial truths, but in their exclusion of all else, and in the imbalance that this represents. It is scientism, not science, that is problematic; and the solution to the current hegemony of scientistic thought does not lie in the repression of science, which would simply replace our existing scientifically based hegemony by one based on the repression of scientific rationality. Framed within a healthier understanding of the cosmos and our place within it, science would become simply one of many ways of relating to the world—one with characteristic uses and limitations—and there would be no confusion between our mapping of the world and reality itself.

The loss of structure that occurs when we define the natural world in terms of abstract categories has something to do with the obliteration of uniqueness and bioregional particularities: a Cascade Lily is a Cascade Lily, a river is a river, and once you've seen one redwood you've seen 'em all. Just as a Beethoven symphony can be summarized in terms of decibels, pitch, and duration, so the world can be quantified in terms of physical characteristics such as board-feet or cubic feet per second—

characteristics that all too easily come to seem fundamental defining attributes. And because this 'subsumption of the particular under the universal"[78] is entangled historically with a particular instrumental vision that is widely accepted as "reality," those sensuous and aesthetic characteristics which have little significance within this instrumental vision, such as smell, texture, or relation to context, become trivial awarenesses with no practical significance. Whereas the reduction of meaning is obvious when applied to Beethoven, it is less so when applied to the natural world, since we have been trained to view this world through the lenses of industrialism since infancy. This reduction in meaning is quite typical of the process of colonization in many of its various forms. Although our discussion has been focused on the colonization of the human psyche by technique, very much the same blindnesses, selectivities, and prejudices are characteristic of colonization in its better-known form involving the imperialistic conquest of one people by another. For example, Murray and Rosalie Wax have drawn a parallel between the "wilderness ideology" of the European settlers of the American continent, which justified them in "populating" the "empty" land, and the "vacuum ideology" that justified the widespread denial of Indian culture and the reeducation of Indian children:

> Confronting a land whose every area was known to and utilised by its native inhabitants, the land-hungry whites had perceived and spoken of "a wilderness."... [J]ust as the wilderness ideology rationalised for the invaders their seizing and occupying of Indian lands, so does the vacuum ideology rationalise for the educators their roles in the schools.... [For] if the child actually had a culture including knowledge and values, then the educators ought properly to learn about these and build upon them, but if, on entering school, he is merely a vacuum, then what need to give attention to his home and community?[79]

Here as in other forms of colonization, the colonizing ideology is blind to those qualities and attributes that are inconsistent with it. The viewpoint of the colonized, in contrast, expresses what is denied: the integration of natural, cultural, and historical structures, and the significance of these in constructing the framework of meaning within which life takes place:

> The White people speak of the country at this period as "a wilderness," as though it was an empty tract without human interest or history. To us Indians it was as clearly

> defined then as it is today; we knew the boundaries of tribal lands, those of our friends and those of our foes; we were familiar with every stream, the contour of every hill, and each peculiar feature of the landscape had its tradition. It was our home, the scene of our history, and we loved it as our country.[80]

In each instance of colonization, the form is identical. The ideological structures that have taken root in the social and psychological life of the invaders are characterized by a projective impetus to expand and to drive out competing structures. Thus an indigenous cultural or ecological system is destroyed and replaced by an exotic which is invariably simpler, less inclusive, and less diverse. The destruction of the natural world, however, is more silent than most; for while native Americans have in the past often been politically disempowered, their poets and writers can speak for their peoples. The nonhuman inhabitants of the natural world, however, possess no poets that most of us can comprehend; having been conceptually destroyed through the "single vision" of what passes for "culture" in the modern world, the holocaust can proceed quietly, unnoticed by many.

While colonizing structures are often abstract and relatively simple (as in the replacement of forest by monoculture) what is lost is often complex and usually unrecognized—at least by the colonizers. Uniformity and standardization, and thus the absence of micro-detail, are essential conditions for the existence of industrialism; and variation and particularities are the "brush" that must be cleared away for the industrial process to proceed smoothly, the diversity that constitutes unwanted deviation, like the unwanted "associations" that interfere with rational thought. Industrialism requires monocultures, not biodiversity—in materials, products, people; and it selects those particular characteristics out of many possible ones that are consistent with its structures, so that these structures will appear as the only possible ones. The price we pay for the products of industrialism thus includes a gross simplification of the most significant structures of our lives, and consequently, an enormous loss of meaning. In contrast, it may be significant that, as we saw in the previous chapter, nonindustrial cultures typically abstain from complex abstract schemes, preferring to emphasize a thing's individuality and uniqueness to a greater extent than we do.[81] For example, Veronica Strang notes that Australian aboriginals, if working on a cattle ranch, "did not count horses as they were brought in, but could tell whether any were missing because they knew them all individually . . . meanwhile, the white stockmen would be trying to count the horses."[82] As we noted before, however, and as a wealth of ethnographic evidence indicates, this attention to what may appear to us as irrelevant detail does not reflect an inability to abstract. For example, Strehlow notes in his monumental study of the Aranda that "[i]t is not as though the natives were incapable of abstract thinking: the ingenious nature of much of their ritual reveals to what extent logical thought has influenced those long-dead forefathers of our Central Australian tribes."[83] In the modern world, however, we have grown comfortable with the illusory wholeness of a reduced and abstracted view of the world, and this preference is reflected in the higher status accorded to those who work with abstractions rather than with specific, concrete physical realities—the theoretician over the technician, for example. We learn in school to prefer purely cognitive approaches to problem solving rather than those that involve counting fingers or beads; a preference that is consistent with the demands of a stratified society where manual labor is the province of the "lower" classes.[84]

As should be clear by now, however, the problematic character of our Western worldview does not rest in abstraction per se, but in our belief that the abstract model we subscribe to somehow reflects an alternative to rather than a complement of the world of down-to-earth embodied experience; and, furthermore, that this abstraction represents a more profound and "real" understanding of the world than the latter. This is what Barfield refers to as an "idolatry" of the scientific worldview[85]—in other words, our mistaking one specific, humanly abstracted model for the reality of the world. While it is true that other cultures may refer to abstract representations of reality, such abstraction tends to be directly related to concrete aspects of the life-world, implying an *elaboration of the world that already exists* rather than an attempt to *replace* this world—a point illustrated in the previous chapter by reference to Thomas Gladwin's research among the Puluwat islanders. Gladwin reports that "there is in Puluwatan navigation a reliance on abstractions," sometimes of "a rather high order." An example is the concept of *etak*, which refers to "a specified but invisible island moving under often invisible navigation stars." But such abstractions, argues Gladwin, are based in a perceptual sensitivity that "we (but not the Puluwatans) would consider extraordinarily acute." In other words, for the Puluwatan navigator, abstraction is grounded in a familiarity with the intimate detail of the world, and "Puluwatan navigation is a system which simultaneously employs fairly high orders of abstraction and yet is pervaded by concrete thinking."[86] Such a system extends knowledge of the world, but also ensures a greater interwovenness with it. For the Puluwatan navigator, taxonomies of wildlife or wave-pattern do not only enable him to travel safely from one place to another, but also express in distilled form an intricate detailed knowledge of the world. It might be said that while the Western navigator travels *through* the world, his Puluwatan counterpart travels *in* it.

The Process of Colonization: Intellectual Development

Ontogeny, so it is said, recapitulates phylogeny; and so if the history of human intelligence can be said to involve an evolutionary process, so this process should also be discernible in the stages of individual development. Is it possible, then, to perceive in the development of intelligence from infancy to adulthood a movement from the undifferentiated, fluid beginnings of representation toward a form of consciousness that is egoic, anthropocentric, and that maintains a consistent separation between self and world?

At first glance, exactly the opposite trend seems to occur. According to theorists such as Piaget, the child moves from an egocentric orientation toward a "decentered" view of the world. But on closer examination, it is not so much that the world of the infant is "egocentric," but rather that the boundaries between the nascent infantile consciousness and the "outside" world are unclear. One can read the child development literature as a narrative of negotiation, involving the child's developing sense of self, the world "outside," and the child's "significant others"—a process that normally, within industrialized societies, results in the emergence of a self that is relatively autonomous and self-directing, and can control and exploit the world for its own ends.[87] This emphasis on control is reproduced in many current theories of child development, which frequently normalize the child's growing willingness and ability to play with and use the world for its own ends while ignoring any possible legitimate structures or interests that the world and its other inhabitants might have. According to Piaget, for example, "every relation between a living being and its environment has this particular characteristic: the former, instead of submitting passively to the latter, modifies it by imposing on it certain structures of its own."[88] True, Piaget also recognizes that the child "accommodates" to the "environment": but as the term "environment" implies, this accommodation is mainly seen as a low-level acceptance of the world's physical characteristics rather than a recognition that it may embody sophisticated structures comparable with the child's own.

In this respect, the child can in some ways be viewed as *less* egocentric than the adult, in that within the nascent infantile ego, intelligence, feeling, and subjectivity are not restricted to the self, but are assumed to be properties that are shared by aspects of the outside world. For example—in the developmental jargon—(s)he may "impute life to inanimate objects," the terminology betraying the ideological assumption that the world *really* consists of "inanimate objects." As Owen Barfield points out, however, the doctrine of animism, according to which "primitive man" had "peopled nature with spirits, [presupposes that] nature must first be devoid of spirit; but this caused the scholars no difficulty, because they never supposed the possibility of any other kind of nature."[89] In certain respects, then, the "egocentricity" that is supposed to characterize infantile experience may in fact imply that the infant has not yet acquiesced to an ideology which limits subjectivity to the individual human being—an ideology which historically and cross-culturally is far from universal.[90] Piaget's view confuses two quite different perspectives: firstly, a diffuse one in which there is no stable point of subjective reference and in which all boundaries are up for negotiation, and secondly, a narrow, genuinely egocentric perspective in which subjectivity is entirely localized. Thus while Piaget's image of the infant's world, as it gradually extends outward to include the infant's own limbs, then objects touched, and finally the world beyond, may be a useful representation of what actually happens in industrial society, there is nothing in this process that implies that an exclusively egoic subjectivity which assimilates the world to an ever more inclusive—one is tempted to say "colonialist"—vision is the only possible one. On the contrary, we have good reason to believe—as, indeed, Piaget's own data suggest—that if they are permitted to, children experience the world empathically, as alive and enspirited; and as Paul Shepard argues, it may be our socialization into an often urban, manufactured environment that gradually teaches us to abandon the notion that the world is alive:

> The absence of numerous nonhuman lives, a variegated plant-studded soil, the nearness of storms, wind, the odors of plants, the fantastic variety of insect forms, the surprise of springs, the mystery of life hidden in water, and the round of seasons and migrations . . . builds in the child the sense that nonlivingness is the normal state of things . . . that the world . . . is not one which feels or thinks or communicates.[91]

What theories such as Piaget's conceal, therefore, are those alternative developmental possibilities that are less consistent with industrialism; and in this respect Piagetian theory closely parallels "modernization" theories of Third World development that perceive a more-or-less "natural" process of development from "primitive" economic conditions to a "mature" stage of steady economic growth.[92]

Moreover, the claim that the mature representation of the world in industrial society can be described as "decentered" is extremely dubious. This representation, as we have outlined above, is the product of historical processes whereby the world has come to be seen as material, passive, and lacking in spirituality and intelligence, by a detached observer who maintains a privileged position in relation to it. We have also seen that the Enlightenment thought which underpins technology is closely associated with the development of linear perspective vision—a style of perception that explicitly distances the sentient, detached observer from a world which is thus viewed "objectively." To the extent that such a representation is shared throughout the industrialized world, it cannot be described as "egocentric"; but equally, to term it "decentered" is to disingenuously conceal its whole ideological history. Perhaps "technocentric" would be an apt descriptor for a view that replaces the impossibly complex mysteries of the organic world by the seductive simplicity and power of post-Enlightenment science. As John Dewey put it, within a genuinely "decentered" viewpoint "neither self nor world, neither soul nor nature . . . is the centre. . . . There is a moving whole of interactive parts"[93]—a view that reminds us of the writings of those pre-Enlightenment philosophers such as Giordano Bruno[94] who had the sometimes fatal courage to challenge the accelerating momentum of Christian/commercial ideological dogma.

Technocentric assumptions are explicit in Piaget's theory—still today the most influential theory of cognitive development—in its portrayal of the growth of intelligence as involving a movement of thought away from the world rather than an engagement with it, and in which "the whole development of mental activity from perception and habit to symbolic behaviour and formal thought is thus a function of [the] gradually increasing distance of interaction . . . [between thought and the world]."[95] This process culminates in the stage of "formal operations," which many children attain during adolescence. "With formal operations," writes Piaget, "there is even more than reality involved, since the world of the possible becomes available for construction and since *thought becomes free from the real world.*"[96]

As we have seen, this autonomy of thought from the phenomenal world can be understood as a defining characteristic of modern industrial society, and its attainment is viewed as a mark of intellectual sophistication. As Susan Buck-Morss puts it:

> For Piaget, the first great cognitive leap is the prototypical experience of alienation. It is the ability of the child to divorce subject from object, hence to grasp the building block of . . . industrial production. . . . With the attainment of object permanency, the idea of an object . . . becomes a substitute for the thing itself, indeed . . . is granted greater cognitive value than the material object, and the child is capable through symbolic play of leaving reality unchanged.[97]

This developing schism between the intellectual and material worlds reflects Piaget's adherence to a dualistic epistemology reminiscent of Kant's prioritization of abstract rationality over concrete particulars. The influence of this epistemology ensures that normative intellectual development is aligned with the requirements of capitalism, so that the detachment of the intellect from the material world, and its justification in terms of allegedly "universal" abilities and developmental trends has become, according to Buck-Morss, "the dominant cognitive structure with the emergence of Western capitalism."[98] This allows the dispassionate categorization, reduction, and destruction of the natural world—processes that are both conceptual and, eventually, physical.

But to conceive of this "prototypical experience of alienation" as involving simply the divorce of the child from the world would be as inaccurate as understanding the environmental crisis as reflecting a simple opposition of humanity and nature. Humanity remains rooted in nature, however politically incorrect it may be to recognize these roots; and any adequate developmental theory needs to explain how an infantile self that is at least potentially whole comes to embody the dissociations implicit in destructive attitudes to the natural world. This dissociative process reflects the interiorization of what Teresa Brennan, following Lacan and Klein, has called the "foundational fantasy"—"a hallucinatory fantasy in which [the ego] conceives itself as the locus of active agency and the environment as passive."[99] This fantasy, according to Brennan, is reinforced by the child's realization of the power of technology (most obviously, through video games and TV), and through their inhabiting a largely built environment that accords with it. These factors reinforce the fantasy, resulting in a dialectic in which nature becomes a passive raw material to be manipulated according to schemes which are largely visual and abstract. This fantasy of power is probably a part of every infant's psyche; but only when it is systematically reinforced by contact with an adult order which assumes it and an environment which responds in accordance with it can it flower into an accepted basis for living. Theories such as Piaget's, then, derive their "accuracy" from their accordance with a particular ideological order and the subset of human potentialities that are consistent with this order. If we are to *transcend* this order, however, we require a theory that can account not merely for the particular developmental course experienced by children who inhabit this specific ideological order, but also for those *alternative* developmental courses that would emerge in *different* cultural settings.

In industrial society, however, the manner in which the infant grows through reference to a constructed world that seems to confirm the power of the ego and the powerlessness of nature has complementary implications for the child's self-concept. The relation to the world that develops is abstract, overwhelmingly visual, and relatively unemotional: these are the necessary conditions that allow the child to conceive of manipulating "objects" in the world in a detached, dispassionate way. This developing configuration of selfhood, as we saw earlier, is that which object relations theorists have described as "schizoid." The price that the infant pays is a loss of the sense of self as constituted in the body—a sense of self that suggests an entirely different orientation to the world. Rather, the child learns to objectify their body primarily as their visual image, during a developmental period that Lacan has described as the "Mirror Stage." The child's body thus becomes something outside the self, something else that it is necessary to have mastery over—a curious situation which comes to be expressed by such paradoxical concepts as "self-control" or "self-discipline," whose implications we referred to in chapter 2.

Bodily feeling is the basis of intimacy between ourselves and the world. Watching a kite soar in the thermals overhead, we feel ourselves soaring with it. Hearing the scream of a hunted animal as it is shot, we feel something of the pain and terror ourselves. And what we experience as we explore some unspoiled area of wilderness draws us into the place in powerfully felt ways that consciousness may be quite unable to categorize or understand. But if our body, and its associated capacity to sense aspects of nature, becomes something *outside* the boundaries of self, then we lose our somatically based abilities to relate to the natural world in this sort of way. This assures the repression of any sensorily based relation to other aspects of the world, and sets the scene for the child's imprisonment within an abstract realm governed by the separation of subject and object, one that is the quite logical fulfilment of Descartes' dream of a world in which sensory and emotional experiences play no part. The repression of nature "out there" is therefore closely related to the repression of our own bodily awarenesses; and these twin repressions are, ironically, nowhere more clearly exemplified than in academic writing about the "social construction" of nature or of the body.

Thus what developmental orthodoxy would conceive of as "educated" thought rejects the phenomenal diversity of the immediate physical world to become organized around certain logical principles of addition, grouping, multiplication, and so on. These logical principles are, correctly enough, presented as reflecting indwelling physical and biological structures. But as we saw earlier, what is problematic about this approach is its incompleteness; and its elevation from a set of pragmatic principles to an all-inclusive account of development leads to a potentially disastrous discrepancy between our assumptions about the world and its actual character. Realizing this, a number of writers have questioned the nature of the relation between operational structures and physical reality. Garfield, for example, asks whether Piaget, "while thinking that he has told us something important about the child's coming to understand reality, . . . has . . . informed us about certain logical categories or formal concepts which he has mapped on to the world of the child." Garfield goes on to consider whether Piaget's approach leans excessively toward idealism; there is a danger, he suggests, that "the world we construct is not a real world at all."[100] In other words, Piaget's theory may be an elaborate projection, an intellectual fantasy, that establishes a comfortable illusion of relation to the world while actually avoiding any real relation.

Moreover, as the part which social and cultural factors play in intellectual development becomes more widely recognized, it is becoming increasingly clear that, in Angus Gellatly's words, "[i]ndividuals do not elaborate, or get greater access to, principles; rather, they learn accepted social practices. They discover what is the accepted way of proceeding in particular circumstances and, maybe, what principles to invoke as justification."[101] In other words, intellectual development cannot be seen merely as the more-or-less successful unfolding of potential toward predetermined and universal logical structures, but is partly a matter of ideologically determined choices among an indeterminate number of possibilities. To recognize that the world allows us to interpret it in a diversity of ways, however, is a long way from claiming that it is "socially constructed"; and a fundamental mistake of developmental theories such as Piaget's is that they argue that characteristics of human intelligence such as the "grouping structures" also accurately reflect the structure of the world itself. Such structures may, it is true, correspond with particular features of the world—indeed, if the theory is to be useful, they *must* do so; but any theory will inevitably omit so much about the world that its correspondence to reality is partial and limited. To recognize the social character of intellectual development is therefore to say little about the reality that such development strives to engage with. Of course, if our chosen form of intelligence were well attuned to the structure of the natural world, these concerns would be merely academic; but the fact that our style of intelligence and the actions that result from it are highly destructive to natural structure indicates that its very success is a matter of concern, for, as Gregory Bateson pithily put it, "the creature that wins against its environment destroys itself."[102]

The neglected concepts of "fluid" and "crystallized" intelligence are helpful in understanding the distinction between two alternative understandings of human ability. "Fluid" intelligence, or "intelligence A," is defined as an innate potential, a capacity for development[103]—a potential that, as a result of experience, becomes transmuted into "intelligence B," or "crystallized intelligence," which is directly related to those forms of behavior and cognition that are valued and practiced within any particular culture. In this process, a flexible, "fluid," undeveloped openness to alternatives, to order as it may present itself, implying a diversity of possible alignments, is replaced by a singular, static, abstract understanding that imposes a single selected order on the world. "Fluid" intelligence—to the rather limited extent that it can be measured—is tapped by test items measuring the ability to perceive pattern in unfamiliar stimuli, or to rearrange elements of a figure in a meaningful way; while "crystallized" intelligence is measured by subtests such as "Information" (general knowledge), vocabulary, and other measures of one's acceptance of a culturally specific knowledge structure. It is not surprising that "fluid" intelligence declines after reaching a peak in the early teens, whereas "crystallized" intelligence has been found to develop into late middle age, reflecting the increasing allegiance to the consensual view of reality and the decreasing awareness of alternatives that accompanies our progress toward "maturity" and old age in the industrialized world. Thus, as cognitive development proceeds, so the individual can operate more and more powerfully within one particular conceptual scheme, at the price of a gradual loss of the sense of openness to alternative schemes—to order as it may spontaneously present itself. In psychoanalytic terms, there is increasing allegiance to conscious, rational, secondary process forms of thought, and a corresponding repression of fluid, symbolic, metaphorical processes, which hereafter must exist mainly within the shadowy realm of the unconscious.

Some would argue that the development of this sort of specialized, literal, singular system of knowledge is an inevitable corollary of growing self-awareness, and that selection and choice are essential aspects of development. But even if we accept, quite realistically, that development cannot simultaneously extend our dexterity within an indefinitely large number of knowledge structures, it is not obvious that we should burn our psychological boats by *repressing* the awareness that the particular "crystallized" form that we choose to emphasize has particular limitations and blind spots, nor that we should abandon the awareness that it is one of many possible systems that are in their way just as valid and successful as our own. In fact, recognizing the limitations of our culturally chosen knowledge system might be expected to have obvious survival value. Technological expertise can and must coexist with an awareness of its own character and boundaries. Choosing to walk down one particular road is no reason for throwing away the map that would allow us to consider alternative paths. Scientism, since it sees the world in terms of a few selected properties, does conceptual—and ultimately physical—violence to the world by ignoring the indefinitely large number of properties that are backgrounded. So although "intelligent" behavior is a technologically powerful system of thought, it is one that "wastes"—in more than one sense—much of the world and our potential richness of relation to it; so whether one should best regard such a system of thought as intelligent or as pathological is debatable. In Paul Shepard's words: "Whether blindness is pathological to those living in a cave depends on whether you think of it in terms of personal adaptability or [in terms of] the inherent potentialities of every member of our species."[104]

Shepard understates the extent of the problem, though: for our technological power gives us the capacity to *externalize* our blindness, destroying those potentially visible aspects of the world that our reduced sensibilities cannot encompass, so forcing the world to accord to this reduced vision. Abstraction, then, needs to be balanced by equally strong abilities to contextualize, to put in place. The Latin roots of these words betray their meaning: "abstract" derives from *trahere*, to draw out, or drag violently. "Context," on the other hand, derives from *texere*, to weave. Following these metaphors, then, abstraction that is not counterbalanced by recontextualization destroys the fabric of the world. And while abstraction, as we have seen, need not imply an alienation from the world so long as there is a fluently articulated relation between the abstract representation and the concrete, phenomenal world, the style of abstraction developed by industrialism is one in which an abstract representation is taken to be *better than*, and a *replacement for*, the phenomenal, concrete world. For example, in the well-known Piagetian task involving a string of wooden beads—mostly brown, a few white—the child is typically asked: "Are there more brown beads or wooden beads?", thus directly counterposing an abstract class with a more perceptually salient one. This apparently innocuous task requires that the child downplay the salience of the colour, prioritizing the more abstract dimension of "woodenness."

In other cases, the separation from the world is even more explicit. Margaret Donaldson[105] quotes an example, from Werner, which is fairly typical of the sort of conversation an anthropologist might have had with an informant until quite recently. The native speaker was asked to translate into his language the sentence: "The white man shot six bears today." The Indian said that it was impossible. The explorer was puzzled, and asked him to explain. "How can I do that?" said the Indian. "No one could shoot six bears in a day." Of course, such a reply is likely to be regarded as "unintelligent" by those who move easily within an abstract "reality" only tenuously connected to a natural world whose limits we are largely indifferent to.

Donaldson criticizes the emphasis on abstract ("disembedded") thought within education, and its separation from the everyday experiential realities lived by children, pointing out that younger children in particular will naturally try to contextualize problems that are presented as purely abstract. What is learned within this sort of educational context is not so much the *ability to abstract*, but rather allegiance to the principle of *abstract formalism*,[106] that is, the structuring of experience according to the separation of form from content. Education within the industrialized world can be seen, in part, as a learning process in which the child is taught to put aside a physical, sensuous intimacy with the world, and instead to exist comfortably within the humanly constructed world of abstraction, scientific rationality, and physical laws. This approach to education is, as Aldo Leopold perceptively put it, "learning to see one thing by going blind to another."[107] The child does not, of course, become oblivious to the phenomenal experience of the world: she will still see, and react to, the colors of leaves and sky, the feel of wind and rain, and the sounds of river and animal, and the states of being that these may induce. However, these qualities, unlike physical attributes such as mass, quantity, or length, will remain relatively unarticulated by the dominant systems of thought available within Western culture, and so will be experienced as relatively trivial—curiously reassuring but ultimately insignificant adornments of more important aspects of our everyday lives. And as our neglected capacities to articulate these qualities shrivel further, we will, like Darwin, come to experience the world in monochrome.

For our purposes, however, the implications of this work go beyond education: for it is a small step from the mentality that can ignore the context of the "six bears" type of problem to that which can comfortably perceive a forest simply as a quantity of lumber. And, incidentally, it is an equally small step to the classification of humans according to race or gender; for prejudices, whether racial, sexual, or ecological, are based on abstractions that implicitly deny individual variation, context, and the potential wholeness of human experiencing. Just as Columbus refused to speak the language or learn the customs of the indigenous peoples he encountered, and just as the experimental psychologist maintains a distant, "objective" nonrelation to the "subject," so abstraction allows us to exploit the world with an untroubled conscience, consciously unaware of the qualities and natural structures we crush in the process.

This chapter has examined the historical emergence of a form of selfhood consistent with industrialism, and also the processes through which we individually acquiesce to this mould in ways that seem "natural" and "intelligent." But grasping the realization that neither the world we inhabit nor our experiential "realities" are inevitable conditions that we have no control over, and that industrialism cannot operate without our collusion—that it is, in Romanyshyn's terms, a symptom—we can glimpse the possibility of recovering those modes of experiencing whose repression is necessary for what passes as maturity in the modern world:

the nonvisual, the arational, what is felt or intuited as well as thought. In this realization, we recognize the inextricable relation between environmental and personal salvation. But I am not about to suggest that environmental problems can be addressed solely through "personal growth," at least as conventionally understood, which would be to fall prey to the individualism I have criticized. Rather, the environmental/ personal salvation I refer to is one that points to and constructs something that bridges the gulf between the modern self and the world, something that is simultaneously an extension of self and a part of the world. I will refer to this "something" as culture, although I hasten to add that this is an inadequate term to describe what I have in mind, so alien is this notion in the modern world. But as we will see in the next chapter, culture has the potential to change both our self-experience and our experience of the world in a way that recovers the wholeness of both.

Part Two

Reintegrating Nature and Psyche

Chapter 4

NATURAL CULTURES, PSYCHIC LANDSCAPES

The world as landscape is now something to look at, something to inspect, and in the face of this world one now sees only one's exclusion, an exclusion which invites one to gaze inside. . . . Modern psychology is the discipline created by this exclusion, a science of the inner self which appears in the midst of an alien world.

—Robert Romanyshyn, *Psychological Life*

The Repression of Culture

Usual understandings of culture, it must be admitted, have not been helpful in comprehending why some societies irreparably damage their natural environments while others seem to achieve more harmonious relations to it. For example, the term has often been taken to mean "high culture"—that is, as a range of more or less esoteric artistic and social activities indulged in by those with sufficient money and leisure; or, for most of us, something to be sampled occasionally, as when we buy a reproduction print to hang on our sitting-room wall. According to this view, culture is strictly an optional extra that has little relevance to the "real" business of life. And even when the term is used—as in Raymond Williams' work—as an attempt to integrate this realm of "arts and learning" with that of "common meanings," the very fact that this attempt has to be made is diagnostic of the rift between them. Secondly, culture has often been understood as superficially adorning and camouflaging a "nature" that is more substantial and more real—an understanding that is symptomatic of the material reductionism of industrial society. In other words, molecules are supposedly more "real" than whole organisms, which in turn are more "real" than cultural or religious structures, so that we assume a sort of ontological gradient in which larger entities ultimately derive whatever validity they possess from their constitution by smaller elements. The long-term attempt of modernity to establish a human cultural realm independent of the natural world therefore coexists in constant tension with a deeper fear that ultimately, culture may collapse back into a state of "nature." Clinging to an interpretation of culture that attempts to subordinate rather than articulate nature, it is not surprising that we sense the eventual victory of the latter; and our awareness of the inevitable flimsiness of an ungrounded cultural realm leads directly to the individualism and material reductionism that are rife within the social sciences. These related understandings of culture converge in a remark made by one of my mature students, a single parent, recently. "I don't think I've ever had a culture," she remarked, "I've just been a mum!" What sort of society is it in which culture is divorced from the central biological processes of our lives?

Both these closely related meanings are quite consistent with the conventional psychological understanding of culture as relatively insignificant in comparison to those processes and mechanisms that reflect "universal" aspects of humanity. In any case, so the argument goes, given that cross-cultural psychology studies the influence of culture, what need is there for any other branch of the discipline to be conversant with it? According to this view, culture can be understood as a sort of socio-psychological custard that is poured over the basic economy-model human being, adding a few colorful extras associated with dress, social behavior, belief systems, and so on. In contrast, allegedly more "basic" functions such as memory, intelligence, and perception are presumed to be rooted in neurophysiological functions that are closer than cultural influences to "real," material explanation. If we were to adopt this view, then any environmental solutions which employed culture as a central concept would necessarily seem superficial and ultimately reducible to biological explanation. Since the "real" character of the human individual, and therefore the basis of our relation to the natural world, is supposed to reside in these allegedly universal biological mechanisms, the most cultural influences could achieve would be some modification of biologically rooted behavior patterns. For example, if we are "basically" preoccupied with satisfying our individual material needs, then a culture of altruism can do no more than cover up and moderate a more basic selfishness.

Material reductionist explanation has not always been regarded as more real than that based on larger structures. In the early Greek world, an atomistic understanding coexisted with one that referred to the equally invisible but radically different world of the gods; and it is only in the modern era that the former type of explanation has almost completely conquered the latter. This victory of substance over form can be traced through the visual arts: as Romanyshyn points out, by the late fifteenth century angels in paintings have to be supported on clouds; and soon after they disappear entirely. Today, the idea of any structure larger than the individual seems flimsy indeed.

The ongoing industrialist demolition of cultural and natural structures makes these assumptions increasingly difficult to challenge, since the acceptable forms of explanation seem to outlaw those larger structures that could define a healthy world. Within such a world, biological, individual, cultural, and natural structures will resonate with each other, mutually constituting each other and so defining a wholeness which each level embodies metonymically. Individual behavior, for example, will accord with and reinforce culture and ecosystem, just as individuality will itself be grounded and confirmed by our participation in these larger structures. Within such an integrated world, culture would not simply be an expendable superficiality, but an essential level within the web of interdependencies that defines individuality. We need to recognize, with James Hillman and Michael Ventura, that "the quality of wholeness is not located within the individual, but in a community which includes the environment."[1]

In the account elaborated in this chapter, I am following leading critics of the conventional psychological view described above, particularly Clifford Geertz, Richard Shweder, and Edward Sampson.[2] But I may seem to some readers to be having my cake and eating it, since in chapter 2 I criticized social constructionists for suggesting that we are culturally constituted. There is, however, a huge difference between their views and mine: for while their emphasis on the cultural construction of individuality virtually eliminates nature as a co-determinant of personhood, I am arguing that personhood is based on the *dialectic between* these two domains, neither of which can be properly understood in the absence of the other. Just as the force that keeps us in our seats during a tedious lecture is a combination of gravity and politeness, as the physicist Richard Feynman put it, so our relation to the natural world has both cultural and physical dimensions; and neither can be reduced to the other. In other words, it is not simply a matter of replacing biological reductionism with cultural reductionism; but rather that the cultural and biological realms are—or rather, should be, in a healthy world—two equally real levels of structure within an oscillating, interdependent, multilayered world. This is a fundamentally different perspective to either the biologistic viewpoint or the social constructionist position, both of which emphasize one domain by suppressing the influence of the other, and therefore collude in the dissociation of the natural and cultural domains. In contrast, I argue that a reintegration of the natural and the social is a prerequisite to any real environmental solution and any adequate theory of our relation to the natural world.

In recent years, a small but growing minority of social scientists, influenced by the work of Geertz, have begun to comprehend humanity in terms of a deep reciprocal permeation of biological and cultural structures. This escape from individualism and material reductionism permits a further, more radical, escape: culture can be a bridge to the natural world, so that subjectivity itself becomes part of nature, not only as an intellectual construct but also as a lived reality. Without such a cultural bridge, talk of our "relation to" nature amounts merely to hot air. Geertz is explicitly critical of the conventional psychological view of culture which holds that the enormous diversity of individual differences and beliefs across the world's cultures disguises the basic and universal similarities in functioning. In contrast, argues Geertz,

> There is no such thing as human nature independent of culture. Men without culture would not be the clever savages of Golding's *Lord of the Flies* thrown back upon the cruel wisdom of their animal instincts; nor would they be the nature's noblemen of Enlightenment primitivism or even, as classical anthropological theory would imply, intrinsically talented apes who had somehow failed to find themselves. They would be unworkable monstrosities with very few useful instincts, fewer recognisable sentiments, and no intellect: mental basket cases. As our central nervous system—and most particularly its crowning curse and glory, the neocortex—grew up in great part in interaction with culture, it is incapable of directing our behaviour or organising our experience without the guidance provided by systems of significant symbols. . . . Such symbols are thus not mere expressions, instrumentalities, or correlates of our biological, psychological, and social existence; they are prerequisites of it.

> Without men, no culture, certainly; but equally, and more significantly, without culture, no men.
>
> We are, in sum, incomplete or unfinished animals who complete or finish ourselves through culture.[3]

Thus culture—which Geertz defines as a "historically transmitted pattern of meanings embodied in symbols, a system of inherited conceptions expressed in symbolic forms by means of which [people] communicate, perpetuate, and develop their knowledge about and attitudes towards life"[4]—contributes essentially to our sense of identity, our ability to live purposefully, and the meaningfulness of our existence. It offers a framework through which we can realize our otherwise latent potentials. Without such a framework, we flounder around rather like runner beans without poles, able to realize our creativity only rarely through heroic individual effort. Culture is also an essential mediator between ourselves and the world—a source of connections that locate us within the universe and give meaning to the whole context of our existence. Spiritual structures, for example, have a particularly central role within many societies, integrating otherwise disparate areas of experience, vitalizing metaphysical concerns that otherwise might become arid and lifeless, and bringing meaning and significance to otherwise trivial details of day-to-day existence. As the theologian Don Cupitt put it, "our religions are produced something like works of folk art. Religion is the heart and centre of culture, and it's through religion that we work out a common vocabulary of rituals and symbols which together make up a kind of house of meaning that we dwell in."[5] This "house of meaning," however, has its foundations in the earth, expressing, in Geertz's words, "the fundamental nature of reality." Culture, then, offers us a sort of symbolic integration of the cosmos, connecting us to what is outside us and so affirming our membership in a scheme much greater than any individual or society.

Spiritual realities tend by their very nature to be difficult to grasp in terms of everyday, rational states of consciousness, especially within an industrialized world in which the very term "spiritual reality" has become something of an oxymoron. But even within those vestigial aboriginal peoples who remain in isolated pockets of the planet, the realities that spiritual experience points toward are not concrete or easily delineated; and systems of religious rituals, artifacts, and beliefs take the form which Sam Gill refers to as "a shadow of a vision yonder."[6] The "thinness" of ritual objects such as a Navaho sandpainting or a Hopi Kachina mask makes it clear that they are necessarily fragile—and, especially in the case of the sandpainting, temporary—indicators or symbols of this "vision yonder." The spiritual integration of the world, then, is not one that can be understood simply in terms of its own internal consistency; nor can it be reduced to a materialist understanding. Rather, it points beyond cognitive and physical frameworks to entities and forms that are not scientifically expressible, and whose presence we may be able to sense but not grasp intellectually.

The ability of cultural forms to integrate what in the industrialized world would be experienced as quite separate aspects of life and the natural world is illustrated by Lauriston Sharp's account of social practices among the Yir Yoronts, an Australian aboriginal group who live in the Cape York Peninsula of northern Queensland. Among this group, stone axes were particularly significant in integrating otherwise disparate aspects of life. Only adult men, for whom axes were important symbols of masculinity, were considered to have sufficient expertise in working the wood, bark, gum, and stone to make them; and although women and children who were related to the axe-owner by particular kin relations could borrow them for a range of everyday activities, only men could use them to make the secret paraphernalia for ceremonies. Axes were therefore important in stabilizing relations between men and women and in the pragmatics of the Yir Yoront's relations to the natural world, and so were considered valuable. Furthermore, the stone used in making the axe could only be found in quarries four hundred miles to the south, reaching the Yir Yoront through complex networks of trading relations in which axes were exchanged for spears; and so the Yir Yoronts' location within the tapestry of aboriginal clans was bound up with these stable trading exchanges. Sharp also points out the religious significance of axes:

> Among the many totems of the Sunlit Cloud Iguana clan, and important among them, was the stone axe. The names of many members of this clan referred to the axe itself, or to activities like trading or wild honey gathering in which the axe played a vital part, or to the clan's mythical ancestors with whom the axe was prominently associated. . . . There was thus in Yir Yoront ideology a nice balance in which the mythical world was adjusted to the real world, the real world in part to the pre-existing mythical world, the adjustments occurring to maintain a fundamental tenet of native faith that the present must be a mirror of the past.[7]

It is hardly surprising, then, that the appearance of missionaries bearing steel axes disrupted virtually every aspect of this highly integrated cultural system. To the missionaries, inhabiting a world in which "things" were clearly differentiated from a largely abstract spirituality, axes were simply technological aids, to be handed out indiscriminately to men, women, and children. From the aboriginal point of view, however, this superficially benign largesse undermined social and gender relations, trading patterns, the grounding of life in the mythological realm of the ancestors, and the linkages between all these and the landscape itself. As Sharp summarizes this dire outcome, the "result was a mental and moral void which foreshadowed the collapse and destruction of all Yir Yoront culture, if not, indeed, the extinction of the biological group itself."[8]

This poignant story illustrates both the ability of culture to stabilize diverse aspects of life within a common universe of meaning, and its vulnerability to unsympathetic external influences. Just as exotic species such as the rubber vine have invaded and choked large areas of their homeland in the Kowanyama area of north Queensland, so the Yir Yoront seemed to be equally vulnerable to introductions which at first seem innocuous.[9] In the cultural world as well as the ecological, changes introduced by those who are insensitive to the invisible structures and symbolic dimensions that connect what to us might be mere "things" are likely to reduce these worlds to those simpler forms of structure that are understandable by science and economics, so apparently confirming the accuracy of this understanding. And if we are surprised by such convergence between ecological and cultural domains, aboriginals would not be; since for them, the cultural realm has long been continuous with the natural/spiritual one. The effects of introducing natural or cultural forms that are inconsistent with these long-established patterns are not predictable by rationality, except in the vague sense that they are likely to be destructive to the original balance.

Of course, to the white Australian cattlemen who have largely replaced the Yir Yoronts, the separation of the ecological and the spiritual is a taken-for-granted fact of life. Veronica Strang found that for the cattlemen,

> Conflict between material and emotional or spiritual needs emerges very clearly in many conversations. Most graziers seem to deal with it by compartmentalising their activities into "work"—highly specialised, focused, quantitative and material—and "rest," characterised by the expression of affective concerns. . . . The land is therefore used in two quite separate ways: as a stage for active and assertive physical and mental challenge, and as a retreat in which emotional or spiritual sustenance can be sought and affective attachments constructed.[10]

The concept of repression, this suggests, only partly expresses the psychological dynamics of the industrialized psyche. Equally essential is the *dissociation* between realms; and this dissociation is reflected back onto the natural world in the form of environmental policies which fragment the world into a patchwork quilt of landscapes that are assigned different "uses." In contrast, the Aboriginal attitude toward the land *integrates* these "separate" realms. The pivotal point about aboriginal land use, argues Strang, is that "economic interactions with country are never wholly divorced from social and spiritual interactions. Land provides a central medium through which all aspects of life are mediated, and economic considerations are merely part of an intimate, immediate, fundamentally wholistic relationship." And this, in turn, challenges our taken-for-granted ontology, since it implies that the experience of personhood involves a flowing outward of the self into the world beyond, and a resonance between "self" and "world" that places a question mark over their separability, reminding us of Jung's remark that "[w]herever there exists some external form, be it an ideal or a ritual, by which all the yearnings and hopes of the soul are adequately expressed—as for instance in a living religion—then we may say that the psyche is outside and that there is no psychic problem, just as there is no unconscious in our sense of the word."[11]

What this suggests, then, is that culture not only integrates us into the landscape; but that it also enables us to be whole. Without adequate cultural structures, the dissociation of life-realms noted by Strang becomes *inevitable*; and we become Geertz's "unworkable monstrosities." This implies that the personhood which we in the industrialized world take for granted as "natural" is a pathological form that not only lacks an adequate relation to the natural world, but that is also marred by dissociations and repressions. For example, Strang argues that white Australian art, since it lacks any work-related or practical relevance,

> Is compartmentalised and invested with . . . "bogus religiosity." In painting, film, photography and literature, landscape is presented either as an aesthetic object or as a more general and romantic concept.
>
> Attachment to land or "home" is thus idealised and sanctified, yet also boundaried and made separate. In cattle station culture, art barely impinges on the daily lives of most people. Their expressions of felt values relating to land are either contained within specialised art forms (such as poetry or story-telling), rationalised as economic or "practical," or so generalised and detached that they are marginalised by their level of abstraction.[12]

The tension between economic imperatives and an affective life driven "inward" toward privatism has become too great, fragmenting the world. Art, unable to bridge the divide between these two realms, tends toward either a frank alignment with industrialism or an impotent sentimentality, paralleling the marginalization of traditional aboriginals in reservations. Furthermore, the natural world seems often to suffer a similar fate, being treated either as a material commodity and a cornucopia of "natural resources," on the one hand, or as an idealized aesthetic object to be kept safe from exploitation, on the other. That alternatives to this fissured psychological and geographical landscape are at least possible is indicated by the aboriginal recognition of patterns that recur in the organization of self, landscape, and everything that these include. Sometimes the subjective continuity of self and world is expressed through what we would term analogy. "For aboriginal people, the landscape is a conscious entity that generates and responds to their actions, creating life with them, nurturing them, grieving with them and sometimes dying with them,"[13] as suggested by one of Strang's informants: "When his . . . father died, Emu waterhole . . . dried out . . . the place died with him . . . If the young people don't come back to this country, the country will feel that."[14]

For aboriginals, meaning is encoded in the land, and this meaning resonates with and structures their psychological and social forms. Language, therefore, is not a primary determinant of meaning; and "linguistic models fail to consider adequately how the physical, visual, and nonverbal symbolic universe contributes to cognitive interaction. . . . Thus the physical and visual worlds, rather than language, are the primary media in which every object or image carries meanings, associations, and values."[15]

Even the concept of time, whose linear flow seems so obvious and unquestionable to us, is patterned by aboriginal groups according to the ebb and flow of recurring natural cycles. We saw in the previous chapter how in the industrialized world a formal and abstract understanding of time that was partly separate from natural rhythms came to replace one that was embedded in the movements of the earth. For the aboriginal, however, time is not a quantitatively invariant flow that nature roughly adheres to, but is rather directly rooted in natural processes. For example, Strehlow documents the extraordinarily detailed Aranda taxonomy for the times of day, although lack of space allows me to reproduce only a small part of this taxonomy:

> *inua topalta*, the time after midnight, when the Milky Way is stretched out across the centre of the sky.
>
> *inuaijinuaia*, the very early hours of the morning when all is still dark.

> *ratajibalelana*, the [hour] which sends the bandicoots back [into their burrows].
>
> *inuupurapara*, the first faint greying of the sky in the east, the time before *lentara*.
>
> *lentara*, the time when the day begins to glimmer, before the first colour comes into the sky.
>
> *altapatara*, twilight; the early morning when the outlines of trees and objects become clearly defined in the growing light.
>
> *inuntinunta*, the time of early or "young" morning; the time when birds begin to sing.[16]

And so on, throughout the day and night.

For the Aranda, then, time is not an abstract pattern that is imposed on the world from an external and autonomous realm of human cognition or scientific necessity, but is rather a relational structure whose applicability lies in its truthfulness to natural cycles and the way it integrates human life within these cycles. In experiencing time as part of nature, we also become part of nature, rather than imposing a mechanical regularity on a nature experienced as *ir*regular. Australian aboriginal cultural traditions are, therefore, quite explicitly grounded in the natural landscape. But this relation between the cultural and physical realms is dialectical in character, for culture is not merely abstracted from the physical world, but also re-applied to it in ways which are, literally, down to earth. David Lewis, in an empirical study of aboriginal navigation in the Western Desert region—an area of several hundred thousand square miles that a European might be forgiven for describing as "featureless"—found that without the help of compasses or other modern techniques, aboriginals could invariably orient themselves with quite extraordinary accuracy. Lewis reports that

> In *physical* orientation the *spiritual* world, manifested in terrestrial sacred sites and Dreaming tracks, would appear to be the primary reference. The emotional attachment, awe, fear and love . . . which link men to the terrestrial features of their Dreamings came to dominate the scene to such an extent that I doubt if ever again I shall be able to look at a hill or rock hole with quite the same eyes as before. . . . The observer might expect to find a clear-cut dichotomy between ecological and spiritual/ritualistic determinants for particular wandering patterns. None are, however, apparent. . . . Ecological and spiritual behavioural determinants became inextricably intermingled into a single spiritual/physical conceptual entity.[17]

Such symbolic realities are clearly recognized in Geertz's writings, even if these writings have on occasion been accused of being rather too cerebral as representations of the embodied spirituality found in many indigenous groups. Meaning all too easily falls into the mental side of the Cartesian mental-physical divide, betraying other dimensions of subjectivity through an understanding that prioritizes intellectual consistency at the expense of other forms of experience such as visual and nonvisual imagery, taste and feeling, muscular action and the embodied recognition of pattern. Nevertheless, Geertz is one of the few social scientists to explicitly challenge commonly held conceptualizations of the "culture-versus-nature" type which are so potentially misleading to environmental theory. He clearly recognizes that if we abandon culture as a source of *connection* to the natural world, then the resulting ethical system will be forced to seek its origins within the arid abstractions of a rational, relatively autonomous individual. This, however, draws us back into the assumptions of individualism, making an exploitative relation to the world almost inevitable. An environmentally adequate understanding of culture regards it not as something external that is "useful" in formulating an environmental ethic; but rather as something that can reconstitute us so fundamentally that an "environmental ethic" becomes an irrelevant abstraction for a self which is *already* part of the natural order.

In the industrialized world, unfortunately, religion—usually Christianity—has tended to suffer a similar fate to art, undergoing a sort of death-by-dissociation in which religious experience and behavior have typically become separated from other experiential domains, so that the power of religion to suffuse our lives with spiritual significance is instead replaced by a narrowly defined and largely impotent dogmatism whose attitude toward both somatic reality and the natural world is often quite negative. The prevalent Christian view of God as transcendent, for example, contrasts with the belief of most tribal cultures that an immanent deity is *part of* the natural world. Among such "state-free" peoples, as Joel Kovel terms them, there is no *supernatural* world: spirit is part of the *natural*.[18] Indeed, "many Indian languages had no word for 'religion'; they expressed the idea by something like the Isleta Pueblo term 'life way' or 'life need.' "[19] Vine Deloria, similarly, argues that while native American spiritual viewpoints typically see creation as an ongoing process that demonstrates the continuing presence of the Great Spirit within the sacred realm of the natural world, in many varieties of Christianity creation is regarded as a specific historical event in the distant past, after which God left humankind to care for a world which is basically a location for the working out of human history.[20] This lack of intimacy between the divine and the natural is unusual even by the impoverished standards of institutionalized religion, and is one of the root conditions of both our technological power and our spiritual impotence.

Many forms of modern Christianity seem to have succumbed to a hardening of the arteries in which an emphasis on conscious understanding and overliteral interpretation superstitiously stand in for a lost symbolic power. A communal meal of bread and wine, for example, was a religious practice centuries before Jesus broke bread and shared it with his friends, the physical nourishment it provided symbolizing a freely given spiritual nourishment. Jesus was therefore tapping into a powerful existing symbol, the "staff of life," when he asked his friends to "take this and eat it," as Elizabeth Rees points out.[21] Today, however, the tasteless, tiny wafers provide scant nourishment, whether spiritual or physical. To take such spiritual atrophy as merely a *religious* issue, however, is to take for granted the absence of spirituality from the rest of life, and so to deny what many indigenous religions affirm: that in a culturally vital world, the spiritual is an intrinsic part both of the most exalted and the most mundane aspects of everyday life and of the world.

In contrast to most forms of Christianity, tribal religions seldom relate to the natural world only in general, abstract terms, preferring a more "bioregional" reference to particular places and areas. For example, as Vine Deloria relates, "[t]he Navaho . . . have sacred mountains where they believe that they rose from the underworld. . . . [T]here is no doubt in any Navaho's mind that these particular mountains are the exact mountains where it all took place."[22] Cultural forms grow out of the natural landscape as readily as other lifeforms if we allow them to do so, as Frederick Turner suggests:

> Every environment encourages a special mythology. Into the sacred narratives, into their ritual enactments, into the personalities of the deities, are filtered and fibered the aerial weather patterns, the size of the sky and what it brings, the shapes of the clouds, the contours of the terrain and its dominant colours, the flora and fauna, the natural rhythms of movement, mating, molting, and perhaps above all human adaptive responses to all this as they develop over time.[23]

Thus an ecology that adequately conveys its implied message of wholeness will be experiential and cultural as well as biological. It makes no sense to speak of a "holistic" ecology if the holism referred to is not complete: that is, if it focuses on biological realities to the exclusion of experiential, cultural, and spiritual dimensions.

Examples such as those above, reminding us of the power of cultural forms to integrate a people with their natural environment, are not intended to imply that we in the industrialized world should imitate or borrow from these cultures in a superficial attempt to restore our relationship to nature. To do so would be to reproduce in the spiritual realm the theft of tribal artifacts by Western explorers that has so often occurred in the past. "New Ageism" and the proliferation of "urban shamans" are embarrassing reminders of our capacity to uproot rituals and traditions from their cultural contexts and assimilate them piecemeal for our own egoic desires. However, other cultural traditions can resonate with dormant aspects of our own being, making us aware of the absence of such traditions in our own lives, and so clearing a space within which authentic spiritual forms can take root. The recognition of an absence is thus an initial step toward allowing that absence to be filled. For this to happen, we first need to recognize the flimsiness of industrialism's claim to completeness—a claim contradicted both by its rapacious need to consume other forms of structure, including the flora, fauna, and cultural traditions of the nonindustrialized world, and by our own underlying sense of restless unsatisfaction.

Culture, Nature, and Psychopathology

A healthy cultural framework, I have suggested, is necessary not only to integrate the human and natural realms into a balanced, mutually sustaining whole but also to allow the self to function effectively. As I write this, I am uncomfortably aware of the difficulty of expressing this idea in a way which does not *assume* the individualism that pervades modern industrial society—that is, which avoids the assumed and prior separation of "psychological," "natural," and "cultural" realms. This problem would not exist among indigenous populations such as the Navaho or Australian aboriginal groups, since psychological breakdown would *intrinsically* be experienced as both cultural and natural, the psyche being experienced as continuous with each of these other realms. In a similar vein, W. E. H. Stanner remarks of the aboriginal peoples of northeastern Australia that "when an aboriginal identifies, say, his clan totem and its sacred site, he is not 'pointing' to 'something' which is 'out there' and external to him, but 'not him'; he is identifying part of his inwardness as a human being, a part of the plan of his life in society, a condition of his placement and activity in . . . a cosmic scheme."[24] As we saw with the Aivilik, *inwardness* for nonindustrialized peoples also refers to the world *outside* the individual. The term "inwardness" is awkward here, however, since to us in the modern world "inwardness" by definition implies a distance and a separateness from what is "out there" in the world. As we move "inward," so we approach a realm that is more and more intensely personal and that has less and less to do with the "external context" of our lives. For us, personal identity resides in this detachment from the world, in the ease with which we can distinguish ourselves from the world. And our way of life reflects this distancing: we strive to control, predict, master what is outside us, minimizing any resonances between the self and natural patterns in the "environment," and downplaying the significance of those patterns that exist in the world. Within this modern configuration, problems in the "environment" are not also immediately "psychological": we can read about and even commit what to many indigenous peoples would be the most devastatingly ecocidal acts while still maintaining at least the outward semblance of personal serenity. Pollution of our drinking water, for example, may eventually be seen by us as problem if it directly affects our health; but it is not an intrinsically psychological problem experienced directly as a pollution of our own selves. In effect, then, it appears that our psychological health is only indirectly affected by the health of the natural world. But this perspective rests on a long-sedimented ontological flaw, that which defines the precondition of modern humanity: the dissociation between self and world. When we recognize this, we are also recognizing that there can be no real health under current eco-social conditions, but only adjustment to these conditions.

Given the large-scale dislocation that many tribal cultures endure today, there is unfortunately no lack of support for a view which more intimately and dialectically relates ecological health with personal well-being. A recent and continuing example involves the forced relocation of large numbers of Navajo from their ancestral homelands, ostensibly because of a land dispute with the Hopi. Since, among the Navaho, culture and psyche are continuous with landscape, removal from one's ancestral dwelling place constitutes a sort of psychological uprooting that is profoundly pathogenic. Edward Casey has written:

> Since [the Navaho] conceive of their land as an ancestral dwelling place and since all significant learning proceeds ultimately from ancestors, culture is almost literally *in the land*. It follows that to learn something is not to learn something entirely new, much less entirely mental; it is to learn how to connect, or more exactly how to reconnect, with one's place. At the same time, to reconnect with that place is to engage in a form of collective memory of one's ancestors; to commemorate them. To be displaced is therefore to incur both culture loss and memory loss

> resulting from the loss of the land itself, each being a symptom of the disorientation wrought by relocation.[25]

"What is striking about the Navajo tragedy," Casey continues, "is the specific acknowledgement by relocated people themselves that the loss of land was the *primary* loss. . . . It follows as a devastating deduction that to take away land is to take away life, that the major cause of illness is not something 'physical' or 'psychological' in the usual bifurcated Cartesian senses of these words but, instead, the loss of landed place itself."[26] If, as Geertz suggests, culture is an essential constituent of selfhood, then it seems that for many indigenous peoples, land is an equally essential constituent, and that the two are, in fact, inseparable. And although there is considerable cultural variation in response to loss of place, among traditional cultures such loss is commonly catastrophic.[27]

A second, and chilling, example of the effects of dislocation is Colin Turnbull's account of life among the Ik—a group of nomadic hunters living in the mountains bordering Sudan and Kenya. Turnbull summarizes the Ik's style of interaction with their natural surroundings in terms that by now will be familiar to the reader:

> In such a fluid society of hunters the environment invariably provides the central theme that holds them together, that gives them a sense of common identity; it is the hub around which their life revolves. It provides all the necessities such as food, shelter, and clothing, and often some kind of spiritual existence of its own is attributed to it. . . . [T]he Ik, in their rocky mountain stronghold, think of the mountains as being peculiarly and specially theirs. People and mountains belong to each other and are inseparable. It is not just that the Ik would not know how to live, to hunt or farm, in the flat, arid plateau below them, because they are as intelligent as any and a great deal sharper, quicker to learn and more adaptable than many. But with regard to the mountains it is different, and their adaptability seems to reach its limits. The Ik, without their mountains, would no longer be the Ik and similarly, they say, the mountains without the Ik would no longer be the same mountains, if indeed they continued to exist at all. . . . And so the two live together, a part of each other.[28]

This complementarity between individual, society, and natural context was broken when the Ik were denied access to the Kidepo Valley, their major hunting territory. With the formation of Kidepo National Park, they were moved from the mountainous regions that were their home to flatter, more arid areas where they were encouraged to adopt an agrarian way of life. Such an intention, like so many decisions made according to the "rationality" of the Western economic system, ignored the complex interconnections between personhood, social structure, and the natural world in which the Ik's lifestyle was rooted. The result was a process of cultural disintegration that fell back into a lifestyle of exaggerated individualism. Selfishness—even to the point of death—viciousness, and a complete lack of mutual caring and affection became the norm, and the lack of cooperation and common goals led to widespread starvation. The very old and the very young were abandoned, and any semblance of family life disappeared. Not only did the Ik not cooperate with each other to ensure their prosperity, or even their survival; they took pleasure in each other's discomfiture and eventual starvation.[29]

The major conclusion that Turnbull draws from his experience of living with the Ik is that "those values that we hold basic to humanity, indispensable for both survival and sanity, . . . are not inherent in humanity at all, they are not a necessary part of human nature. Those values that we cherish so highly and that some use to point to our infinite superiority over other forms of animal life may indeed be basic to human society, but not to humanity."[30] Turnbull's perspective, based on two years of living among the Ik, suggests an extended Geertzian viewpoint: a healthy psychological and social life depends, at least for many peoples, not simply on healthy cultural frameworks, but also on the integration with the natural world that such frameworks allow.

Cultural breakdown, of course, may the result of causes other than relocation. We have already seen that—as in case of the Yir Yoronts—quite subtle interference with the cultural system may be enough to cause a severe disturbance. The basic point is that culture, place, and psyche, if not distorted by some other influence, tend to coalesce to form a mutually sustaining whole, and that a disturbance anywhere within this whole will cause repercussions throughout the system. These secondary disturbances can thus be read as symptoms of a cause that may not be apparent to consciousness, since according to our cultural logic, causes must lie in the same domain as effects. As we saw in chapter 2, for example, depression is usually ascribed to (genetic or other) weaknesses within the individual psyche or, at best, to immediately proximate "social stressors." Thus the possible causal links between depression and more distant disturbances of the cultural/ecological fabric remain unexplored. There must be many psychotherapy clients who have had experiences similar to that of Joanna Macy, who relates that when she told her psychotherapist of her outrage over the destruction of old-growth forest, the response was that her distress sprang from fear of her own libido, as symbolised by the bulldozers.[31] As another perceptive social commentator, David Smail, remarks, "individual, so-called 'psychopathology' cannot be understood out of the environmental context in which it occurs, and indeed cannot be attributed to any pathological process *inside* people. It is, in fact, more correctly characterised as a pathology of the environment."[32] Conversely, there are few studies that relate environmental destruction to the individualism and narcissism that plague the modern psyche, although the recent emergence of "ecopsychology" is an important beginning in reintegrating these conventionally separate spheres.[33]

The Colonizing Impulse

The embrace of cultural structure, which integrates diverse elements by expressing symbolic resonances between them, is radically different from that of the colonialist impulse, which mutilates otherness so that it conforms to existing structure. For example, if the idea "tree" is allowed its symbolic aspects, it expresses its interconnection with a diversity of things: it can symbolize the earth's fertility (as in the English maypole), birth and rebirth, the home of spirits—especially female earth deities—the axis of the world, and various oracles (although no doubt Newton would not have understood his apple tree in this way); and, of course, each specific species of tree has its own symbolic referents.[34] These referents, however, have gradually been pruned in the modern world as the meaning of "tree" has come to conjure up a more limited range of meanings such as "timber," "aesthetic object," and so on. Today, the poverty of our common understandings of "treeness" is well expressed by those bare trunks that sometimes line our urban streets, their limbs having been repeatedly amputated as too dangerous or inconvenient—a mutilation which is redolent of our suspicion toward those dimensions of the world that cannot be contained within rational understanding.

This impulse to eliminate those dimensions of being which are refractory to the spirit of rationalism is perhaps the defining motif of industrial civilization. One of the tragedies of the discovery and rape of the "New World" over the past several centuries, for example, is that the opportunity for regenerating the culture of the old world, with its fossilized religious dogmas, isolation from the natural world, and aggressive ideology, has usually been overshadowed by the impulse to assimilate the new to the old. Although many of the early adventurers and colonists sent back enthusiastic and admiring reports both of the flora and fauna of the newly discovered lands and of the peoples who lived there, the possibility that the Europeans could learn something from these native inhabitants and their cultures was never considered seriously except by a deviant few. Columbus exemplifies the pattern, writing that the Arawak islanders were

> So affectionate and have so little greed and are in all ways so amenable that I assure your Highnesses that there is in my opinion no better people and no better land in the world. They love their neighbours as themselves and their way of speaking is the sweetest in the world, always gentle and smiling. . . . [Y]our Highnesses must believe me when I say that their behaviour to one another is very good and their king keeps a marvellous state, yet with a certain kind of modesty that is a pleasure to behold, as is everything else here.[35]

But if Columbus felt that the behavior of the native peoples he encountered might hold lessons for his own civilization, he gives no indication of it; for his intentions are quite clear. The Spanish cleric and historian Las Casas, who accompanied several Spanish expeditions to the New World and transcribed the log books, tells that one of Columbus's first acts on landing was to call together others of his group and demand that they should bear witness that he had taken possession of the island for his sovereigns and masters, the King and Queen. Furthermore, Columbus reports in his first entry after landing that the Arawaks

> Should be good servants and very intelligent, for I have observed that they soon repeat anything that is said to them, and I believe that they would easily be made Christians, for they appear to me to have no religion.

Two days later he wrote:

> These people are very unskilled in arms, as your majesties will discover from the seven whom I have caused to be taken and brought aboard so that they may learn our language and return. However, should your Highnesses command it all the inhabitants could be taken away to Castile or held as slaves on the islands, for with fifty men we could subjugate them all and make them do whatever we wish.[36]

Such attitudes were almost unquestioned among the early colonizers. Almost, because a few, such as Las Casas, were able to criticize the assumption of the God-given right to colonize, pillage, exploit. Las Casas' eye-witness account[37] of the destruction of the indigenous peoples of the Americas is disturbing to us not only because it lays bare the extreme violence on which the modern world is historically based, but also because this violence echoes metaphorically much of the violence that is subtly implied by our own lifestyles and ways of relating to the arational. As Frederick Turner remarks:

> Las Casas had come to see the New World for what it should have been—itself—and to recognise it as an enormous spiritual opportunity. Just as he had come to understand that enslavement is destructive of *any* race of human beings, so he gradually moved away from feeling that the New World was an enormous evangelical opportunity toward the conviction that it was a mysteriously granted gift through which Christianity could recover its primitive vigour. Whereas he had once pitied the Indians in their gentleness and simplicity, increasingly he was drawn to an admiration for the *rightness* of their cultures. . . . Though he never could bring himself to admit that there was something radically wrong with Christianity itself, Las Casas became adamant in his conviction that it was surely diseased in its New World practice. . . . Though he could not admit that something in the history of this religion made men callously commit great cruelties in its name, he could name those cruelties and he came very close to discerning one of their causes: the divorce of faith from the human body. The Christians despise these natives, he wrote, because they are in doubt as to whether they are animals or beings with souls. Such a distinction is foreign to the world of living myth, and if Las Casas could never make his way back to that, he was nevertheless quite certain that for all their attachment to nature, to the earth, these people had immortal souls.[38]

The implication of Turner's remarks is that the Indians and their culture represented that conjugation of the natural and the cultural which was so energetically being dissolved in Europe, and that this embodiment of a taboo association was the principle sin that justified their violent extermination. Las Casas, in challenging this interpretation of sin, was a rarity in a Europe where the separation of the human from the natural was fast becoming a measure of humanity's "ascent" from the "lower animals." Very few Europeans possessed the cultural imagination and undefensiveness that were required to subject their own way of life to critical examination; but another of these rare souls was Montaigne, who was contemptuous of the invaders' preoccupation with material acquisition and conquest, recognizing, like Las Casas, that contact with "minds so pure and new" offered regenerative opportunities to the Europeans' jaded spirituality that were potentially far more significant than "the traffick of Pearles and Pepper." "We have made use of their ignorance and inexperience," wrote Montaigne, "to drawe them more easily unto treason, fraude, luxurie, avarice and all manner of inhumanity and cruelty, by the example of our life and patterne of our customes."[39] As a result of their fateful, if unwitting, decision in favor of plundering the Americas rather than opening themselves to the new social and spiritual possibilities it offered, the invading Europeans ensured that the New World would become an extension of the Old World. "America once held out the promise of a land on which to base a new ethos or mode of dwelling on the earth," Robert Pogue Harrison insightfully remarked. "No such luck. The oceans that separate also unite, and Thoreau's nation as a whole now swims in the streams of opinion, delusion, and the old ways."[40]

This indicates a defining aspect of the relation between colonizing structures and those that are benign: the remarkable hostility of the former toward the latter. While the Indians typically welcomed the Spanish colonists with a good deal of generosity and courtesy[41]—at least until the murderous character of their mission became clear—Christian dogma experienced any form of otherness as a threat and a challenge; and so the terrorization of the native inhabitants of the Indies was matched by the equally brutal suppression of those traditions and practices which challenged Christian orthodoxy at home, beginning with the massacre of the Cathars in the thirteenth century, through the long years of the Inquisition, and continuing in the torture and burnings of "witches" that persisted into the eighteenth century. Christianity had become a cornerstone of the new anthropocentric order that counterposed the natural to the "spiritual." In embodying this nascent order with its inbuilt conflict, the Christian psyche also embodied its tenuous fragility, and so needed to engage in strenuous psychological efforts to defend it against the otherness of those symbolic awarenesses which were increasingly fading into the unconscious.

The churches' antagonism toward embodied religious experience suggests that the disentangling of the soul from the body was a key element of the new order. Christian mysticism, for example, as "a series of solitary attempts to find some way out of the prison of the soul that Christianity had become, back to the primitive joys of the sacred period,"[42] was a clear threat to the dogma of the mainstream Christian churches, since it insisted on the contemporary reality of *spiritual experience and revelation* rather than on an abstract "belief" in religious events that were distanced from people by space and time. The "victory" of Christian dogma over traditions such as Catharism within Europe and over indigenous spirituality abroad can be seen as the victory of dogmatic ritual over spiritual renewal, of intellect over the body, and of dissociation over integration, proclaiming the dualistic values of a totalitarian ideological system that was, and is, profoundly hostile to the natural order.

The Repression of Otherness

If culture, as I have argued, is such an essential integrating force within a world that is otherwise prone to forms of fragmentation and collapse which are simultaneously environmental, social, and psychological, why has this not been generally recognized; and why, in fact, have the various disciplines constituted themselves so as to deny it?

The widespread academic denial of culture, viewed as a fundamental integrative medium, is less puzzling if we recognize that this denial is a founding characteristic of the social milieu we inhabit. If culture, understood in a Geertzian sense, *integrates* the human and the natural, then the emergence of a "human" realm that attempts to *sever* its symbolic connections with the natural is a process which is fundamentally hostile to both culture and nature, reducing the first to a stockpile of social variation and the second to biology. Thus "culture" has often been co-opted as part of the "human" realm that explicitly distances itself from "nature," reflecting the curious taxonomy of modern experience generated by industrialism; and within this taxonomy, "culture" tends to be *dis*integrative as much as integrative. Consider, for example, the ways social groups may be distinguished according to their artistic preferences for, say, rap music or the Beethoven sonatas; and the way both forms are used to distinguish "human" activities from those of "animals." We will only be able to rediscover the integrative power of culture if we are prepared to discern the cadence of a more distant drummer above the more strident rhythms that structure most academic activity—the slow, subtle rhythms of the natural order. Conventional understandings of culture and cultural differences have been heavily demarcated by the fault line between the "natural" and the "human" that characterizes modernity, and even the insights of Geertz refer mostly to a type of "meaning" in which the natural world provides the content of mythology rather than influencing its underlying form. What we need, then, is a description of the world which, in recognizing the full significance of the "natural," is "thicker" than even Geertz intended, and in which symbolism does not imply the arid intellectual abstractions of Lévi-Strauss but rather one in which human patterns find their felt analogies in natural processes.

Anthropology, of course, has not merely documented cultural differences, but, at a deeper level, has embodied them as well. It is a well-founded truism that ethnographies have—often startlingly—revealed as much about the ethnographer's background as they have about the ethnic group studied. To this extent, they are diagnostic of the reactions of the modern ego when faced with real otherness, and, writ large, of tendencies within the modern world to assimilate or otherwise neutralize challenges to its own configuration. In present-day ethnographies, these characteristics are more deeply concealed as we consciously strive to avoid denigrating otherness and exalting our own cultural preferences, often disguising them under a pervasive cultural relativism or, alternately, retreating into an assumed objectivity. The ambiguities contained in both these stances and the lack of any clear alternatives to them has provoked a confusion that as much as anything else is symptomatic of the moral impasse affecting not only anthropology, but also the modern world as a whole; for our claimed allegiance to diversity, choice, and freedom all too often takes place against a subtle but all-encompassing alignment with industrialism.

The reactions of the European colonialist mind when faced by structures that are at variance with its own are well illustrated by the problems experienced by colonialists as they attempted—often with difficulty—to inculcate a sense of personal possession among tribal peoples. For aboriginal groups, most of the items used in everyday life are as much a part of the preexisting fabric of the world as they are personal possessions. The Yir Yoront axes referred to earlier in this chapter, for example, were located within a complex and sophisticated cultural scheme that related to spiritual and social realities as much as to individual needs; so what we translate as "ownership" may be more an expression of the membership of the individual within this larger scheme than it has to do with individual "rights" to do with something as one pleases. In contrast, the industrialist notion of "ownership" emphasizes only the latter aspect, referring to a fundamentally *economic* "reality." The contrast between these two conceptions of ownership is diagnostic of a difference between a world whose structure permeates one's own life and behaviour, and one which consists of "things" that can be manipulated or bartered. Given these radical differences, it is hardly surprising that trading relations between tribal peoples and European colonialists have been the origin of some fundamental misunderstandings. For example, the French trader Tabeau, who established a trading post among the Arikara of the upper Missouri during the early years of the nineteenth century, remarked contemptuously:

> Their minds not grasping our ideas of interest and acquisition beyond what is necessary, it is a principle with them that he who has divides with him who has not. "You are foolish," said one of the most intelligent seriously to me. "Why do you wish to make all this powder and these balls since you do not hunt? Of what use are all these knives to you? Is not one enough to cut the meat? It is only your wicked heart that prevents you from giving them to us. Do you not see that the village has none? I will give you a robe myself, when you want it, but you already have more robes than are necessary to cover you.[43]

This being the regrettable situation, not surprisingly efforts were made to educate Indians so that they could take their place within the fold of Western civilization. As one white trader put it: "We need to awaken in him wants. In his dull savagery he must be touched by the wings of the divine angel of discontent. . . . Discontent with the tepee and the starving rations of the Indian camp in Winter is needed to get the Indian out of the blanket and into trousers—and trousers with a pocket in them, and with a pocket that aches to be filled with dollars."[44] By such schemes was the Indian wrenched out of his natural context and pressured into feeding cannibalistically off it.

But if statements such as this seem often to be the unorganized and intuitively apprehended expression of an ideology whose agents are themselves uncertain of its purpose and direction, on occasion we find the colonialist process driven by a frank awareness of its character. Frank Waters tells us that after the Plains tribes were forcibly moved to reservations,

> The Cherokees, now settled in their new home, made such progress that Senator Henry L. Dawes of Massachusetts paid them a visit. He reported: "The head chief told us that there was not a family in that whole nation that had not a home of its own. There was not a pauper in that nation, and the nation did not owe one dollar. Yet the defect of the system was apparent. They have got as far as they can go, because they own their land in common. *There is no selfishness, which is at the bottom of civilisation.* Till the peoples will consent to give up their lands, and divide them amongst their citizens so that each can own the land he cultivates, they will not make much more progress."

> Wherefore in 1887, the "Dawes Act" or "General Allotment Act" was passed. Each Indian was to be allotted a piece of reservation under a fee simple title."[45]

Similar scenarios were also frequent in South America during the nineteenth century, where colonization was backed by a degree of violence remarkable even by the standards of that barbaric century. The indigenous populations of Colombia, for example, were forced into servitude on the rubber stations by means of floggings, torture, and exemplary mass murder; for the opinion expressed by one plantation overseer that if "Indians were not flogged they would not bring in rubber"[46] was generally accepted among colonists and exporters. And such views were very likely realistic; for the Indians showed little inclination toward sharing the whites' obsession with amassing material wealth. Jules Crévaulx, who explored the Putumayo River in 1879, reported that

> [S]ometimes these children of nature enter into relationship with somebody searching for sarsaparilla or cacao, but this does not last long. As soon as they have exchanged their stone axe for a knife or for a machete, they find the connection with the Whiteman insupportable and they isolate themselves in the forest. The problem in civilising the Indians of South America is that they lack ambition. The Indian that has one knife will give nothing, absolutely nothing, to possess another.[47]

Such stories reveal starkly contrasting ideologies, of course; but they leave unexplained the reasons for the colonialists' fervent advocacy of an economic system whose destructive effects must have already been apparent to them. We have already noted Frederick Turner's answer to this question, suggesting that those whose beliefs have lost their relation to the grounded realities of their lives are easily threatened by alterity, and so punish those competitors whose essence is threatening to their own. And Turner's answer is consistent with that of another great commentator of the American West, Roy Harvey Pearce:

> What generally emerges [from white writings about Indians] is a simple and clear demonstration . . . of a proposition which Americans, in their feelings of pity and censure over the fate of the Indians, needed desperately to believe: that men in becoming civilised had gained much more than they had lost; and that civilisation, the

> act of civilising, for all its destruction of primitive virtues, put something higher and greater in their place.[48]

The actions of the colonizers, then, may be directed not only against those others whom they sought to assimilate, but also against those residual aspects of themselves that the Indians' lifestyles so poignantly evoked in them. In this respect, they remind us of the desperate technological optimism of our own era. The nascent industrialist structures which parasitically occupied the colonialist psyche were profoundly threatened by, and so demanded the repression of, competing structures which resonated more intimately with the natural order; and it is therefore not surprising that colonialism elevated this repression to the status of a moral imperative. In the words of Colonel Seth Eastman, "The Indian . . . is yet ignorant of the greatest victory of which man is capable—the conquering of oneself."[49] Alterity is not a psychological problem if the other is also part of the same shared cosmos; but in a fragmented cosmos in which the relation to the other has been lost, alterity is feared, so that native populations becomes "savages," animals become machines or "organisms," nature becomes an assortment of "things," and feelings become the sometimes incomprehensible impediments to rationality. In each case, depending on its capabilities, the other becomes either a threat to be destroyed or a resource to be exploited.

It is in the nature of such dynamics that they express themselves symptomatically in a diversity of apparently unconnected forms. Some of their least sophisticated manifestations simply represent the other as deficient, as we have already seen, and as documented in classic studies of prejudice. Such views, of course, are both politically incorrect and inconsistent with the humanistic claims of modern industrial society, and as a result their straightforward expression may be becoming less common. But it is characteristic of industrialism that the colonizing process takes progressively more covert forms, and even apparently enlightened attitudes can serve a latent purpose of assimilating that which they claim to defend. Within academia, for example, where crude prejudice has long been unacceptable, there may be a reactive fetishism of colonized peoples and their art, so that overt prejudice is replaced by an anxious condescension. And in a later stage of this process, what often emerges is a kind of detached pseudo-objectivity that is the final outcome of the ebb and flow of prejudice and antiprejudice. Thus if the initial colonizing impulse (which Pearce calls "savagism") may be covered up by the opposite impulse to glorify the indigenous (which he refers to as "primitivism"), "objectivity," thereafter, resides in a sort of rhetorical Punch-and-Judy show of argument and counterargument. A final twist to this dialectic occurs with the pronouncement that whatever reality the contestants were debating is actually "socially constructed" through the dialectical process of debate—which, by that stage, is probably a fairly accurate assessment! This whole approach, then, is one that serves to distance us from that *initial* reality which was certainly *not* an artifact either of social consciousness or of the dialectical process of debate.

Pearce himself is not immune to this problem. His tendency to see all firsthand accounts of relations between white Americans and Indians in terms of the dialectic of savagism and primitivism precludes the possibility that these accounts might contain any accurate information, and even those firsthand accounts that Pearce regards as relatively reliable are assimilated to this savagism-primitivism dialectic rather than accepted as possible windows into the realities of Indian life. Benjamin Franklin, for example, whose integrity Pearce acknowledges, wrote that "it has appeared to me that almost every War between the Indian and Whites has been occasion'd by some Injustice of the latter towards the former," a verdict that has been generally confirmed by historical studies. Nevertheless, Pearce dismisses Franklin's views, arguing that he "could avoid the facts and, in the finest primitivistic fashion [i.e., one that glorifies the 'primitive'], use the Indians to criticise the excesses, and possible excesses, of his own civilised society . . . constructing a conventional eulogy of Indian politeness, Indian 'Conversation,' Indian religion, and Indian hospitality and honor. None of it is to be taken seriously."[50]

Pearce is equally dismissive of other firsthand accounts of Indian life, such as Crevecoeur's. "There must be something more congenial to our native dispositions [in Indian life] than [in] the fictitious society in which we live," wrote Crevecoeur, "or else why should children, and even grown persons, become in a short time so invincibly attached to it?"[51] William Bartram, too, found that Indians received moral guidance from "a more divine and powerful preceptor, who, on these occasions, instantly inspires them, and as with a ray of divine light, points out to them at once the dignity, propriety, and beauty of virtue."[52]

Pearce offers many more such examples which allegedly demonstrate the unreality of white images of native American life, invariably assessing them as examples of either savagism or primitivism. The possibility that such writings might indicate in an acceptably truthful fashion actual qualities of Indian life is not one that he is prepared to consider. Rather, such accounts become part of a rhetorical game involving competing and equally illusory texts, existing within an academic world that seems safely insulated from the real history of colonization. They are emotionally detached from the tragedy of the destruction of American tribal life, whether joyful or tragic, cruel or humane, benign or destructive, existing instead within a narrowly academic world of deconstructive analyses whose power or motivation to address political inequalities, whether present or past, is flabby or nonexistent. The implicit and curiously nonreflexive assertion that all white writing is ideological and reflects nothing worthwhile of the reality of Indian life amounts to a denial both of the native experience and of the capacity of white writers to engage passionately with it. It is a stance which, rather than sifting the biased from the authentic, perceives the former as inevitable and the latter as impossible in principle. The original realities of Indian life therefore disappear in the face of the alleged universality of one or other species of ideological bias; and the focus of discussion becomes this bias rather than the original realities. Thus both the "reality" being studied and the processes through which such study takes place are assimilated to the same, rhetorical world; and the action that results from this stance is also contained within this world of words and ideas which have lost their grounding in physical, cultural, and historical reality. Vine Deloria, for example, observes that we periodically "rediscover" the genocide on which modern industrial society is partly based, and go through a spasm of rhetorical hand-wringing before it is again forgotten. Such writing is a sort of verbal academic sport rather than a passionate and embodied engagement with the ecological, cultural, and political realities that writing can *refer to.*

Stances such as Pearce's therefore unwittingly collude in the denial of otherness and its assimilation to a separate, "human," realm. All too often, recognizing that our ways of empathizing with realities removed from us by time or culture are susceptible to various prejudices and distortions has led to a defensive focus on these prejudices and distortions rather than on the realities themselves—a tactic that camouflages the realities even more effectively than the distortions it is designed to reveal. If all writing that argues for the value of native American cultures is glossed as reflecting "primitivism," and all writing that reports less appealing aspects of these cultures is depicted as exemplifying "savagism," then the cultural realities are entirely assimilated to this dualistic scheme. Thus "as one academic field after another falls into the paralysing coils of obsession with language and communication,"[53] the real substance of Indian life and traditions recedes into the background, to be replaced by the discursive concerns of the academic world. John Ellis has noted the generality of this trend within academia, pointing out that "reading the classics with an interest in what they have to say to us (rather than diagnosing their race-gender-class attitudes) is described scornfully by many [academics]."[54] A rigid and exclusive emphasis on deconstruction is in danger of rendering Indian life and history itself as illusory—as if an analysis of white attitudes were all that is required to right past (and present) injustices. As a result, material situations of continuing oppression and poverty becomes secondary to this analysis;[55] and cultures that could challenge the assumed inevitability and progressive nature of modern industrial society are made to appear illusory and insubstantial. As Ellis concludes, "[t]here is a real danger, then, that prevailing practices within academia may actually further the colonialist aims which they claim to be critiquing."

If this danger is realized, we are left with modern industrial society as the *only* possible reality, anything else being dismissed as the product of white imagination and ideology; and a potentially critical comparative social science is reduced to the study of competing accounts and desiccated historical fragments, distanced from us by time, place, and apparent irrelevance to the urgent questions of the contemporary world. These historical fragments therefore become the "ore," the "raw material," that is ground up and reconstituted to form the commodities of the academic world: they are assimilated to current understanding rather than being allowed to *challenge* that understanding.[56] As Peter Mason argues, "[u]nderstanding the other reduces the otherness of the other. . . . To understand the other by comprehension is to reduce the other to self. It is to deprive the other precisely of the very alterity by which the other *is* other."[57] Such understanding, although extending the scope and reach of Western ideology, reflects a reordering internal to the closed world of that ideology, and any radical potential that it might possess is therefore all too easily undermined by its acquiescence to the values of an academic world which denies realities that are not intellectual or discursive. As such, it is ideologically contiguous with earlier, cruder techniques by which otherness has been assimilated to modernistic understandings, such as those employed by European commentators regarding the "New World." As Mason remarks:

> The way in which observers of America resorted to the world that was familiar to them is a timeless response by self when faced with the challenge of the other. In using the elements familiar to them, they were in fact engaged in a double process of reduction and construction. In *constructing* the New World, resemblance was linked with imagination to avoid the endless monotony of the same. The result is a continuing process of construction and reconstruction of a world, which we may therefore call an imaginary world. . . .
>
> The other side of this process is its *reductive* aspect. The perception of the other was not limited to observation from a distance. It was coupled with violence, and the violence carried out against the other was an attempt to *reduce* what was refractory to the bounds of self.[58]

Thus, says Mason, an "exegetic apparatus is imposed on the New World to assimilate it to the closed circle of words and things."[59]

Unfortunately, Mason himself, although recognizing how such reductive and constructive processes led to the formation of an "imaginary world," adopts the fashionable view that the reality so constructed is the only valid reality. "Mythology does not consist of imaginary accretions as functions of some underlying (historical) reality. . . . On the contrary, there is much to be said for the suggestion that 'reality' itself is a product of the activity of our imagination."[60] Once again, then, the meaning of historical realities is confined within the human psyche. We do not have to remain within this literary prison, however: the text, like Sam Gill's "shadow of a vision yonder," can be an attempt to model, to point toward, to empathize with a reality that exists beyond its own boundaries; and even if it is not "the mirror of nature," it can at least be a signpost to it. We need, therefore, to reach beyond deconstruction's focus on the discourses that describe particular events or natures, to these events or natures themselves, drawing on our embodiment in the world as the vehicle of our empathy and comprehension.

In view of the taboo on embodied personal engagement with realities that extend beyond literary or academic boundaries, however, it is perhaps unsurprising that those few social scientists who *have* had the courage to engage with other cultures have often been criticized for a lack of "balance"; in other words, their work, lacking the wishy-washy relativism, earnest triviality, and emotional disengagement which characterizes most social science, is experienced as threatening, and so has to be defended against. The failure of nerve that underlies the unwillingness to seriously consider radical cultural critiques is particularly apparent in some of the reaction to Colin Turnbull's work. A case in point is Marcus and Fischer's criticism that Turnbull relies on "static us-them juxtaposition to deliver criticism of American (and Western) society. The cultural other becomes chauvinistically valued to the point of unrelenting pessimism about the conditions of American society in comparison"[61]—a criticism that is particularly misplaced given that Turnbull's book about the Ik was also, as Robert Edgerton relates, "severely criticised for painting too *dismal* a picture of Ik culture."[62] Here Turnbull's writings are depicted, respectively, as "primitivistic" and "savagistic." It is hard to see, however, just what form of critical cultural comparison might be acceptable to Marcus and Fischer unless it becomes so vague, defensively qualified, and abstracted from political realities as to lose all impact.[63]

But Marcus and Fischer offer plenty of indications as to the sort of "cultural critique" that *they* would accept as valid. Such a critique, they argue, would play down anthropology's traditional interest in "Africans, Indians, and Pacific Islanders," and focus instead on "domestic" topics.[64]

This "repatriation" of anthropology recognizes that "distinctive cultural variation is where you find it, and is often more important to document at home than abroad."[65] Such a stance runs the risk of realigning anthropology entirely within the orbit of the industrialized world; and our apprehension about this increases when we read that "all other cultural worlds have been penetrated by aspects of modern life," so that "what matters . . . is not ideal life elsewhere, or in another time, but the discovery of new recombinant possibilities and meanings in the process of daily living anywhere. . . . Alternatives . . . must be suggested within the bounds of the situations and lifestyles that are the objects of cultural criticism."[66] The "traditional . . . strategies of the cultural critic" are "increasingly easy to dismiss" because they are either "thoroughly pessimistic" or "thoroughly idealist or romantic." After all, if "the exotic cultures that remain are increasingly marginal in a world that appears to be homogenising, then what relevance do their isolated realities and experience have for modern life?"[67]

Any real alterity is therefore discounted, and cultural criticism is limited to those alternatives that exist *within* the orbit of the industrial world. This stance serves a similar ideological purpose to claims that nature is already "artefact and habitat"[68] rather than a source of *alternatives to* the industrialized world: in both cases, there is an attempt to assimilate to industrialism what is at least partly separate from it, and so to facilitate the physical assimilation of the other. Thus today, "it is commonly thought that with advances in communications and technology, the world is becoming a more homogenous, integrated, and interdependent place, and with this process, the truly exotic, and the vision of difference it held out, is disappearing."[69] This, regrettably, is largely accurate; but the question is whether we are prepared to relinquish the "exotic" and the diverse without a whimper of protest. "For a long time," argue Marcus and Fischer, "the primitive other—a vision of Eden, where the problems of the West were absent or solved—was a very powerful image that served cultural criticism"; but today, "to invoke another culture . . . is to locate it in a time and space contemporaneous with our own, and thus to see it as part of our world, rather than as a mirror or alternative to ourselves."[70] What we are being asked to accept here is not only the reduction of natural and cultural diversity toward the monoculture of industrialism, but also the inconceivability of any alternative to this monoculture. This is rather like the approach of a recent BBC wildlife documentary, which seemed to suggest that although species were disappearing at an alarming rate in the wild, this doesn't really matter, since other forms of wildlife are flourishing by the verges of motorways! Similarly, Marcus and Fischer argue that by de-emphasizing the "exotic other," anthropology can instead focus on "more important" issues at home, such as "providing ethnographic data for administrative policy."[71] Ethnography, according to Marcus and Fischer, "explores possibilities that are strictly within the conditions of life represented, rather than beyond them in some other time or place," locating alternatives "by unearthing . . . multiple possibilities as they exist in reality."[72] Thus the poor *favela* dwellers of the Brazilian northeast described by Nancy Scheper-Hughes, whom we will discuss shortly, should presumably be content to live within the possibilities defined for them by a "reality" dominated by rich plantation owners, rather than playing with "idealist or romantic" ideas about redistributing land ownership or lobbying for an unpolluted water supply; and we could similarly argue that environmentalists should focus on the diversity of animal life existing in urban areas and abandoned factories rather than on romantic conceptions of wilderness, grizzly bear habitat, and so on. So it is that anthropology follows the lead of psychology in becoming the servant of industrialism rather than its critic.

The political conservatism of Marcus and Fischer's approach is obvious; but it is a conservatism which, in a more subtle form, pervades much of academia's preference for textual analysis over interaction with the real world. This, in turn, is a part of a larger tendency within industrialism: the distancing of humanity from the natural world, and the replacement of nature by an alternative "reality" that aspires to autonomy from its natural roots. The unintended irony of Marcus and Fischer's approach is that the "reality" they prefer to "romantic and idealistic" notions is actually an aspect of the grand illusion of industrialism, the fantasy that industrialism can develop independently of the natural order. The "sense of impending loss [which] is still poignant in ethnographic writing"[73] is misleading, they argue, since "there [is] no real indication that anthropologists [are] running out of subjects"—as if anthropologists' supply of "subjects" were the important issue! By denying that anything has been lost, we adjust ourselves psychologically to its absence; and the discrepancy between present conditions and our painful ability to envision healthier alternatives is dealt with by adjusting the psyche to fit these reduced conditions rather than by fighting to retain cultures, species, and wilderness in physical reality.

Part of this psychological adjustment involves the dissociation of values and feelings from the empirical description of the world. Thus for Marcus and Fischer, "the statement and assertion of values are not the aim of ethnographic cultural critique."[74] Rather, anthropology should stand back from any assertion of values, and instead adopt an "engaged discourse" with its subjects that "fulfils in practice the idea of relativism."[75] Specifically, "art and philosophy are the domains in which values, aesthetics, and epistemology have been systematically debated, but these discourses thrive on a self-conscious detachment from the world to see their issues clearly." Values, in other words, should be detached from physical or political realities. Values, however, according to the perspective suggested in this book, are the indirect articulation of an embodied rootedness in the natural world, and so the distancing of values from anthropological and ecological realities can be seen as another manifestation of the culture-nature divide which, I have argued, is central to the project of modernity. And this is the central issue that divides cultural critiques such as Turnbull's from that of Marcus and Fischer: while the former begins from the embodied participation in a reality outside industrialism, the latter *accepts* industrialism as the *only* reality and dismisses anything beyond this reality as "unreal." But this rejection of anything beyond industrialist reality as "romantic and idealist" leaves Marcus and Fischer with no basis for *any* radical critique of modern industrial society, and so they have to be content with the politically impotent comparison of particular variations *within* it. Since all these variations share common ideological properties, including an anthropocentric understanding of nature, such a "critique" becomes in practice an *affirmation* of the values of industrialism and an acquiescence to the safely descriptive, business-as-usual approach to social science. The featureless ethical landscape advocated by Marcus and Fischer, which "fulfils in practice the idea of relativism," is one that has long ago lost any sense of a grounded morality. Such a relativism, in the words of Stanley Diamond, "is in accord with the spirit of its times, a perspective congenial in an imperial civilisation convinced of its power." Cultural "criticism" which begins from a mindset that is unchallengeable is thus a colonialist practice that can only, in Diamond's words, "convert the experience of other cultures into a kind of sport, just as Veblen's modern hunter mimics and trivialises what was once a way of life. Relativism is the bad faith of the conqueror, who has become secure enough to become a tourist."[76]

It is this transgression against the academic taboo on experiential wholeness and embodied involvement which accounts for the threat that more conventional academic writers have perceived in Turnbull's writings,[77] since conventional anthropology, in Johannes Fabian's words, defends the idea that "the Other, as an object of knowledge, must be separate, distinct, and preferably distant from the knower."[78] Such work as Turnbull's keeps alive the possibility of encountering the other on the same experiential terrain—a possibility that otherwise remains at the level of a mute, inarticulate yearning for wholeness. Tribal traditions and mythologies represent a genuine "otherness" for us, an otherness that we desperately need as a rootstock for the regeneration of our own culture; for the alternatives which are available within the almost all-encompassing logic of the technological worldview are inevitably ideologically determined, and so lead back to the same destructive "solutions." As Ulrich Beck has pointed out, "modernisation has consumed and lost its other,"[79] psychologically, culturally, and—almost—ecologically, trapping us within a spiral of increasing alienation from the natural world; and this is an entrapment which environmental theory must recognize if we are to generate solutions that are more than variations on the industrialist theme.

Integrity and Adaptation in the Modern World

Culture, I have argued, has the ability to articulate and develop our basic biological potentials in ways that extend the already wonderful diversity of biological evolution itself. Modern industrial society, however, has come to regard culture as *opposing* nature rather than articulating it; and this situation has led many to look toward the patterns embodied in indigenous societies in their search to reconnect the human and the natural. As we have seen, there are indeed indigenous societies that have this potential; but equally, it is quite clear that some indigenous societies can be taken as models neither of ecological wisdom nor, for that matter, of social harmony, as Robert Edgerton has made clear.[80] However, the blanket dismissal of indigenous societies as sources of environmentally relevant principles is as mistaken as their unreflective acceptance: clearly, each society needs to be considered on its merits.

The notion of "ecological wisdom" is one that is intrinsically difficult to define in the reductionist terms of scientific discourse, since it implies both an internal cohesion and a fit with its environment, so that "nature within" is harmoniously integrated with nature in the "outside" world. A problem that immediately appears is that societies which have possessed these characteristics have often been conquered by those which are more violent and expansionist, so that the characteristics that are associated with ecological wisdom may also, under current political conditions, be those that make a society less likely to survive in the long term. We can understand this in terms of a previously well-adapted society becoming ill-adapted due to changed, more hostile, environmental conditions that select for advantage in warfare rather than a benign attitude toward the natural environment. This argument can be generalized to include economic "warfare": in other words, the conditions of global economic competition imply a quite different environment to that of a world consisting of widely separated societies with relatively few communicational links and limited military hardware. This suggests that the conditions of modern life may be intrinsically hostile to "ecological wisdom," so that searching for possibilities of environmental health *within* these conditions is inherently paradoxical. It may be that an ecologically "healthy" society can only survive in a world wherein the grounding conditions are such as to favor the existence of diversity rather than selecting only for the most economically and militarily viable options.

To blame an indigenous society for its military inadequacy in the face of a more powerful and technologically advanced foe is rather like blaming the Great Auk—a flightless seabird that once lived on European coasts—for not evolving body armor which would have protected it from the clubs of avaricious seamen. In other words, in order to survive, the Auk would have to have become *what it was not*. The natural order, however, is not given to this sort of adaptive regression, which would constitute an accommodation to industrialism. Societies, too, have often been unwilling to accommodate to more aggressive nations, as Edgerton relates:

> From the rebellion led by Spartacus against Roman slavery to thousands of similarly desperate uprisings by African slaves in the New World, people have chosen to die in a quest for freedom rather than live in bondage. So it was among several North American Indian tribes such as the Southern Cheyenne and the Nez Percé, many of whom lost their lives in their attempts to escape confinement on reservations. The Yahi died because they chose not to submit in the first place. And Jewish epic history records the decision to die made by nearly 1,000 men, women, and children who defended the rocky plateau fortress known as Masada against a Roman army in the first century AD.[81]

The choice between survival or adaptation made by a cultural system in the face of overwhelming odds may also be a more subtle affair. Gregory Bateson tells how the anthropologist Sol Tax attended the National Convention of the Native American Church, whose central sacrament involves the use of peyote:

> The church was under attack for using what would be called a drug; and it occurred to Sol Tax . . . that he would be helping these people if he made a film of the convention and of the very impressive rituals which would go with it. . . . He therefore dashed back to Chicago . . . and was able to get a movie truck and some technicians and a stock of film and cameras. He told his people to wait in Iowa City while he went and talked to the Indians to get their approval of the project. In the discussion that ensued between the anthropologist and the Indians, it gradually became clear to Tax that they could not picture themselves engaged in the very personal matter of prayer in front of

> a camera. . . . As he [Sol Tax] . . . listened with fascination to the speeches, gradually the realisation came that they were choosing their integrity over their existence.[82]

As Bateson relates, the "curious paradox in this story is that the truly religious nature of the peyote sacrament was proven by the leaders' refusal to accept the pragmatic compromise of having their church validated by a method alien to the reverence in which they held it."[83]

Species and individuals, too, can in effect choose between integrity and adaptation, although the former lack chroniclers who could record their decisions, preferring often to disappear without apparent protest from a world that is no longer congenial to them. Over the millennia, many tribes and societies have been conquered, exterminated, or incorporated into more aggressive societies; and it is these more aggressive societies that have tended to survive and therefore to be better known. They are also very likely closer in structure to, and therefore more compatible with, the competitive organization of the modern world. For example, societies such as the Zulu under King Shaka, the Asante empire of what is now Ghana, the Aztecs, and the white European colonizers of the New World successfully assimilated those tribes who were willing or able to adapt to the conditions that they imposed, and exterminated those that were not. Each of these cases, and many others, illustrates not only the conquest of a weaker nation by a more powerful; but also the imposition of a logic which insists on the primacy of aggression and military force over more tolerant and empathic attitudes to their surroundings. This is a logic, of course, that is easily recognized in the modern world, since the principle of imposition of a human order on what is not human is, as we have seen, widely accepted.

It appears, then, that as the industrialist order sweeps aside those species for which it has no use, it similarly extinguishes those cultures that are equally inconsistent with the new order. *Some* of these extinguished societies may embody what I have called "ecological wisdom," which we could learn from; which is not to say, of course, that we should slavishly imitate their practices or beliefs. However, we should be aware that the psychological, cultural, and ecological structures which might exist in a healthy world will necessarily be judged as impractical, unrealistic, and irrational under current social conditions; and that those solutions which appear practical, realistic, and rational are very likely already colonized by the industrialist ideology that makes these judgments.

Multiple Pathologies in a Fragmented World

If psyche, nature, and culture within aboriginal societies often seem to be interwoven in a way that ensures that the life-world is stable and coherent, it might still be objected that this has little or no relevance outside these aboriginal societies, since the industrialized world operates in radically different ways, and science and liberal humanism have rendered obsolete our dependence on the cultural articulation of nature typical of indigenous societies. However, I will argue that this dependence has been distanced from awareness rather than functionally eliminated, and that the chronic defensive adaptations that our separation from the natural realm demand are damaging both to selfhood and to nature. If the technologically colonized self operates according to laws and principles which are increasingly disconnected from those of the natural world, then it is hard to see how this emerging disparity can have any outcome other than disastrous conflict between the natural and cultural realms.

Researchers who have examined cultural breakdown in aboriginal societies have often drawn parallels between the symptoms that are associated with the influence of industrialization in these societies and those that are common in our own social context. Colin Turnbull, drawing on his experiences of living among several societies, arrives at conclusions about the importance of culture which are strongly consistent with Geertz's views, suggesting that the integrity of the psyche will inevitably suffer whether the atrophy of culture is acute or chronic.[84] He finds a number of striking similarities between the fragmented culture of the Ik and our own society, arguing that the signs of psychological and social disintegration that were so prevalent among the Ik are now recognizable within modern industrial society. In particular, he points to the loss of family structures and the "cutthroat economics, where almost any kind of exploitation and degradation of others . . . is justified in terms of an expanding economy and the consequent confinement of the world's riches in the pockets of the few."[85]

In a similar vein, Cisco Lassiter argues that the sort of cultural and personal disintegration that he discusses in relation to those Navaho who have been relocated is, in a more chronic, subtle form, detectable within white culture.

> We, too, are now finding ourselves increasingly vulnerable to the kind of "psychopathology" experienced by the Dineh relocatees: homelessness, disorientation, rootlessness, alienation, loneliness, depression, and despair. In a society driven by the pursuit of an ever-increasing material standard of living, often at the expense of home, rootedness, and membership in the biotic community, these forms of suffering are probably inevitable. Many of the common illnesses of contemporary society suggest experiences which parallel the relocation of Dinehs from their homeland.[86]

The symptoms which Lassiter notes among relocated Navaho are acute responses to disruptions of an ongoing relationship to culture and to the land. Such acute responses are commonly found among refugees or otherwise displaced peoples.[87] The situation among the stable industrialized populations is likely to be somewhat different, however, since it involves a chronically evolving and centuries-long dialectic between personhood and the social, economic, and technological environment rather than an acute response to sudden upheaval. In chronic situations like this, psychological problems will reflect particular character structures, embodying adaptations and compensations that are deeply sedimented within the social and psychological realities shared by both researchers and "subjects." Acute symptoms, in this context, would be found only when an individual's adaptive capacity is stretched beyond endurance by other factors, in which case these other factors rather than the latently pathological social context will be identified as "causing" the symptoms—in much the same way that the causes of death among a population weakened by starvation may be identified as a range of diseases and infections. Thus the "background" factors reflecting chronic cultural changes and tensions will be included in the assumptive context of almost all studies and so will effectively be invisible to the methodologies available to us. As we noted in chapter 2, the more obvious psychiatric problems identifiable within the industrialized world, such as schizophrenia, depression, anxiety, phobias, and so on will appear to be due to specific factors involving genetics, stress, family communication, or whatever; but the underlying cultural factors which move the population as a whole toward a predisposition to these problems will be common to every person, and so will not usually appear as causally related to psychological distress. Guntrip has suggested that beneath all the so-called neuroses, psychoses, and character disorders lies what is "*always* the ultimate underlying problem": schizoid dissociation of mind from body and intellect from feeling;[88] and so the discernible causes of these disorders are in fact only the final straws for a psyche already weakened and fragmented by underlying and almost universally accepted social conditions. Thus the cultural ideology that leads to the demolition of the natural world also covertly undermines our own ontological security.

Of course, we might compare the incidence of psychopathology within industrialized societies and those that are not yet industrialized—an increasingly difficult task given the global reach of industrialization. Moreover, the methodological problems involved in such cross-cultural epidemiology are notorious. There are considerable qualitative differences in the expression of human misery in different cultures, and little agreement on how to categorize psychopathology.[89] In addition, the stresses involved simply in surviving in, say, a Third World slum area would almost certainly dwarf those that might be apparent in an affluent European context; and so differences in affluence, to name only one factor, would probably overwhelm the more subtle background influences associated with industrialized lifestyles. There are, however, a few careful studies which have yielded suggestive results. In a review of the area, Arthur Kleinman found that anxiety disorders are diagnosed at a rate of between 12 and 27 cases per thousand population across a range of cultures, with the notable exception of Australian aboriginals, among whom only a single case was found among 2,360 individuals.[90] There is quite convincing evidence that both the incidence and course of schizophrenia is more favorable in less developed parts of the world.[91] A few types of psychopathology, such as anorexia nervosa, are so specific to the industrialized world that they are virtually culture-specific syndromes. Overall, however, Kleinman's review of relevant studies suggests that while there are culture-specific differences in the *form* psychopathology may take, any underlying differences between industrialized and nonindustrialized societies in the incidence of psychopathology are concealed by a mass of other factors, such as differences in material standards of living, stresses due to cultural change, and the specifics of individual life-situations. Such factors, of course, may themselves often be related to industrialization; but the paths of influence are complex and indirect, and so usually beyond the reach of currently available methodology. In addition, even those societies that at first glance we might classify as "nonindustrialized" are, on closer examination, often deeply affected by modernizing influences. Furthermore, the cross-cultural application of psychiatric categories derived from European groups is, to put it mildly, dubious. All in all, cross-cultural comparison of psychopathology is something of a methodological minefield.

If, however, industrialisation fragments the life-world in a fashion that affects all the allegedly separate domains of life that result from this fragmentation, perhaps we can identify the effects of this process by exploring parts of the world that are in the "acute" stage of industrialization, and where the defensive gloss which industrialism assumes in its more advanced stages has yet to disguise the less acceptable evidence of its "progress." The Third World, regrettably, includes many such places. One that can serve as an example is "Bom Jesus," the Brazilian shantytown described in Nancy Scheper-Hughes' *Death Without Weeping*, and set in the northeast of the country in an area once covered by ancient forest and inhabited by Tabajara and Caetés Indians. In this century, however, most of the lush forest has been cleared, the Indians are long gone, and the landscape is dominated by sugar plantations. This domestication of the once wild landscape is the setting for, and continuous with, the social and cultural changes that followed. Today, Scheper-Hughes reports, even the peasants' subsistence gardens have disappeared under sugarcane, making them even more dependent on the inadequate wages paid by the sugar companies. The sugar workers are effectively serfs within a feudal system that is violently enforced, suffering chronic malnutrition and weakened by diseases once thought to be things of the past—typhoid, dengue, malaria, Chagas' disease, TB, and many more. As Scheper-Hughes summarizes the situation, "[t]he history of the Nordestino sugar plantation is a history of violence and destruction planted in the ruthless occupation of lands and bodies."[92]

If this Third World scenario seems a million miles from our own political experience, perhaps we should remember that life in early modern Europe was in many respects similar to that in the Brazilian *nordestino* today. Furthermore, such otherwise disparate parts of the world are today joined by their complementary roles in the global economic system, which ensure that the violence which is displaced from the affluent world surfaces elsewhere. Today, much of the overt brutality of industrialization in its acute stages has moved to the Third World, and as we buy our air-freighted vegetables from the supermarket we remain oblivious to the exported violence on which this commercial arrangement depends. Equally, our own European landscape has long since been "pacified," its native large animals mostly exterminated, its brutal history covered up by the appearance of rural tranquillity in farming communities. In areas such as modern Europe, then, the violence of industrialism is largely implicit within the organizational principles of taken-for-granted bureaucratic, economic, and conceptual structures, together with the anthropocentric value systems which these imply; or else is displaced to Third World shanty-towns such as Bom Jesus, where the character of the destruction of the natural, the impoverishment of the people, and the domination of all by economics is more starkly felt.

In this situation of acute industrialization, how does the replacement of a pre-industrial order by an industrialized one manifest itself? Conventional social theory might understand this situation as the exploitation of one group or class by another, or in terms of the mysterious workings of "power," or through the lens of an economic analysis. Each of these viewpoints has something to offer. However, each of them omits something important; and I will argue that one of the important dimensions that is missing from all of them is the repressive dissociation of the "human" from the "natural." The indigenous Indians who once inhabited the Brazilian northeast had a relation to the world which can be summarized in the term "organic," in that they were extensively integrated into the natural landscapes, rhythms, and processes that surrounded them and provided the pattern for their own, human, structures. In contrast, for the Brazilian *nordestinos*, every aspect of their lives is marked by a pervasive alienation from the natural, and its replacement by a reality imposed by industrialism.

This alienation is reflected at the most basic level in the monotonous and impoverished diet of the cane workers and their families, and in the polluted water which they carry home in five-gallon tin cans. This is consistent with Mark Cohen's findings that the transition from a hunter-gatherer existence to more sedentary modes of living involving agriculture has invariably "been accompanied by a reduction in the proportion of animal products in the diet, a reduction in the proportion of foods eaten fresh, a reduction in dietary variety, and, consequently, a decline in the overall quality of the diet."[93] One of the fundamental principles of industrialism is that the basics of life—food, water, clothes—are not the fruits of one's relation to the natural world, but must be bought through one's involvement in the *economic* system. Secondly, and equally clearly, the industrialist pattern determines the lives of the cane workers through their crippling daily toil in the fields or in the sugar factory.

However, there are other, more subtle alienations; and one of these is that which makes one a stranger to one's own body. Scheper-Hughes puts forward a thorough and convincing argument that the workers' political helplessness is inscribed in their bodies, in the breathlessness, lack of balance, and weakness that are encapsulated by the term *nervos* in a similar manner to that in which an affluent European might talk of "stress." The body expresses symbolically as well as literally the failure of the patterns which surround them to resonate with their character and needs. Hunger and malnutrition may be the most obvious aspects of this failure; but the victory of the industrialist order over a human order which is also potentially natural is expressed more profoundly through the inhabitants' acceptance of a medicalization of their situation that hides its physical realities. Thus, as Scheper-Hughes shows, hunger is redefined as *nervos*, a medical condition of the nerves; and the malnourished inhabitants of the Alto do Cruzeiro seek drugs rather than food to alleviate their suffering.[94]

It is, therefore, not simply that the oppressed workers of the Northeast are being denied what they need to live adequately; but also that their assimilation by an economic system that occludes the patterns of the natural order denies them the power to comprehend their situation or to relate to more constructive patterns. *Nervos*, says Scheper-Hughes, speaks to a profound sort of mind-body alienation in which hunger is redefined as an individual psychological problem. Unable to express their suffering and rage directly, they manifest a "dissociation from reality, a kind of collective psychosis"[95] in which they accept politically motivated redefinitions of the relations between self and an imposed environment that is hostile, dangerous, and insufficiently providential rather than to face this environment as a whole person who is nevertheless powerless to change anything. As Scheper-Hughes puts it:

> The recent history of the persecution of the Peasant Leagues and the rural labour movement . . . has impressed on rural workers the political reality in which they live. If it is dangerous to engage in political protest, and if it is . . . pointless to "complain to, or argue with, God" (and it would seem so), hungry and frustrated people are left with the possibility of transforming angry and nervous hunger into an illness, covertly expressing their disallowed feelings and sensations through the idiom of *nervos*, now cast as a "mental" problem. When they do so, the health care system, the pharmaceutical industry, commerce, and the political machinery of the community are fully prepared to back them up in their unhappy and anything but free "choice" of symptoms.[96]

Political and economic oppression, then, are redefined as *internal* problems, aligning behavior and personality with an apparently unchallengeable economic system. The loss of the natural in the external world may either precede or follow its loss in the human psyche; but either way it reflects the fragmentation of an ecological whole by a hostile order. A natural reality which extended from the "disappeared" forests to the complementary somatic structures of the indigenous peoples who inhabited them has been replaced by a radically different one that forcibly restructures selfhood to accord with it. Within this involuntary alliance of psyche and commercialism, the body is not only denied, but is in many ways replaced by a mechanized body whose movements and desires reflect industrial rather than natural processes, representing in microcosm a more general replacement of the natural by the commercial order. And this self, wrenched painfully from the natural order, is the nascent form of the autonomous Western self, the schizoid self which is taken by psychology as its assumed subject, a self which understands the natural order only as a source of raw materials.

Perhaps, too, Scheper-Hughes' ascription of *nervos* mainly to physical malnutrition and political impotence is incomplete; for hunger can be more than merely literal, and the destruction of the natural also obliterates the meaning structures through which resistance might occur. Extending a Geertzian understanding of this situation, if we are reliant for our adequate functioning on our patterned interrelatedness with the cultural and natural realms, then the demolition of these realms might in itself be expected to produce a pathological redefinition of self that defensively emphasizes its own autonomy in the face of an unreceptive "environment." And this, perhaps, is the fundamental link between the otherwise contrasting situations of the poor of Bom Jesus and the more affluent peoples of the industrialized world; for the disruption of the self-world *gestalt* that occurs in the *nordestino* can be understood as the acute phase of a situation which we take for granted. We, too, are out of touch with our bodies and with nature "outside"; only we can possess all the commodities and accoutrements that industrialism offers as substitutes for what we have lost. And if *nervos* reflects the mapping of a fundamental disruption of our embodiment as natural beings within a context of poverty, perhaps eating disorders are symptomatic of the same disruption as it appears in more affluent areas of the industrialized world.

Part of the reason why the global spread of industrialism has not usually been understood in these terms may be that academia, as we have already seen, has itself mostly accepted the industrialist redefinition of the world and is therefore blind to the character of this redefinition. For example, our awareness of the experienced, embodied realities of political repression and physical malnutrition often fades within an academic perspective as situations are redefined in terms of "slow development," "lack of investment," or the inadequacy of "education" or "communication." This perspective, of course, is not without some validity; but our concern needs to go beyond this in recognizing that even an industrial society that in its own terms is functioning effectively accepts as a *fait accompli* huge disparities in income, the dissolution of integrative cultural structures, and the alienation of entire populations from the natural world. To the extent that academia accepts an economistic frame of reference, so it will be unable to recognize those dimensions of suffering that are a "structural" part of industrialism. Scheper-Hughes points to anthropology's frequent denial of the reality of hunger, as indicated, for example, by suggestions that small size or "stunting" is adaptive because it allows a greater number of adults to survive on the available food, or that "toddler malnutrition was a biosocially adaptive mechanism for population 'pruning'," or that hunger is simply a metaphorical expression of the inevitable conflict between individual desires and collective social needs.[97] Anthropologists who have frankly depicted hunger and suffering, such as Colin Turnbull, have been "largely ignored and discredited" by their peers, since such depictions break through the defensive intellectualizations that convert anthropology into a comfortable spectator sport. "More than any other ethnographer," states Scheper-Hughes, "Turnbull broke the taboo of silence on hunger, and Turnbull and his book suffered the consequences."[98] But while Scheper-Hughes succeeds in her insistence on portraying hunger as lived experience rather than simply as adaptive mechanism or as expressing inevitable human conflicts, she nonetheless understates its character as part of the larger tragedy of the destruction of the natural order. Just as the bread and wine of the Christian sacrament symbolically expresses other forms of nourishment, so hunger expresses an impoverishment of relation to the world that is simultaneously physical and more than physical. If *nervos* is an "oblique but nonetheless critical reflection by the poor on their bodies and on the work that has sapped their force and their vitality, leaving them dizzy, unbalanced, and, as it were, without 'a leg to stand on,' "[99] it can also be understood as expressing metonymically their ungroundedness, their lack of support in any solid ontological structure with which their own personhood could resonate, and the absence of any meaning-structure that is embodied as well as conceptual.

If this analysis of the acute effects of industrialization can be said to have any generality beyond the particular context of the Brazilian northeast, then other areas, too, should demonstrate similar effects. We would expect the symptoms to vary from place to place depending on the detailed accommodations of existing social traditions to capitalism; but we should nevertheless be able to trace the origins of these symptoms to the fragmentation of a natural order that included humanity and its replacement by an industrialist understanding that reduces the world to "resources." And that this generality does, indeed, hold is suggested by ethnographies relating to many areas of the world. For example, let us move to the other side of the South American continent to Colombia, where the introduction of a capitalist economy has been associated mainly with the growth of the sugar industry. Colombia, to an even greater extent than most other South American countries, has a long history of violence and exploitation; but traditional cultural and spiritual beliefs among the indigenous and peasant populations have shown a remarkable resilience in the face of the invasion of a market morality.

The differing styles of relation to the natural world embodied in peasant farming, on the one hand, and wage labor in the fields or mines, on the other, are striking. Here, I will draw heavily on Michael Taussig's description, in his *The Devil and Commodity Fetishism in South America*.[100] Peasant farming, argues Taussig, is in many ways more efficient than large-scale agribusiness:

> The main tasks in peasant agriculture are the harvesting, which occurs every two weeks, and the weeding, which is done once or twice a year. Both tasks are light and require little time. Around two hectares cultivated in this way provide a subsistence living for the peasant household and demand no more than one hundred labour days a year. . . . Firewood, house-building materials, cordage,

> wrapping leaves, packing, gourds, a little corn and manioc, and many medicinal plants are also obtained from the plot, on which poultry and pigs are maintained as well. Commercial as it is, this type of agriculture preserves most of the preexisting ecosystem in its vast diversity of cultigens, and the soil is constantly nourished by a compost of the fallen leaves.[101]

This system is consistent with a variety of Andean folk beliefs, all of which emphasize and embody the *unification* of experience. "To the Andean Indians nature is animated, and persons and nature form an intricately organised unity. . . . The conception of nature and society as fused into the one organism is here most explicit. The land is understood in terms of the human body, and the human body is understood in terms of a culturally perceived configuration of the land . . . the analogies between the human body, the social body, and nature form a cultural system that is like a language with its own autonomy and integrity."[102]

The "unification of experience" that Taussig describes does not, however, imply "the mushy totality that phrases like 'the oneness of the universe' or 'the unity of all' might suggest." On the contrary, "this unity is composed of a highly differentiated system of dualities." But the "dualism inherent in this scheme bears no resemblance to Cartesian dualism," Taussig argues, "but is ontologically and epistemologically opposite."[103] That is, the dualities which structure the lives of the Aymara Indians, for example, together form an essentially monistic ontology in which patterns are repeated across what might otherwise be separate social, psychological, and natural realms. Thus "shrines on Mount Kaata are divided into those associated with death and those associated with life; lineages are made whole by division into male and female kin groups; shrines are served in pairs—male/female, young/old, mountain/lake," and so on. The contrast between these Aymara distinctions and those that govern our lives in the modern world has something to do with our loss of a sense of integration: whereas in their world, distinctions also imply a complementarity that points toward a larger whole, our distinctions—between, say, human and nonhuman, or between social classes or racial groups, or between male and female—seem to have gradually lost some of this sense of a counterbalancing wholeness.

This, in turn, may be related to our reliance on a form of cognition which emphasizes the reality of "things" while finding a good deal of difficulty in articulating the *relations* between them. We have been schooled toward literal and unambiguous definition, and to favor analysis and the reduction of complexities into their component parts—powerful and successful techniques which, nevertheless, need to be balanced by holistic vision and a sense of integration. And this integration, for the Andean Indians as among many other such groups, is provided by their sense of the analogies between the social, natural, and psychological realms—in particular, the analogy between the physical body and that of the earth. Thus "in the fertility rite of the New Earth in Kaata, llama fat and blood are circulated from the centre of the mountain body to its extremities. Community life and the energy present in all the parts have to be circulated and shared out. Political authority lies in the system of parts and whole and not in a leader, man, god, or thing."[104] This suggests a symbolic rather than a logical form of relation:

> The analogical mode of reasoning is compelling . . . because things are not seen as their self-constituents but as the embodiments of relational networks. Things interact because of the meanings they carry—sensuous, interactive, animate meanings . . . and not because of meanings of physical force locked in the privatised cell of self-enclosed thinghood . . . in the peasant and working-class epistemology individual terms or things are conceptualised as are Hegel's "moments": each expresses the totality of which it is the manifestation. Things contain the totality within themselves, so to speak . . . they are of interest . . . primarily as ciphers that echo the meaning of the system that society forms with them.[105]

Analogy, then, seems to facilitate a sense of meaning based in *relationality* and *metonymy* (the use of a small part or attribute to denote a larger whole) rather than in the detached significance of individual items. Thus a "thing," rather than possessing its own, discrete meaning, derives meaning from its relation to other such things, and ultimately to the whole. Meaning becomes a nexus of relationality, not in the arid intellectual sense of poststructuralist linguistics, but rather as an inherent property of a world that is simultaneously relational and physical. Within such a world, a "thing" is defined by what it symbolizes as much as by its characteristics as an isolated entity. And individual persons, too, derive their sense of identity from their awareness that they are part of a much greater social/natural whole.

By now, the reader will probably have a sense of déjà vu, as this Andean ontology recalls the Australian aboriginal and Navaho realities that I discussed earlier in this chapter. Like them, Andean cosmology connects the social and natural worlds through symbolism and metaphor, defining a life-world which is both differentiated and highly integrated.

The wholeness that we find in Andean communities, then, is not just a social wholeness that is defined in contradistinction to a wild, undomesticated world; but one that reflects the natural order and echoes its patterns and rhythms. The integration of these otherwise separate realms does not result in a whole that is fixed, unmoving, and static, but one that reflects the ever-changing character of the natural world itself in its vitality and responsiveness to the pressures of change. As Taussig puts it, "like language, culture changes systematically . . . in this dialectical manner the system of analogies obtaining between the human body, the social body, and nature is never completely fixed or isomorphic."[106]

There is a danger here of falling back into a dualistic style of thinking that polarizes the modern and the traditional Andean worlds so as to deny the overlap between them and the exceptions to what I have argued. Nevertheless, the picture that emerges is a fairly clear one, and, moreover, one that can in its essentials be replicated in many other areas of the world. But our story does not end here; for there is another way in which Andean village life reflects in microcosm more widespread processes, and that is in its exposure to the expansion of the "market economy." The precursor of this economy was the removal in the early decades of the twentieth century of many peasant smallholders from their lands by rich landowners—an assault that "lives on in popular legend as a holocaust."[107] And the more recent impacts of capitalism on the working classes and remaining peasants of Colombia are just as far-reaching, if more subtle and sometimes less obviously violent in the forms they take. However, in this acute phase of industrialization, the continued influence of traditional cultural frameworks offers insights into economic processes which we in the already modernized world find hard to articulate. Thus "peasants represent as vividly unnatural, even as evil, practices that most of us in commodity-based societies have come to accept as natural,"[108] and in a resistive accommodation to the Christianity that has in the past often been forced on them, it is ironic but perhaps not surprising that industrialism, in the tin mines of Bolivia as well as in the sugar cane fields and factories of Colombia, has come to be symbolized by the figure of the Devil.[109] This reaction of indifference or hostility to an ideology experienced as oppressive and destructive, which we also noted earlier, differs only in detail from that found by observers in a wide variety of places; and Malinowski was only one of many who chronicled natives' (in this case, Trobrianders') rejection of and contempt toward the Europeans' acquisitiveness.[110] Perhaps, however, this widespread rejection of the "market economy" is less surprising than our easy acceptance of it—an acceptance that is most convincingly explained by the cultural and epistemological changes, occurring over many centuries, that predated or accompanied it, and that underpin the "social phantasy system" which is the psychocultural dimension of industrialism. There are good reasons, as we saw earlier, for seeing a scientistic frame of reference as in many ways consistent with capitalism, and our acceptance of one implies a lowered resistance to the other. As Taussig expresses this historical consistency,

> Reality and the mode of apprehending it became defined in commodity terms that are based on the epistemological canons of atomistic materialism. Man is individualised, as are all things, and organic wholes are broken into their supposed material constituents. Irreducible atoms related to one another through their intrinsic force and causal laws expressed as mathematical relationships form the basis of this cosmology, and in so doing, embody and sustain the commodity fiction of social reality. The mechanical and atomistic view of reality, the basis of which was outlined in the works of Descartes and Galileo, found its most perfect expression in the physics and metaphysics of Isaac Newton, who may with all justice be regarded as the father of modern science and as the man who gave to capitalist apprehension the legitimising and final smack of approval that only science can now endow.[111]

If many of us in the already industrialized world are too extensively colonized by its assumptions to acquire much in the way of critical leverage over it, then to the Andean peasants its disastrous effects are apparent in every arena of life. Most obviously, the replacement of a relatively autonomous existence as smallholder by that of the wage slavery of work in the sugar factory or the cane fields is experienced as humiliating and debilitating, leading to hunger, ill-health, and premature death. Like the Brazilian *nordestinos*, the Colombian fieldworkers recognize that "plantation work makes people thin and prematurely old in comparison with even the least remunerative peasant occupation. They [describe the sugarcane] as a plant that dries or eats one up."[112] And if the change from peasant farming to wage labor is conceptualized as a Faustian bargain, as "Devil work" that rejects an accordance with the natural order by its allegiance to a system that is hostile to it, it is also a social and psychological change which has the most far-reaching effects. In Taussig's terms, "the meaning of personhood and thinghood is at issue as capitalist development reworks the basis of social interaction and subjugates that interaction to the fantastic form of relations between things . . . the advance of market organisation . . . also tears asunder a way of seeing. A change in the mode of production is also a change in the mode of perception. The organisation of human sense perception is determined by historical as well as natural circumstances."[113] Any analysis which comprehends the change from smallholding to wage labor simply in pragmatic terms, therefore, assumes and incorporates the fragmentation of the life-world that is a taken-for-granted fact of life in already industrialized areas. We are not, therefore, talking simply of a change of occupation; but also of the abandonment of one cosmology in favor of an entirely incompatible one. The alienation from the natural order that accompanies this change is not *just* an alienation from the natural order; but also represents the collapse and fragmentation of an articulation of nature that is social, psychological, spiritual, and epistemological.

Ethnographies such as Taussig's, which take a holistic view of ecocultural issues are, unfortunately, rare; but if anthropology has usually been reluctant to engage with the realities of hunger and other aspects of our alienation from the natural, other social sciences have usually ignored them altogether. This willful neglect is accomplished in part by selectiveness: each discipline chooses its subject matter, so that anthropology, for example, concerns itself with the varying symbolic organization of different societies, psychology with the "interior" structure of individuality, sociology with conflict or integration between different social groups, and so on. In this way, each discipline restricts its scope to a single part of a "human" realm covertly defined by that larger and entirely overlooked dissociation of the "human" from the "natural"; and it is the recognition of this larger fragmentation that is fundamental to an adequate appreciation of "environmental," "psychological," or "sociological" realities. If we are to overcome the blinkered focus on a "human" or "cultural" realm which is torn from its immersion in the natural, we will need to look outside the self-imposed boundaries that keep each discipline in a state of comfortable ignorance of its larger context, and seek connections which challenge the accepted alignments of these smug academic enclaves. One tradition that challenges these accepted alignments is that of critical psychoanalysis; and within this tradition, writers such as Marcuse and Lasch have attempted to use psychoanalysis as a tool of cultural criticism rather than as theory of the individual within a taken-for-granted social context. As we have seen, cultural theory opens up fundamental questions about the constitution of self as alienated from the world; and critical psychoanalysis can help us to explore these questions without losing sight of the whole—an exploration we undertake in the next chapter.

Chapter 5

THE PSYCHODYNAMICS OF SELF-WORLD RELATIONS

A man lives in things and things are moving. He stands apart in such a temporary way that it is hardly worth speaking of. If that perception dims egocentrism, that illusion of what man is, then it also enlarges his self, that multiple yet whole part which he has been, will be, is. Ego, craving distinction, belongs to the narrowness of now; but self, looking for union, belongs to the past and the future, to the continuum, to the outside. Of all the visions of the Grandfathers the greatest is this: To seek the high concord, a man looks not deeper within—he reaches farther out.

—William Least Heat-Moon,
Blue Highways: A Journey into America

The Psychodynamics of Modern Selfhood

The field of psychoanalysis, although concerned with intrapsychic processes, also patrols one of the great taboo areas of modern civilization—that uneasy twilight zone between what is considered to be "within" the self, and what is "outside" it; and like most other forms of understanding in the industrialized world, it serves the latent purpose of legitimizing our experience of "innerness" and "outerness," adjusting and filtering it to fit the assumptions of industrialism. In most of its manifestations, then, psychoanalysis reinforces the concept of an atomistic self with clear boundaries—a very different conception to the notion of self as a nexus of relations, extended and articulated into the world through culture, that we explored in the previous chapter. Among nonindustrialized groups such as Australian aboriginals or Andean Indians, a sophisticated interweaving of individuality and the natural world implicitly questions the idea of self as discrete and separable; but in the industrialized world, this interweaving is mostly discouraged, and its remnants are "explained" in terms of such "mechanisms" as projection or introjection—a form of "explanation" that allows us to throw any phenomenon into one or other of the baskets labelled "self" or "not-self."

In spite of this basically conservative intent, psychoanalysis conveys a subversive power that derives directly from the nature of the materials it deals with. The relation between a psychoanalyst and the unconscious is in this respect rather like that between dam-builder and river: in both cases, there is an ever-present threat that one's carefully built construction will crumble in the face of the power one is attempting to understand and manipulate. And the free-flowing unconscious, like the free-flowing river, is not necessarily the destructive thing that consciousness fears, although the order which each embodies may not be one that would be consciously recognized. Most forms of psychoanalysis, then, are "colonialist" in the sense that they impose on unconscious material a conscious rationality, as Freud himself recognized in his famous statement that "where id was, there shall ego be," and in his comparison of this process to the draining of the Zuider Zee. Such remarks express much the same spirit as that suggested by Columbus's assimilation of the lands and inhabitants of the Americas, as does Freud's even more explicit admission that "I am by temperament nothing but a conquistador."[1] Nevertheless the unconscious, like the free-flowing river, is potentially consistent with a healthier lifestyle than that which we achieve through its subjugation; and its potential destructiveness stems mainly from its confinement by rationality.

This, however, is not a view that would be welcomed by most traditional psychoanalysts, for whom the conventional opposition between inner and outer, as enshrined in Freud's "reality principle," is as basic as it is for any scientist. Environmental theory, too, has tended to assume styles of subjectivity and selfhood that seldom stray far from those that are conventionally accepted. In turn, such assumptions subtly constrain and influence environmental analyses in ways which prevent them from escaping the gravitational pull of the same common sense assumptions that underpin the "business as usual" mentality. Hardin's discussion of the "tragedy of the commons,"[2] for example, while a fine contribution to our understanding of environmental *problems*, can offer us only authoritarian solutions because of its tacit assumption that current forms of selfhood are the only possible ones. Personhood and industrial society form a *whole*; and we cannot envisage alternatives to this whole if we begin by assuming one of its key components. In the same way as a growing crystal enlarges itself by constantly reproducing the same molecular pattern, so environmental theory, if it begins by assuming any part of the edifice of industrialism, will find itself reconstructing the whole. Consequently, if we begin by assuming a self who is detached and alienated from the world, then the world will necessarily take a complementary form, appearing distant and uninvolving. Approaches such as Hardin's take for granted what is better regarded as one particular historical alternative—and one whose reconsideration is basic to environmental progress.

Industrialist consciousness and the fabricated environment that is replacing the natural world come into being together, through a series of dialectical processes which repress those uncategorizable forms that cannot easily be assimilated to this dualistic scheme. Historically, changes in consciousness evolved together with changes in methods of subsistence, initially through the use of tools and new farming methods, and more recently through the power of a technological system whose character is still largely mysterious to us. Nevertheless, we experience ourselves as outside this system, separate from it, and, to a great extent, in control of it—illusions that disguise our extensive colonization by it. Such seductive beliefs are supported by a mental health industry that would regard any claim to be controlled by "the system" as paranoid, although it is difficult to see this sort of claim as being any more out of touch with reality than the reverse notion that "we" are in control of the direction of industrialism. The conscious experience of being "in control" is basic to most definitions of mental health; and most of us experience ourselves as separate and autonomous beings, content in our unawareness of the extent to which we are formed to be consistent with the requirements of commerce and technology.

This process of colonization, for a number of reasons, cannot be regarded as a healthy one, although its pathological character may not be palpable from within the integrated system of consciousness and technique. Within this constructed "reality," the fissures, dissociations, and literalizations of modern life are accepted as "normal" and legitimated by the available forms of discourse; and challenging them therefore demands that we somehow identify structures that exist *outside* this constructed reality. To some extent, this is implicit in environmentalists' emphasis on notions such as biodiversity and ecological integration, which hint at the existence of a more inclusive world than that envisaged by science and industrialism. This alternative world, however, is so far one from which subjectivity continues to be excluded; so the Cartesian divorce between the ensouled and the soul-less parts of the world remains. An adequate environmentalism cannot be one in which we simply "think differently" *about* or "behave differently" *toward* the world; rather, behavior and thought have to relearn their membership and participation *in* the world. To take a fairly general example, to empathize with something "out there" is both to discover something about the world and to extend the quality of our own awareness in a way that goes beyond the simplistic rationality of class inclusion and exclusion. As a result, the notion that an animal is either "like us" and so worthy of moral consideration, or "unlike us" and so simply a resource can be replaced by a sophisticated web of meaning that weaves together complex strands of both difference and similarity. What potentially grows up between self and world is this resonance between them, extending subjectivity as it constructs a cultural realm which is both "self" and "not self." This is very different to the regressive loss of ego boundaries that occurs in identification, and that reflects a sort of amorphous fusion rather than relationality.

Freud, while sharing the general presupposition of his times that the development of a science-based civilization was both desirable and inevitable, also recognized the particular stresses that modern society imposed on individuals. The most important source of such stresses, according to Freud, is the suppression of sexuality, which he believed to be the cause of widespread neurosis and unhappiness. As Freud conceptualized this situation, there is a fundamental conflict between the instinctual drives of the id and the requirements of civilization; and this conflict becomes internalized as the superego incorporates society's demands. He maintains that this conflict within the personality can be to some extent alleviated if these id-derived instincts are diverted, controlled, and harnessed according to the "reality principle" of the ego and the moral constraints of the superego. "We shall never completely master nature"[3] he states in *Civilisation and Its Discontents*, so our suffering is to that extent inevitable. But by our partial victories over nature, he suggests, we can reduce this suffering. As we saw in chapter 2, Freud advocates "going over to the attack against nature and subjecting her to human will"[4]—words that are strikingly reminiscent of Bacon's writings. This "attack against nature" increasingly enlists the help of the ego as the child grows up, developing the intellect in a way that meshes with the defense mechanisms to subdue and redirect "wild," emotional, and instinctual aspects of self. Freud, in his unswerving allegiance to the development of "civilization," saw this separation of the intellect from emotion as essential to progress, as we saw in chapter 3. It is quite clear, then, that Freud's allegiance was to a "rational" civilisation whose values derive from science. It is equally clear, moreover, that the "attack against nature" that Freud has in mind is directed not only against the urges of the id, but also against *external* nature, for a few pages later he argues that

> [In] countries which have attained a high level of civilisation . . . we find that everything which can assist in the exploitation of the earth by man and in his protection against the forces of nature . . . is attended to and effectively carried out. In such countries rivers which threaten to flood the land are regulated in their flow, and their water is directed through canals to places where there is a shortage of it. The soil is carefully cultivated and planted with the vegetation which it is suited to support; and the mineral wealth below ground is assiduously brought to the surface and fashioned into the required implements and utensils. . . . Wild and dangerous animals have been exterminated, and the breeding of domestic animals flourishes.[5]

This domestication of nature is not only advocated by Freud's writings; it is also embodied in them. For "nature" in Freud's work is a nature reduced to its scientific image, a biologistic, mechanistic simulacrum that has already been assimilated to the post-Renaissance reductions of the scientific establishment. Freud continues the Enlightenment tradition which was earlier expressed in Bacon's call to "hound," "mould," and "enslave" nature, and in Locke's remark that "[t]he negation of nature is the way to happiness."[6] That Freud's theories are both biologistic and nature-denying is only superficially paradoxical; for biologism requires the mapping of nature onto a mechanistic template. It is clear, then, that Freud identifies with and writes from the standpoint of rational consciousness, perceiving the boundaries of the ego as protecting what is most valuable about the self. Anything outside these boundaries needs to be acted on by the ego and its attendant rationality for the "benefit" of the human individual or for the "welfare" of civilization as a whole—these being defined, of course, from the perspective of the ego. Any aspect of human personality that is not ego-syntonic is cast outside the city walls of civilization into a "natural" domain that is then re-assimilated and reformed to be rationally understandable; and there are strong parallels between the way Freud hunted down the "meaning" of unconsciously motivated behavior and Bacon's attempts to "put nature on the rack" in order to torture her into divulging her secrets. In each case, nature is dragged into an arena predefined through egoic consciousness, reduced to "raw material," and forced to conform to the laws of the ego; and in this colonialist encounter, egoic structures remain essentially unchanged. Nature, thus viewed, whether it exists "within" the person or in the "outside" world, needs to be "mastered"; and the success of this project, according to Freud, determines the extent of human happiness—a view which suggests that the "self" whose happiness Freud seeks to maximize is one from which "nature" has already been excluded. Within this scheme, civilization, rather than articulating nature, is at war with it; and an opposition is built into the relation between the warring factions that becomes sedimented into both psyche and world.

Whatever accuracy Freud's tripartite approach to personality may possess therefore derives from its embodiment of the "deep structure" of the civilization that he inhabited and the way it succeeded in mapping the geography of this structure as it permeated the psychological realm. This "deep structure," therefore, is not *primarily* psychological, since it reflects a deeper, industrialist, ideological patterning; and because of this, the Freudian personality is neither universal nor inevitable. In fact, as Geertz has put it, "[t]he Western conception of the person as a bounded, unique, more or less integrated motivational and cognitive universe, a dynamic centre of awareness, emotion, judgement and action organised into a distinctive whole and set contrastively both against other such wholes and against its natural and social background, is, however incorrigible it may seem to us, a rather peculiar idea within the context of the world's cultures."[7] Nineteenth-century Europe was struggling to deny its natural roots, and Freud's theories reproduced this struggle and presented it as psychological necessity, polarizing aspects of self that may potentially be intertwined in more sophisticated ways. It is not difficult to see that Freud's formal separation of the rational and the arational reproduces a social reality that had been developing at least since the Renaissance; and Freud himself, for all the acuity of his detailed insights, seems to have had little ability to contextualize his findings in terms of large-scale historical processes. Consequently, while he pointed to the possibility that civilization itself had become "neurotic," he failed to follow up this idea, retreating instead to a concept of neurosis as defined in contrast to the apparent "normality" of the majority of the population.[8] This stance, of course, denies the possibility of a "sociosis," a *general* pathology that affects the overall direction of the modern psyche. In the final analysis—as always—Freud allies himself with "civilized" values, even if these values necessarily imply that the most basic parts of the internally divided personality he proposes are at war with society. Thus although his remarks that the apparent happiness and fulfilment of "primitive" peoples were due to "the bounty of nature and the ease with which human needs were satisfied,"[9] he nowhere follows up the alternatives that such remarks hint at.

From Freud's viewpoint, whatever degree of neurosis may exist within the Western world is due to the inherent character of biological factors intrinsic to the human condition, and their inevitable conflict with social organization. This view freezes any possible dialectic between nature and culture, so that the opposition between them becomes reified as a psychological and social inevitability. Both destructive and integrative tendencies are viewed as instinctual, so that "beside the instinct to preserve living substance and join it into ever larger units, there must exist another, contrary instinct seeking to dissolve those units and to bring them back to their primeval, inorganic state. That is to say, as well as Eros there was an instinct of death."[10] Even though the latter causes conflict within human society and within the individual as well as "the destruction of its own organic home,"[11] Freud's biologism prevents him from seeing this "instinct" as originating in any organizational principle *outside* human biological organization. Rather, both instincts are assumed to be intrinsic aspects of the human personality; and the organization of society and of our relation to nature are understood as *following from* this assumed instinctual structure.

But suppose, instead, that our most basic instinct is a *relational*, life-affirming one; and that destructive, narcissistic tendencies originate not only within our instinctual structure, but also from outside. Suppose, moreover, that it is the frustration of our relational tendencies which causes that pathological turning inward of the libido which Freud labelled narcissism.[12] In other words, if we find ourselves in a fragmented world that negates our "erotic" tendencies toward relationality, the only option open to us is to prioritize our own egoic needs in a survivalist way, abandoning an empathic reaching out toward the world and instead experiencing the world as resource and "object." From this perspective, our need to control and exploit the world is not due to an inbuilt destructive "instinct," but is, rather, a defensive reaction to finding ourselves in a world that seems unresponsive to our relational needs. If we cannot fulfil ourselves through relation to the world, then the course of action open to the ego is one of "narcissistic enjoyment," involving "a fulfilment of the latter's old wishes for omnipotence . . . [and] control over nature."[13] In other words, if our potential for relationality does not result in an interweaving of self and world so that the wholeness and integrity of the world is maintained, then we narcissistically prioritise egoic structures by "feeding off" the world in a "colonialist" fashion. This process will be a dialectical one, since the narcissistic self that is the outcome of this process will energetically set about *constructing* the sort of structureless, relationless world that is its complement. In effect, then, a radical restructuring of the world occurs: the natural order, in all its diversity, is replaced by a technological order whose agent is the narcissistic ego.

Destructive behavior, if we understand it this way, is less the result of a "death instinct" than of the frustration of our *integrative* tendencies. This, however, is not a possibility which Freud's biologism allows him to recognize. Today, too, we are more aware than Freud that in many cultures, and especially traditional foraging societies, individuality does not usually take the form of a narcissistic abandonment of relationality and its replacement by an exploitative individualism. Freud's biologistic reductionism, then, ensures his faithfulness to the basic project of industrialism: the fragmentation of the world and the denial of integrative natural structures. It is not only the natural order that Freud cannot see, but also the order of industrialism that is destroying it, so that the struggle between these two—the defining struggle of the modern era—is simply invisible to him.

As Frank Sulloway argues, Freud was highly ambivalent about the rootedness of his ideas in biology, and his writings can be partly understood as his struggle to come to terms with humanity's (and his own) natural inheritance by transcending or even denying this inheritance. In this context, Sulloway's suggestion that once Freud "had finally achieved his revolutionary synthesis of psychology and biology, [he] actively sought to camouflage the biological side of this creative union"[14] is readily understandable. In part, this camouflage was probably due to Freud's desire to present psychoanalysis as an autonomous science; but we can also understand it as expressing a modernistic desire to locate psychoanalysis firmly in a "human" realm that is explicitly distanced from the "primitive" world of the natural. It is as if Freud, not content with reducing our relationality with the rest of the natural world to the partial and mechanistic understandings of biology, needed to relinquish even this connection by translating the quasibiological concepts of psychoanalysis into a more exclusively "human" language. In this case, then, the "human" becomes a disguised, and tragic, reflection of the natural rather than an *articulation* of it into the cultural sphere; and Freud's inability to recognize that culture can be *facilitative* as well as repressive ensures that his vision of the human is one in which we strive to control and distance ourselves from nature while ultimately being drawn back into it. Thus the dialectic between nature and culture—or more accurately, as Freud saw it, the struggle between them—is not only the subject matter of his writings; it is also what fatefully determines their form and direction. His insights into our natural being were from the standpoint of an intellect which was quintessentially civilized, rational, and detached, rather than consistent with the natural order. This being so, Freud epitomized the modernistic zeitgeist of his era, and his attitude toward the natural world was one of distance, separation, mastery, and control.

If Freud himself seems more to exemplify the rationalistic impasse than to bring to light any potential environmentalist solution, his writings nevertheless, and almost in spite of themselves, imply the extent of our immersion within the natural domain; and the implicit tensions that result from this underlie the widely diverging interpretations of his work. Thus, notwithstanding his allegiance to "civilisation," Freud's work often seems to embody an ambivalence—almost a nostalgia—for a lost world that he cannot be part of. Creativity, religion, enjoyment of nature—these are all seen as comforting illusions, sublimations of basic instincts, as if Freud were struggling within himself against more exalted interpretations. Of course, Freud only recorded one side of this conversation, but he lets drop more than a few hints that there was a very different viewpoint struggling to emerge. Reading Freud, one gets the impression that a Damascene conversion is not all that far beyond the realms of possibility.

This concealed ambivalence emerges more clearly in the writings of some of Freud's followers. The modern individual, according to Erich Fromm, enjoys freedom from the oppression and the material needs that his ancestors suffered; but at the same time, "this growing individuation means growing isolation, insecurity, and thereby growing doubt concerning one's own role in the universe, the meaning of one's life, and with all that a growing feeling of one's own powerlessness and insignificance as an individual."[15] Fromm's reservations concerning individuality in the modern world stem from the way he glimpses the paradoxical character of modern conceptions of "freedom": in other words, he comes close to recognizing that a truly positive freedom is not one that allows a sort of unrestrained expression of our desires, unmodified and unmediated by any external structures, but rather one that incorporates our need to relate to some source of structure and meaning outside ourselves. Without such a structure, we are prone to the feelings of powerlessness and insignificance which he describes; and in the face of such feelings, he argues, we may be attracted to any form of social organization which offers us a degree of certainty and meaning, including fascist and authoritarian political dogmas. The simplistic power structures embodied by such political systems serve the purpose of enabling us to orient ourselves more clearly, according to Fromm; hence the attraction to individuals who otherwise might feel lost and insecure. Fromm's argument implies what many theorists of the 1960s forgot—that "liberation" from any sort of cultural mediation of our lives is likely to be repressive rather than emancipatory; for without appropriate cultural structures we cannot articulate our relation to the world outside, and so selfhood remains mute and undeveloped. In this respect, Fromm's concept of "freedom to" foreshadows Geertz's emphasis on the need for cultural structures which allow us to make sense of the world and our place within it.

Nevertheless, the sort of "reconciliation" between nature and culture which Fromm hints at is one that, as Marcuse has pointed out, requires us to jettison much of the radical potential that is almost reluctantly embodied within Freud's thought. The ideal of health envisaged by Fromm suggests not so much a type of cultural organization which fully recognizes and accepts the archetypal patterns and forms provided by millions of years of evolution, but instead conceptualizes our "instinctual" endowment, if at all, in a way that is laundered and domesticated so as to allow its superficial rapprochement with existing social "reality." Indeed, our grounding in nature is recognized by Fromm mainly as something to be left behind as we strive to become "fully human." He is quite explicit on this point:

> Man's evolution is based on the fact that he has lost his original home, nature—and that he can never return to it, can never become an animal again. There is only one way he can take: to emerge fully from his natural home, to find a new home—one which he creates, by making the world a human one and by becoming truly human himself.[16]

Thus while Freud at least acknowledged natural realities and the implicit challenge they pose to the ego and to civilization—if only, finally, to collude in their repression—Fromm begins by denying these realities and the conflicts they necessitate. In effect, while Freud assumes the inevitability of the conflict between "nature" and "culture," Fromm takes the repression of nature a stage further, assuming the possibility of a "human world" that has somehow "emerged fully" from our "original home, nature." Fromm, like Freud, is in this respect caught up in the great delusion of industrialism: that a human realm independent of nature is practically realizable and sustainable, and that "[man] must proceed to develop his reason until he becomes the master of nature, and of himself."[17]

According to the viewpoint developed in this book, however, a reason which makes itself the enemy of nature rather than its ally is a suicidal reason which attempts to flourish by consuming its own roots; and to the extent that both Freud and Fromm advocate such a reason they are unable to generate real solutions to those environmental and psychological problems that characterize our age. A truly unrepressive form of social organization is not achieved simply by reinterpreting nature according to the requirements of an alienated rationality, but is one that can recognize and incorporate natural patterns, growing out of them and extending them into what today might be understood as social, spiritual, political, and intellectual realms as well as into the more pragmatic organization of our everyday lives. Within such a genuinely nonrepressive cultural sphere, the conflict between the social and the natural disappears, and in its place there emerges a full spectrum of life which is simultaneously social and natural.

This is not to deny, of course, that natural forms may be capable of constructive extension by social factors; nor, even more importantly, is it to imply that the influence of the natural order is a rigidly deterministic one. Rather, the natural order can be understood as providing basic patterns (not unlike the Jungian archetypes) that can form the basis of a range of social and psychological possibilities, only a few of which will be able to flourish within the overall ecology of natural, social, and cultural organization. Within this "thick" ecology—to borrow Geertz's term for an understanding that integrates micro- and macro-worlds, and social and individual realities—some forms of social organization (we discussed several in chapter 4) will be more consistent with natural structures than others; a point that has often been lost within current, politically correct social perspectives which somehow seem to float freely above an increasingly unreal natural world. While Fromm's approach is less extreme than these current trends in the extent of its failure to recognize the natural basis of being, it can be seen as a step in a direction that becomes explicit in more recent, postmodern understandings of social being. Such views as Fromm's cannot articulate the nature-culture dialectic, since they begin by reconfiguring the natural to make it compatible with the existing social world: the "natural" is in effect replaced by a socially derived fantasy of the natural. Such a nature-denying model, in spite of its radical pretensions, is consistent with an economic system that increasingly assimilates all else to it.

Fromm, I emphasize, does not go as far as many postmodernists in his denial of the natural; but within his work, nevertheless, even Freud's biologistic understanding of the natural fades, becoming a mere shadowy presence devoid of real form, something that is left behind in the development of civilization. Only in our "historical infancy" are we "rooted in nature"; whereas modernity is characterized by "the decisive step to emerge fully from nature," creating "a definite demarcation line between [ourselves] and the animal world."[18] Freud, as we have seen, can only envisage forms of culture that subdue a nature misunderstood as crude and primitive; but in Fromm, nature as an order alternative to the constructed world of civilization disappears entirely.

Thus although their "solutions" to the problems of modern society differ, both approaches are fatally flawed by their assumption that human welfare demands the conquest and assimilation of nature. If we accept this, then the only possible environmentalism is that of softening our conquest of the natural world by doing "less of the same." Industry and commerce will be more muted, more "sustainable" according to this line of reasoning; and we will drive smaller cars, recycle more of our rubbish, eat free-range chickens, and so on. In other words, such views lead to reform environmentalism which, while no doubt preferable to the free-market consumeristic orgy tacitly promoted by most commercial enterprises, merely dampens the symptoms of commercialism without solving any of its basic problems. Such an agenda leads us to a gray place indeed—a muted, guilty form of life in which potentially full and vibrant forms of living must constantly be renounced in the quest for less damaging lifestyles.

The alternative is one that Freud does not consider, and Fromm only glimpses in his concept of "freedom to." Marcuse, however, suggests the possibility of a nonrepressive interaction of culture and nature, referring to Margaret Mead's discussion of Arapesh culture as an example:

> To the Arapesh, the world is a garden that must be tilled, not for one's self, not in pride and boasting, not for hoarding or usury, but that the yams and the dogs and the pigs and most of all the children may grow. From this whole attitude flow many of the other Arapesh traits, the lack of conflict between old and young, the lack of any expectation of jealousy or envy, the emphasis upon co-operation.[19]

"Foremost in this description," Marcuse argues, "appears the fundamentally different experience of the world: nature is taken, not as an object of domination and exploitation, but as a 'garden' which can grow while making human beings grow. It is the attitude that experiences man and nature as joined in a nonrepressive and still-functioning order."[20]

While the "revisionists" criticized by Marcuse—and he is principally referring to Erich Fromm—allow the idea of a natural order with some undeniable structure to fade away in their search for "adjustment," Marcuse himself advances his idea of "nature" well beyond Freud's conception of a "biological drive." "Instinctual liberation," according to Marcuse, "involves not simply a release but a *transformation* of the libido . . . [leading to] erotisation of the entire personality."[21] This, however, could be said to underplay the extent of the transformation required; for eroticism is a potential characteristic not only of the personality, but of the rest of the world, too! In other words, eroticism can be understood not as a property of a particular form of personhood, but as a particular style of *relation*. In the absence of such relationality, eroticism becomes the disguised phantasy of a wholeness denied in reality—an attempt by the individual to reach out into the world to fulfil an unrequited relationality. Viewing instinctual liberation as a purely individual process is as paradoxical as attempting to "save" a wild animal by removing it from its habitat: in both cases, foregrounding the individual and ignoring the context ensures that the liberation is a pseudoliberation into an industrialist arena rather than an authentic articulation of individual structures into a sympathetic natural space. Individual and context *together* define a whole, the characteristics of which are not entirely predictable from a familiarity with their alienated forms. It is only within a society that is fundamentally "anerotic" that sexuality will remain an "individual" and "instinctual" characteristic rather than a feature of our multiple relationships with the world beyond our physical boundaries. In anything approaching its mature form, then, individuality should be the offspring of a nature and a culture that mutually inform and extend each other.

It is not so much a question, then,—as Marcuse himself points out[22]—of envisioning a "nonrepressive" society within which instinct could be fulfilled, for such straightforward "fulfilment" will be oddly unsatisfying; but rather one in which instinct and culture can engage in a conversation, or a dance, which extrapolates from natural structure in a way that remains faithful to it. To view adult sexual organization, for example, as "instinctive" is to reify the repressive social organization which individualizes sexuality and which prevents the evolution of cultural structures that could embody a genuinely erotic reality. To view this stunted form of sexuality as natural is rather like mistaking the seed for the tree: just as the seed can only grow into a tree within a sympathetic context, so a sexuality that is allowed to extend into a sympathetically structured world will be radically transformed by its reintegration into the realm of the cultural. An authentic eroticism, in Marcuse's words, would "minimize the manifestations of mere sexuality by integrating them into a far larger order. . . . In this context, sexuality tends to its own sublimation."[23] This is not to imply some form of social experiment like that carried out in China during the Maoist era, in which love relationships between individuals were proscribed and punished on the collective farms; for whether eroticism is expressed individually or collectively, repression is no alternative to its constructive articulation. Rather, I am suggesting an extension of eroticism into a wider realm, so that intensely personal relationships will coexist with, and metonymically resonate with, the infusion of passion into the world as a whole. Similarly, although Marcuse does not explain the "far larger order" that he refers to, it can only be one in which the "human" realm is culturally realigned with the natural.

But the liberation of sexuality will be repressively structured by industrialism rather than by any more authentic natural pattern so long as it is understood as an *individual* liberation rather than as one that reunites the individual with what is outside the individual. This individualist form of "liberation" imposes individual structures and desires onto an other that is outside the individual and therefore an object; and so this is a fundamentally "colonialist" style of relation. An *authentic* form of liberation would be one in which a genuine erotic resonance occurs, so transcending the categories of subject and object and defining a new whole which is not dominated by either partner. We therefore see that just as the repressive forms of self and world, and subject and object, arise conjointly, so their liberation is also, necessarily, a joint liberation. Nature in any form cannot authentically be liberated into a context that remains dominated by industrialism, as John Rodman points out:

> The buffalo herd in Stanley Kramer's film, "Bless the Beasts and Children," thunders out of the pen, released by the daring efforts of a group of heroic boys, only to stop and graze peacefully on a nearby hill, allowing themselves to be rounded up and imprisoned again. Elsa, the Adamsons' pet lioness, "born free" and then tamed, must be laboriously trained (sic!) to become a wild predator before she can be safely released.[24]

Our concepts of the "natural" and of "sexuality" have been formed by centuries of repression and distortion, so that, in Rodman's terms, "our domestication of nonhuman animals and our division of ourselves into a 'human' part that rules or ought to rule and a 'bestial' part that is ruled or ought to be ruled are by now so hopelessly intertwined that it seems doubtful that we could significantly change the one without changing the other."[25]

Marcuse implies that the separation of instinct from what it relates to is itself part of a repressive social organisation, and that instinct in its reified, static form is an artifact of this organization, its apparent separateness being due to the failure of an anerotic environment to offer a matrix within which libido could extend and develop. Thus the central conflict within the modern world is not intrinsically, as Freud suggested, that between nature and culture, but rather that between the natural and industrialist orders; and it is this latter order that isolates nature, drives it into discrete cages such as that labeled "sexuality," and orchestrates its reluctant opposition to civilisation. In contrast to such already reified understandings of sexuality, an alternative, reintegrative understanding of sexuality would be in terms of a reaching out, a relational drive which constantly attempts to transcend the narrowly individual form which it takes within modern industrial society. In this alternative viewpoint, "instinct" as realized in adulthood would transcend any distinguishable spheres of "nature" and "culture," forming a whole that would be the culmination of the ecological project.

Suggestions such as this take us beyond the industrialist sphere of influence that structures much psychoanalytic theory. As Marcuse points out, for example, Freud's insistence that the only possible paradises are remembered ones makes utopianism necessarily regressive. In this, Freud complies with a reality principle within which the only "practical" realities are those which exist within the confines of current rationality: anything else is seen as unrealistic, regressive, or infantile. In this he is in plentiful company. Marcus and Fischer are only two of the many social scientists who criticize any attempt to diverge from current rationalities, labeling a radical stance as "idealist or romantic," as we saw earlier. Such a stance blocks the potential for rejuvenation that lies within the diversity of experience, and any experience which lies outside currently accepted, waking, rationality is defined as childish or even deranged—a destructive definition that today is usually accepted without question. For example, empathic relational feelings toward aspects of the natural world are often denigrated as "animism" or "anthropomorphism"; and spiritual experiences, such as the "oceanic" feeling that Freud views in terms of the persistence of childhood or "primitive" experiences, tend to blow away like smoke in the breeze because of the absence of any cultural structures that could stabilize them or lend them permanence. Such denigratory definitions are defensive maneuvers designed to protect the dissociations and prejudices which define a basically colonialist mentality—one that can assimilate but not learn from alternative structures. It is characteristic of these dissociations and prejudices that they can only exist within that particular state of alert "rational" consciousness that is encouraged in the modern world; and industrialism invalidates and pathologizes alternative states such as those that are commonplace among nonindustrialised cultures, together with the forms of knowledge that are associated with them. Consider, for example, the definition of "ecstasy" given in *Black's Medical Dictionary:*

> Ecstasy is a term applied to a morbid mental condition in which the mind is entirely absorbed in the contemplation of one dominant idea or object, and loses for a time its normal self-control. This condition usually presents itself as a kind of temporary religious insanity.[26]

Such views are consistent with a "reality principle" that speciously claims a monopoly on possible social and political forms, and that defends itself against the threat of a whole world. Those psychologists who have dared to express a contrary view have been rare indeed. One such was William James, who claimed that "[W]e pass into mystical states from out of ordinary consciousness as from a less into a more, as from a smallness into a vastness, and at the same time as from an unrest to a rest. We feel them as reconciling, unifying states."[27]

This is no mere incidental point, but is rather one which is crucial to the enterprise of developing an environmental subjectivity. Simply put, we cannot effectively defend the natural order if we rely solely on a conscious rationality structured by the dualistic splits that implicitly deny this order. This is not, however, to argue that egoic consciousness should be replaced by a new, "environmentally aware," form; but rather that both, or all, forms of awareness can exist within a wide spectrum which therefore contains its own relativistic humility. *Any* form of consciousness, if taken as the exclusive representation of reality, will be misleading, and will eventually collide with its own limitations in the real world. Consciousness therefore needs to recognize its own limitations, partiality, and inaccuracy, constantly checking its conclusions by reference to a broader organic awareness of reality, and recognizing the persistence of a natural world that is essentially unfathomable and mysterious. Egoic consciousness, therefore, is not in itself the problem. Rather, the problem is that we assume egoic consciousness to be capable of offering an accurate and complete description of reality, and so lose sight of any broader frame that could remind us of its limitations.

Healthy subjectivity, then, will include an egoic component; and this amounts to a temporary prioritization of a particular type of individual awareness. But this is only one of many possible states of being; and consciousness will move fluently between states that foreground individual structure, those that are essentially relational, and those that reflect an identification with a much larger system. Subjectivity, in embodying these varied states of being, will be able to remain closer to the natural world, reflecting its variations, changes, and multifaceted character. This is in contrast to a colonialist consciousness that defines itself in contradistinction to "external" structures, such as that of nature, which it feeds off and dissolves. We will also need to relinquish the striving for complete consistency among the parts of our worldview, recognizing that some ambiguity and uncertainty will inevitably accompany any relation to the world that transcends individual structure. Resonance with the natural world will be considered a more profound value than internal consistency; for internal consistency, if attained at the price of fidelity to the phenomena we are open to, is a paranoid stance that merely disguises the inaccuracy of our standpoint. This fidelity to the natural world implies a "bioregional" rootedness to the localities we inhabit, a sensitivity to changing landscapes, so that our ways of being and experiencing are sympathetic to and are supported by local features of the landscape, ensuring that our behavior is in accord with the forms which the land offers us.

An environmental awareness, like any other form of organic structure, can ultimately only flourish within certain types of habitat—within a sympathetic ecology. Put differently, an environmental subjectivity is part of the landscape as much as it is part of the human mind. To argue otherwise is like attempting to ensure the survival of a particular species while allowing its forest environment to disappear: species and environment jointly define each other, and the animal in a zoo is not the same animal as that which exists in the wild. One part implies and requires the others; and the destruction of one part imperils those with which it resonates.

Nature, however, is more resilient than this description would imply, and this resilience becomes more crucial as the modern landscape becomes progressively more urbanized and domesticated. Like any other system, nature embodies homeostatic tendencies, and implicitly recognizes a loss of diversity and integration as a need for restitution. Subjectivity, as part of nature, can recognize the sense of loss resulting from its imprisonment within the human mind due to the absence of a sympathetic environment, even if this recognition is somatically sensed rather than consciously identified. Our awareness of this sense of loss, however, does not imply that we can consciously envision the type of change that would allow the needed fulfilment, since fulfilment and restitution occur through processes whose span exceeds that of consciousness. However, an awareness of our loss, and of the part that existing forms of consciousness have played in bringing it about, can enable us to open ourselves to what is beyond consciousness, and so loosen consciousness' claimed monopoly of meaning. Thus if consciousness itself is of little direct use in our search for alternatives, consciousness' learned awareness of its own limitations can guide us in the recovery of our wholeness.

Whatever environment we inhabit will be sympathetic to certain types of subjectivity—currently, those consistent with a commercial/technological orientation—and hostile to others. What we might call the "ecological tendency" of systems ensures that thought processes and physical surroundings, at least in a facilitative ideological environment, tend toward consistency with each other; and aspects of subjectivity that are inconsistent with this evolving gestalt will be driven further from consciousness. For this reason, it is probably no coincidence that much radical environmentalism seems to originate in those parts of the world where wilderness survives, such as the western United States and Tasmania, which can still nurture a conscious environmental awareness. Those of us who inhabit more urbanized areas find it difficult to develop the environmentally aware states of consciousness that are sensitively attuned to the "more-than-human" world, and so may be more prone to developing variants of environmentalism that are derived reactively from the patterns of industrialism. The final victory of the ego will occur when and if there is no longer any felt contradiction between the largely economic and material "reality principle" which forms an increasingly central aspect of subjectivity among many in the industrialized world, and the physical character of the world we inhabit. Within this scenario, children will inherit a world which contains little that is wild, spiritually pregnant, or mysterious; indeed, even what is "natural" about it is likely to be genetically engineered so as to maximize its usefulness to humanity. What inchoate yearnings and fantasies, one wonders, might our children's children experience, if the landscape they survey is one that is entirely domesticated, neutered, and rationally ordered?

Alternatives to the Narcissistic Self

Environmental awareness can only begin from a frank recognition of our present psychological situation. This recognition, however, cannot be *only* a psychological one; since, as I argued above, we need to see our situation in ecological terms—"ecological" in this case indicating a recognition that the structures we inhabit are simultaneously social, economic, and cultural, as well as "ecological" in the narrower sense. The fragmentation of understanding into a large number of specialisms and disciplines mirrors the reduction of the world to the "bits" that are the raw materials of industrialism; and only a relatively integrated and holistic methodology can begin to recognize those elusive systemic properties which characterize both the natural order and the industrial order which is replacing it.

Bearing this in mind, let us begin by taking seriously the question that Freud glossed over, concerning the possibility that civilisation is "neurotic"—a question that is explored by Christopher Lasch in *The Culture of Narcissism* and *The Minimal Self*, building on Freud's theory of narcissism.[28] According to Freud, all children pass through the phase of "primary narcissism," in which the boundaries between the child and the world have not yet been established, and in which the fulfilment derived from the mother's attentions are experienced as due to the omnipotence of the self. As the child learns to distinguish herself from a world experienced as "outside," so she will seek to cathect objects outside the self—in other words, to establish satisfying relationships with aspects of the external world. If this attempt is frustrated by an unresponsive environment, however, then there is a withdrawal of libido back toward the ego. As Freud puts it, "through internalisation the patient seeks to create a wished for love relationship which may once have existed and simultaneously to annul the anxiety and guilt aroused by aggressive drives directed against the frustrating and disappointing object."[29] *Secondary* narcissism, then, is a reaction to the frustration of relational needs, so that individuals are forced into "seeking *themselves* as a love-object, and . . . exhibiting a type of object-choice which must be termed 'narcissistic'."[30] In short, if the child fails to find structures in the outside world through which successful relationships can develop, then there is a turning inward of the libido so that a fantasy relation is established as a substitute for the aborted real relationship, and the world outside is experienced as radically alien. There is an important distinction here between the world experienced as *other*, as distinct from self, in which case I can have a relationship with this other; and the world experienced as *alien*, as unresponsive to me, so that it becomes psychologically meaningless. A healthy self will complement the world outside itself so that their integration forms a relatively harmonious whole; while a narcissistic self will lack this complementarity, seeing the other only to the extent that it can somehow be assimilated to its own concerns. In the former case, the world will be experienced as full of meaning, originality, and diversity; while in the latter, the only meaning will be that which I impose on the world. If the "health" of a culture can in some respects be measured by its ability to integrate individual and "environment" within a harmoniously functioning whole, then a "narcissistic" culture is clearly an unhealthy one.

Lasch argues that the modern world is characterized by the absence of relational possibilities which would make healthy psychological functioning possible, and that the self therefore has good reason to substitute a set of fantasy relations for the external object-relational network it strives toward. In its mature form, this fantasy world is catered to by television, shopping, sport, video games, the stock exchange, the worship of cult heroes, and other adjuncts of the "culture of narcissism." From this viewpoint, there is an unhealthy collusion between a self that is vulnerable, needy, and grandiose, and that attempts through lifestyle and possessions to fulfil fantasies of power and perfection, and a society that provides an endless flow of consumer goods while ensuring the neediness of the population by subverting those traditional cultural structures around which the self could stabilize. It doesn't require much imagination to recognize that this scenario is environmentally disastrous.

If the character of modern society is, as Lasch and others have argued, such as to deny the relational connections which would root us meaningfully into a cultural, and ultimately (as I would argue, but Lasch doesn't), a natural, context, then we necessarily become "narcissistic," adopting an attitude of distance, attempted domination, and even fear or revulsion toward anything whose essence seems different to our own. In effect, the resonances between structures within the self and those outside become stilled, resulting in an exaggerated separation between them, and in turn facilitating the dualistic mapping of reality that so profoundly pervades the modern psyche. We maintain a "survivalist" attitude, shrinking into ourselves and relating to otherness only in a limited and often exploitative way.

As we have already seen, however, Lasch's analysis doesn't go far enough; for the universe within which his analysis is played out is one that is *already* separate from the natural world. Highly perceptive though his diagnosis of the modern psyche is, it nevertheless itself embodies that split which most significantly underlies the industrialist psyche: that between the "human" and the "natural"; and the person finds herself within a purely social, humanly constructed realm with only the most meager connections to the natural world. This socially constructed realm thus remains ungrounded, free floating within a universe of its own making, its links to the natural world abandoned and forgotten. For example, although arguing that a "genuine affirmation of the self . . . insists on a core of selfhood not subject to environmental determination," Lasch cannot locate this "core" within the natural order, preferring to see it as historically determined by "an older conception of personality, rooted in Judaeo-Christian traditions."[31] This "older conception of personality" is itself, presumably, rooted in an "even older" conception of personality, the origins of which remain mysterious. It is also significant that the Judeo-Christian tradition is one which itself strenuously divides the natural from the human, so cementing the repression of the natural within Lasch's work. Thus personality is grounded only in a social world that is presented as historically unconnected with nature. The long-standing dialectic within which self and nature emerge as separate is not recognized; nor are the implications of this separation in the genesis of narcissism. Lasch's work, although importantly connecting selfhood and social trends, therefore perpetuates the illusion of our inevitable separateness from the natural. *Why*, we need to ask, is it that traditional cultural forms have crumbled, leaving the self vulnerable and narcissistic? Only an analysis that recognizes the dependence of culture on the patterns of the natural world can successfully address this question.

The artificial separation of humanity from nature deprives us of the most basic source of meaning; and the subtle, multifaceted, and complex array of relations that exist within the natural world is replaced by an artificial and narrow "rationality" that offers few metaphoric overtones by which we might relate ourselves to a meaningful context. More concretely, our frequently urban lifestyles fail to offer the opportunities for interaction with a natural world; and Paul Shepard, in his last published work, explicitly related this situation to the growth of narcissism.[32] Complementarily, he also recognized that a healthy relation to what is outside us facilitates the integrity of a world that includes the self. Thus "the perception of a flower, seed, plant, garden, or prairie spontaneously refers us to fugitive aspects of the self . . . these interspecies interactions between ourselves and plants are ecological as well as psychological."[33] For children to grow up to become mature, fulfilled individuals who live purposefully and constructively, they need to inhabit a context that they can recognize as meaningful, that resonates to and can articulate their feelings, instincts, intuitions. Unfortunately, the spiritual, ontological, and aesthetic barrenness of much of the modern landscape offers an environment that is largely bereft of these nurturant qualities, implying a reduction of the world from a rich source of diversity, beauty, and meaning, to a simplified landscape defined primarily in economic terms. As a result, growing up in modern industrial society is all too often, in Charlene Spretnak's words, a "passage into emptiness."[34] Given this situation, it is unsurprising that the psychological condition of many young people in the industrialized world is similarly barren, and that the "adulthood" which awaits them contains the regressive elements described by Lasch. This personality configuration is reinforced by commercial interests that feed the addictions to fantasies of power, sex, and violence, easy satisfaction of artificially stimulated "needs," and instant oral gratification in the form of sugary foods. So it is that an exploitative relation to a world experienced as meaningless other than as a source of "things" to be used originates in infancy and childhood; and the mature form of this exploitative relation is exemplified in the "objective" attitude of the scientist, in the colonialist exploits of European explorers, and in the neediness of the modern individual. What is happening here is that the alienation from the world that occurs in infancy facilitates a fantasy of the world as alien, harsh, and threatening; and this fantasy, in turn, motivates the development of an attitude that is controlling and ultimately destructive.

A recognition that our "instinctual" makeup is orchestrated and sometimes distorted by social realities defines a range of approaches within critical psychoanalysis; and here Lasch's work has a good deal in common with that of Marcuse. For example, it is characteristic of modern life that libidinal energy is withdrawn from cultural frameworks wherein it might be sublimated, and used in the service of commercial interests, a process that Marcuse refers to as "repressive desublimation." This overt use of sexuality—most obviously, in advertising—has often been presented by the media as a sort of liberation from repressive constraints.

However, this superficial viewpoint illustrates the difficulties with the Freudian theory of sublimation, which portrays the direct expression of sexuality as most satisfying, and any cultural mediation of this expression as necessarily repressive. Sublimation, Freud implies, is always preceded by repression; and so spiritual experience and the enjoyment of the natural world are seen, and experienced, as poor substitutes for straightforward sexual experience. This viewpoint assumes that culture necessarily opposes the aims of instinct: "On the one hand, love comes into opposition to the interests of civilisation; on the other, civilisation threatens love with substantial restrictions."[35] Freud was clearly unhappy about a cultural situation in which "the life of present-day civilised people leaves no room for the simple natural love of two human beings;"[36] but reluctantly, he believed this situation to be inevitable, and his "reality principle" reflected this belief.

Freud's sexual theories undoubtedly had a degree of accuracy when applied to a Victorian Europe in which the repression of sexuality was accomplished through quasimoral injunctions forbidding particular types of sexual activity or relationship. Today, however, the sexual issues that we face are most usefully viewed not simply in terms of the prohibition of sexuality, but rather as involving its decontextualization and commodification, which are precursors to its assimilation and exploitation by the interests of power. In this, I am in some respects closer to Foucault[37] than to Freud. While Freud's view of social reality involves a relatively straightforward opposition between human instinctual needs and a social structure that heavily constrains direct sexual satisfaction, Foucault's influential view sees sexuality as defined by and diffused through social structure in a way that allows certain possibilities while failing to articulate others.

A third alternative, and one that avoids either the biological determinism of Freud or the social determinism of Foucault, is that of Geertz, in which culture is seen as potentially completing humans and relating us to the external world, thus complementing our "instinctual" structure rather than repressing it. It is only under certain social conditions—when, for example, potentially facilitative cultural structures are perverted or overwhelmed by the interests of power or commerce—that culture can be understood as "repressive." If we follow this line of argument, then the "direct satisfaction" which Freud saw as the most fulfilling form of sexual experience may represent a sort of glorified masturbation, a *forme fruste* of an act which could relate our deepest subjectivity to the cultural and natural worlds, and so integrate our universe of meaning. Is it possible that sublimation, instead of representing a less satisfactory avenue for "instinctual urges," may actually be experienced within a healthy culture as an individually fulfilling stance? Such a view would be consistent with that of Lasch, who also sees a totally desublimated form of sexuality as socially and psychologically destructive, arguing that "sex valued for its own sake loses all reference to the future and brings no hope of permanent relationships."[38]

However, to conceptualize the concept of sublimation as an *alternative* to direct instinctual satisfaction reproduces the dualistic polarization of nature and culture that will by now be familiar to the reader, reflecting an ideology in which individual and social expressions of any human propensity are necessarily experienced as mutually exclusive. Within such a polarized universe, sublimation can only take a repressive form, since to the extent that instinct is expressed in a way that is consistent with social structure, it will to that extent be unfulfilling as individual expression. In this case, as Norman O. Brown has noted, "sublimation is life entering consciousness on condition that it is denied."[39] However, if we reject the precondition that the instinctual and the social are inevitably opposed, then we can abandon the concept of sublimation entirely, and envisage a spectrum which, to a greater or lesser extent, integrates both "instinctual" and cultural elements, and within which erotic elements permeate the whole of the life-world just as cultural forms articulate and enable the expression of "natural" propensities. Put differently, the clear distinction between "direct" and "sublimated" forms of expression would vanish in a social context in which culture was understood as *articulating* rather than *opposing* "instinct." Such an arrangement would only be possible within a culture which was genuinely in synchrony with human predispositions.

Marcuse is one of several theorists who have glimpsed this possibility, envisioning "a civilisation very different from that derived from repressive sublimation, namely, civilisation evolving from and sustained by free libidinal relations."[40] Now, in spite of Marcuse's loyalty to Freud, this argument implies a very un-Freudian form of instinct—namely, one which is structured coherently enough for a civilization to be based on it. This, however, is hardly an outlandish possibility, since the natural world is, in general, copiously structured, in stark contrast to the Freudian belief—which, we might note in passing, is recapitulated by many postmodern theorists—that any structure is necessarily *imposed* by culture. Marcuse also points to Ferenczi's concept of a "genitofugal libido" in which eroticism diffuses throughout the entire organism, supporting the view that a narrowly sexual interpretation of libido that places it in direct opposition to culture may be only one possible, culturally specific, arrangement out of many.

The idea that the id is a blind, structureless entity is itself the product of forms of thinking, virtually universal in the modern world, which accept science's claim to a monopoly on structure. If theory is shaped by dualisms such as those that contrast "civilization" and "nature," then it is easy to recognize the structural properties of the former while regarding the latter as the manifestation of chaos. However, Freud's thinking was challenged on this point by object relations theorists such as Fairbairn, who argued that:

> The Ego is . . . conceived by Freud as a structure; but the Id is described in a manner which implies that it is essentially structureless and is merely a reservoir of instinctual energy. . . . [However], the Id must be regarded not simply as a source of instinctive energy, but as an inherently dynamic structure.[41]

A society that is nonrepressive and ecologically healthy cannot be based on the same, repressive form of libidinal organization as ours, according privileged structure to certain aspects of the world while denying it to others. Rather, we need to look forward to the possibility of a libidinal "ecology" that avoids the destructive splits between body and mind, thinking and feeling, self and world, structure and energy, so that feeling becomes diffused through a network of relations rather than located within a narrow egoic consciousness. Freud's view—which undoubtedly expressed the experience of many in Victorian society, as it still would today—was that "when a love-relationship is at its height there is no room for any interest in the environment."[42] But under the more favorable conditions which Marcuse envisages, the "biological drive becomes a cultural drive"; and the id, instead of seeming to be blind, primitive, and unintelligent, takes an articulated form such as the "Superid" envisaged by Charles Odier.[43] These possibilities imply a rejuvenated dialectic between culture and nature: nature would be articulated culturally, and culture would derive its ultimate meaning from its rootedness in a natural world viewed as structured and intelligent. The differences between such a society and our own, therefore, would go far beyond a mere "liberation" of sexuality in its existing forms, and the resulting configuration of nature and culture would reach toward a redefinition of both. Recognizing that the "less preferred," dualistically oppressed aspects of existence have their own structure and intelligence would imply some truly radical changes sexually, socially, and environmentally, some of which I will explore later.

The withdrawal of meaning from the world which occurs through the denial of its structure is, however, not restricted to those areas traditionally seen as relating to "biological needs." There is also a more general literalization and a loss of the metaphoric sense; and this impoverishes our lives, bringing everyday consciousness into line with a reductive scientistic vision. Paul Shepard has argued that we can see the beginnings of this process in the transition from a hunting-gathering existence to settlement in villages, involving a focus on the "mother earth" metaphor and its adoption as a stable reference point in determining our sense of relation to the earth. According to Shepard, the widespread acceptance of this metaphor led to the exclusion of other possible metaphors, leading to an attitude toward the natural world that draws on the image of a sometimes nurturant, sometimes harsh, mother. Among the effects of a lifelong subordination to this mother image, argues Shepard, are "resentment and masked retaliation, displaced acts of violence, and the consequent guilt." Nevertheless, the same metaphor "enables emotions and bonds of kinship, compassion and responsibility to be felt not only within the human group but to be directed to the earth,"[44] so the effects of this dominant metaphor were positive as well as negative. All the same, if we follow Shepard's argument, the focus on this metaphor represents a preliminary narrowing-down of the meanings associated with nature. Today, of course, we have gone much further than this, replacing the maternal metaphor by a mechanistic understanding whose reification is indicated by its acceptance as literal truth.

Colin Turnbull's work among the Mbuti offers a telling comparison between a cultural viewpoint in which each person's erotic nature is intertwined in a sort of "metaphoric free play" with everyday life and the world around them, on the one hand, and one in which sexuality is a matter of individual urges, isolated from context and therefore essentially meaningless, on the other. The Mbuti address the forest as "father" or "mother," says Turnbull, for it offers food, shelter, clothing, warmth, and affection:

> The word that I translate as "affection" is kondi, which may equally be used to mean love and need, between which the Mbuti seldom differentiate when discussing human relationships. . . . It is clear that on occasion the emotion is one of sexual love, for the sexual nature of the relationship between an Mbuti man or woman, boy or girl, and the forest is sometimes demonstrated overtly enough by an erotic gesture of the body, in imitation of the act of copulation. Playful youths may even specify verbally that they want to copulate with the forest, and if this wish is accompanied by well-executed body movements, it is sure to give rise to mirth among the youth's companions. But as a motive, that hardly obtains when a youth behaves like this in privacy and solitude, as I have often seen. Then, at least, it is done for something

> other than the approval and laughter of others; it is more in the nature of a spiritual, if sexual, communion with the forest.
>
> On other occasions the emotion is sometimes more one for which I can only use the word "adoration." I use the word without shame, rather with the joy felt by Teleabo Kengé when he slipped into the bopi (children's playground) one moonlit night. He was adorned with a forest flower in his hair and with forest leaves in his belt of vines and his loincloth of forest bark. And with his inner world he danced and sang in evident ecstasy. And in answer to my question, he said, "me bi ndura, me bi na songe"—"I am dancing with the forest, dancing with the moon." It is reasonable to assume that the Mbuti child, growing up, sees all this and much more and is transformed accordingly.[45]

It is a measure of our own alienation that this passage may be experienced as fanciful, childish, or overly-romanticized. Such reactions serve to hide the pain of our own loss, deeply recognized and defended against, for our own early erotic experiences, unless we are unusually fortunate, are unlikely to have found so receptive a context. Turnbull himself compares his own experiences at an English public school, where

> The boys who prided themselves as being above . . . homosexual activity used to compete with each other to see who could splatter his sperm highest up the white tiled wall. To me the emptiness of that act was more debasing than anything else that went on, symbolic of another emptiness that permeated our adolescence. If two boys formed a liaison, either for mutual pleasure or protection, they were criticised for exclusivity, for selfishness, and lack of "team spirit." But far worse, if two boys formed a liaison because of mutual affection and respect, whether such a liaison was accompanied by sexual interaction or not, the two were condemned publicly and accused of all manner of perversions. Yet gang rape or splattering sperm on walls was just "good clean fun." In all these ways our first sexual experiences were systematically divorced from normal human relationships and set against the concept of sociality. Far from being acts of creation, even in our minds they were acts of destruction; in place of beauty there was ugliness.[46]

Such learning processes reinforce the dissociation of "biology" from "sociality."

Focusing on one particular, historically constructed way of articulating eroticism as "normal," as a "biological given," fixes and reifies its metaphorical play into a single, literal form. Viewing human sexuality as an "individual urge" is a particularly unconstructive form of this reification, since it alienates the individual from structures in the outside world, caging the expansiveness of eroticism and denying relational possibilities between ourselves and the rest of nature. In a healthy culture, "sexuality" can be a power that connects, that relates, that permeates both us and the world in which we live, a world that is therefore spiritually and erotically vital. Complementarily, Eros need not be experienced as the merely physical impulse portrayed by Freudian accounts of repression, but also as one which integrates body and spirit. It is only two thousand years of theology, says Marcuse, that disguises the fact that "Eros and Agape may be one and the same."[47] This suggestion is consistent with our historical knowledge of the somatic dimensions of early Christianity, involving bodily awareness and ecstatic experience, long since lost in the evolution of a narrower, more dogmatic, and more cerebral Christianity.[48] Complementarily, we need to connect body and spirit from the other direction as well. As James Hillman has argued, religion can be understood as a form of instinct, and one that is "as basic to psychological life as the so-called more organic, physical urges."[49] The fundamental aim of the ecological project must be to connect, or reconnect, nature and psyche in all their forms, so that potentialities present as bodily awarenesses or other natural structures engage in a mutually respectful dialogue with the ideational structures that might articulate them.

The capacity of a spiritual/erotic "instinct" to engage us in the ebb and flow of life that is the biosphere, however, is reliant on the cultural structures which relate individual awareness to the natural order. The current absence of such structures suggests that we inhabit what Romanyshyn terms "an age which in offering no mirrors for the reflection of an authentic sexuality reflects in its place a way of living one's sexuality, and one's passions, as only a hidden and forbidden wish of a helpless child."[50] What we may be seeing in the development of Eros from an infantile, "polymorphously perverse" form toward what we know as adult sexuality is not so much the mature flowering of our biological potential, but rather a destructive narrowing and literalization whose effects are most far-reaching not in the domain of sexual relations, but in those other areas of life which are *deprived* of their erotic dimension. That is, we "develop" in terms of our power to act in the world; but our actions retain the immaturity of the child who enjoys his power without recognizing a responsibility and a deeper involvement with the world. Our toys become bigger and more powerful as we grow; but they remain toys rather than the vehicles of a more mature relation to what is outside us. The development of "intelligence," which we discussed earlier, for example, illustrates how human experience can be narrowed and deprived of any erotic relevance. The squeezing of eroticism into the narrow domain of human "sexuality" is therefore directly related to the absence of eroticism from the rest of our lives and from the world outside. James Hillman's conclusion that "romantic love keeps the world dead"[51] may be overstated; but it is nevertheless essentially accurate. While we have become ever more emotionally dependent on loving another individual, we have lost the ability to reach out into the world in a manner that Robinson Jeffers described as "falling in love outwards."

Few psychodynamically oriented theorists have recognized the need for a reconceptualization of "instinct" as clearly as Fairbairn, whose work has been extended by Guntrip. More recent object relations theory has tended to lose sight of the poignant and theoretically pregnant disparity between the full and integrative relationality sought for by the infantile psyche and the stunted forms of relationality that are actually possible within the "culture of narcissism" described by Lasch. Increasingly, object relations theory has lost the critical edge that derives from the awareness of this disparity, often suffering a similar fate to the post-Freudian "revisionism" criticized by Marcuse. For this reason, I will focus on those critical insights which the early object relations theorists imply but never quite manage to develop into a fully critical social theory.

Fairbairn, while recognizing that superficial pleasure-seeking behavior occurs, asserts that "explicit pleasure seeking represents a deterioration of behaviour"[52] that occurs in order to relieve the tension which has built up as a result of failure to achieve some object relationship. The ego, according to Fairbairn, is fundamentally object seeking rather than pleasure seeking, and the impulses concerned are part of the ego process rather than originating within a separate id-like structure. According to this view, sexuality aims for emotionally significant relationships with objects in the outside world rather than simply a sort of hydraulic release of pent-up libido; and only when such relationships are blocked do we find the individualistic, crudely pleasure-seeking attitudes which Freud interpreted as reflecting the natural configuration of human needs. Fairbairn suggests that "explicit pleasure-seeking has as its essential aim the relieving of the tension of libidinal need for the mere sake of relieving this tension. Such a process does, of course, occur commonly enough; but since libidinal need is object need, simple tension-relieving implies some failure of object-relationships. The fact is that simple tension-relieving is really a safety-valve process. It is thus not a means of achieving libidinal aims, but a means of mitigating the failure of these aims."[53] The near universality, in the industrialized world, of the individualized and literalized forms of sexuality to which Fairbairn refers does not make them any healthier or more "natural." Rather, he implies that a healthy erotic life is something which can arise only out of a full relation with the outside world; and, furthermore, that this implies a reconfiguration of selfhood. "It is impossible to gain any adequate conception of the nature of an individual organism if it is considered apart from its relationships to its natural objects," he suggests, "for it is only in its relationships to these objects that its true nature is displayed."[54] Of course, I am here arguing for a wider understanding of "natural objects" than Fairbairn would have countenanced.

Fairbairn's argument that the ego is fundamentally object seeking—that is, seeking to establish relationships with the outside world—is consistent with the views of Geertz and others that a healthy culture complements and articulates individual propensities, seeking to construct larger wholes. In a sense, the Freudian id is *created* by the inability of society to articulate our relational needs, leaving behind a reservoir of those needs so that they appear to be in principle merely "instinctual" drives which are necessarily in conflict with social structures. Such a concept would be entirely superfluous within a culture that effectively articulated the biological potentials of its members.

Given that the approach suggested here emphasizes the potential interwovenness of individuality and cultural structure rather than the conflict between them, a possible criticism is that we are in danger of losing a site for protest. If protest against the social determination of our lives depends on our ability to distinguish between cultural structures and "natural," "instinctual" ones such as the "id," then a complete alignment between the natural and cultural realms will remove the distinctions on which protest depends. Within a hypothetical society of the future which approached the ideal of a cultural realm that fully articulated the natural, individual and society would exist within the same expansive universe of meaning, so that dissent would be difficult to envisage. How, then, does the approach suggested here differ from the social constructionism criticized earlier for its political impotence? In both cases, we see the alignment of the "human" with the "social." A crucial difference, however, is that whereas social constructionism is based on a reinterpretation of nature as a *social* construct, I am accepting nature as having its own structures regardless of their consistency or inconsistency with the social realm. Nature, in other words, is the basic grounding of life; and the need for, and possibility of, protest depends on whether the social realm is broadly consistent with this frame, or whether—as in industrialism—it attempts to become independent of it. To the extent that protest can be understood as originating in a misalignment of the social realm with the natural, then the elimination of such misalignments will to this extent render protest redundant and incoherent. Clearly, other types of protest are conceivable: for example, conflicts of interest are clearly possible within the social sphere even if it is aligned with the natural realm, since the natural may be socially articulated in a diversity of sometimes inconsistent ways. However, within a society in which the human and the natural mutually permeated each other, and in which culture was the medium of this permeation, the purpose of and need for protest would be much more limited, and perhaps in such circumstances the idea itself would be conceptually incoherent. Similarly, individual "creativity," which like the "freedom" to protest can be seen as the necessarily individual expression of aspects of subjectivity repressed within a society that denies the natural, would in part give way to forms of action more consistent with the cultural context. Such a situation is quite different to one involving totalitarian political regimes in which subjectivity both fails to be socially articulated and is also repressed at an individual level. In a sense, the adequate articulation of individuality would allow alternatives to be expressed *through* the medium of culture rather than as protest *against* it. In practice, of course, this "adequate articulation" will remain an ideal to be aimed at rather than a practical possibility.

The vision of a cultural arrangement in which individual, social, and natural orders are more or less in harmony with each other, thus reducing the necessity for protest, is therefore quite different to that offered by those theorists who perceive existing forms of selfhood as being entirely socially and historically constructed. Within this latter, socially deterministic vision, the construction of individuality through the interweaving of cultural and natural factors is denied as completely as is the case with the biologistic theories which until recently dominated psychology, as we saw in chapter 2. Whatever its radical pretensions or emancipatory intent, it is a vision which is politically impotent and fundamentally nihilistic; for it systematically denies the existence of those aspects of selfhood that remain unexpressed within this cosy collusion of socially determined consciousness and social structure, perpetuating the vision of a human realm that is set above and apart from the nonhuman. In contrast, the views developed here are more sympathetic toward Joel Kovel's suggestion that "if we are to be true to people—and grant them a dimension in which the administered world of political economy is only partial—we need to retain an 'instinct like' concept; and this concept cannot be divorced from the realm of nature or the trans-historical."[55] Nevertheless, the "realm of nature" that Kovel seems to have in mind is *internal* nature—in other words, something close to Freud's original concept of instinct. However, Kovel's use of the term "instinct" is potentially misleading, fragmenting a relational ecology into desiring subject and desired object. What we need is a concept which retains its independence from social determination, while including the relation-seeking property emphasized by Fairbairn: in other words, that recognizes that nature "within" reaches out toward integration with "external" nature, and that this integration would be an essential aspect of healthy development. Jung is one of the rather few social scientists to have recognized this need for alignment between natural structures that are both within us and outside us:

> Deep inside us is a wilderness. We call it the unconscious because we can't control it fully, so we can't will to create what we want from it. The collective unconscious is a great wild region where we can get in touch with the sources of life.[56]

Within present-day Western cultures, however, this recognition is stifled, and therefore instinctual cravings will appear as strictly "individual" drives. What appears as biological necessity, then, can be recognized as a reflection of the ideological manipulation of experience. As Eugene Gendlin has put it: "The self's new intricacy seems only inner because the external controls prevent it from being lived out. Therefore it can be lived only in private self responding. But if the intricacy is accepted as inherently only something inner, then the social controls are accepted without having been noticed. What prevents one's outward efficacy is masked and unseen."[57]

Winnicott's "true self" or Fairbairn's "libidinal ego"—that "inner" part of us which embodies our deepest and most poignant hopes, yearnings, and feelings of selfness—may thus be the residue left by the frustration of our relationality to the world "outside," and an index of the failure of our culture. A fundamental part of the self therefore develops in isolation, remaining needy and unfulfilled, unable to take part in the movements and flows of a world that comes to seem increasingly remote and separate from ourselves. Our behavior toward the natural world is likely to be strongly influenced by this "inner" self as it seeks to find some form of substitute fulfilment through, for example, domination, identification, consumerism, or compulsive sex—those neurotic substitutes for relation adopted by a self which unconsciously recognises its disconnection from its natural context. There is, therefore, an evolving and destructive dialectic between a form of selfhood made needy by its failure to find relation, and a constructed substitute world of things and experiences to be bought—a dialectic which becomes more vicious as our opportunities for authentic relation are steadily reduced by the physical destruction of the natural world. In effect, then, our natural, bodily relation to the world has been suppressed, leaving a form of inner experience that feels aimless and isolated.

However, within a cultural context in which the individual, social, and natural realms were harmoniously interwoven, the solipsistic character of our innerness would disappear, to be replaced by forms of relation that would often be beyond our current repertoires. In other words, the cravings, hopes, and needs that have been ascribed individualistically to "instinct" may be viewed as stunted forms of relational desires which are unrequited by existing, narrowly social, opportunities. Fairbairn may have been correct in recognizing the relation-seeking nature of the ego, but too narrow in seeing the existing social context as potentially "good enough" for the consummation of this need. The cravings and dissatisfactions which stem from existing social reality are compounded by the apparent impossibility of alternatives: an impossibility that originates in the specious hegemony defined by the collusion between an acquisitive, colonialist self and an industrial world which offers the illusions of fulfilment. As Kovel has written, "The self does not arise prior to the transformation of the world, but in the transformation of the world. As the object is made, so is the human subject;"[58] and it is this complementarity between subject and object which suppresses dissent. Seen from this perspective, the dissociation and repression that Fairbairn came to understand as typifying the "normally schizoid" character organization of our times is entirely consistent with the ongoing "constructive technological destruction"[59] that obliterates nature in the *external* world.

Toward the end of his life, Fairbairn began to move away from conventional psychodynamic assumptions involving the separateness of internal and external reality and the limitation of environmental influences to parent-child relations. Winnicott, too, began to criticize "the tendency to dwell either on a person's life as it relates to objects or else on the inner life of the individual,"[60] recognizing the existence and significance of a "third area"—"an intermediate area of experiencing to which inner reality and external life both contribute."[61] He saw this third area as developing out of the "transitional phenomena" of childhood, by which he meant the borderline realm in which self was gradually crystallizing out from not-self, and which involved such ambiguous "transitional objects" as teddy bears. The fate of such transitional objects is

> To be gradually allowed to be decathected, so that in the course of years it becomes not so much forgotten as relegated to limbo. By this I mean that in health the transitional object does not "go inside" nor does the feeling about it necessarily undergo repression. . . . It loses mean-

> ing, and this is because the transitional phenomena have become diffused, have become spread out over the whole intermediate territory between "inner psychic reality" and "the external world as perceived by two persons in common," that is to say, over *the whole cultural field*.[62]

Winnicott is here beginning to recognize the potential importance of that ideologically occluded space between the individual and the world that is a blank space on the map of industrialized society; and it is this space that is the key to revivifying the world and our place in it, if we can allow ourselves to recognize it. What if teddy bears, rather than being understood as temporary and expendable toys, were seen as residual attempts to reach out into the natural world, the pathetic substitutes for—or, more optimistically, the precursors of—a real relation with the wild? In this case, the loss of meaning referred to by Winnicott is the essence of a tragedy repeated in the socialization of every child in the modern world; and what is needed is not so much the "loss of meaning" that occurs as feeling is "diffused" throughout the cultural field, but its increasingly mature *articulation*. Among the Tikopia of Polynesia, for example, the child learns to experience herself as part of a social nexus, a widening circle of "meaningful others," beginning with the mother who suckles the child, followed by aunts and other relatives, other households, and eventually the land itself, to the extent that the name given to man and wife is determined by the place where they live.[63] The extent to which self is culturally interwoven with landscape is indicated by the fact that Tikopia separated from their land often simply die.[64]

Harry Guntrip also saw culture as essential to a healthy subjectivity, emphasizing its capacities to articulate what cannot be expressed literally or scientifically, and to "move us profoundly to a deeper experiencing of human living."[65] Object relations theory, then, hints at the natural development of the individual as involving an extension of selfhood into the outside world, so that there is a mutual exchange of meaning between one's pregiven makeup and the natural structures one inhabits; and through the mediation of culture, these basic beginnings can flourish into a mutually expressive whole. The third area which Winnicott referred to is not, therefore, merely an "illusion," but is essential in linking the individual with external reality, so that a creative, playful approach to life might exist. "It is creative apperception more than anything else that makes the individual feel that life is worth living," he wrote. "Contrasted to this is a relationship to external reality which is one of compliance, the world and its details being recognised but only as something to be fitted in with or demanding adaptation."[66]

Winnicott seems here to be struggling to elaborate the importance of culture as an area which is neither exclusively "individual" nor exclusively "external," while, in the main, attempting to retain a Cartesian separation between the active, intelligent, developing child and a world that is formulated in vague terms such as the "facilitating environment." In effect, he almost reluctantly implies a redefinition of selfhood, and of the relation between self and world. While the influence of Freud is evident in Winnicott's assertion that "illusion . . . is inherent in art and religion,"[67] he nevertheless recognizes that the "place where cultural experience is located is in the potential space between the individual and the environment."[68] Within this potential space "there develops a use of symbols that stand at one and the same time for external world phenomena and for phenomena of the individual person."[69] Winnicott's concept of the "potential space" within which individual and world are integrated has profound environmental significance; for he is sketching out the conditions under which a genuine environmental subjectivity could emerge. He is clear, however, about the gap between the potential self which could exist in a healthy cultural milieu and the stunted form of individuality common under present conditions:

> There is for many a poverty of play and cultural life because, although the person had a place for erudition, there was a relative failure on the part of those who constitute the child's world of persons to introduce cultural elements.[70]

There is little doubt, Winnicott continues, that "the philosophic attitude of the age" contributes to this problem.

Could one not envisage, however, a Tikopia-like environment in which meaning derived not only from one or two individuals, but from a broader context, including the natural world, and involving a large number of individuals? In this way, the potential space that is so often left empty could be filled with a rich matrix of relations which could connect the individual in a secure and meaningful fashion to the world outside. I am reminded of a Navaho student of mine who, shortly after the birth of her child, told me of a ceremony that she and her daughter had taken part in to welcome the child into the community and the world, attended by 163 "relatives"—blood relatives and ceremonial ones. For a child, such an introduction into a rich nurturant context surely bodes well for her future. Could it be that our present childcare situation, in which the child's identification is limited almost entirely to the parents, and the external world is seen as largely devoid of emotional significance, is a diminished form of a more broadly inclusive developmental situation in which the parents would be merely among the more important mediators of a relationship between the child and the cultural and natural worlds? As we saw in chapter 3, the schizoid personality, which is closely related to the development of scientific "objectivity," approximates our "normal" personality configuration. Fairbairn argued exactly this, suggesting that "[e]veryone, without exception, is schizoid at the deeper levels." Guntrip agrees, commenting that the schizoid condition is

> In varying degrees universal. No human being ever has perfect mental health. Instead of saying that there is in health a situation of this kind, analogous to that found in pathological conditions, it would seem that this radical ego-split is actually a universal phenomenon, present in all of us without exception, not intrinsically or theoretically inevitable, but practically inescapable; . . . the schizoid problem in the above sense is the ultimate problem.[72]

But if this schizoid splitting of the self is "not intrinsically or theoretically inevitable," then under what conditions could the self be whole? Presumably, given that the split-off parts of the self are those potentially relational qualities that are incompatible with the individualism of our time, these conditions would be those that would allow selfhood to become part of the cultural and natural fabric of the world. Possibilities such as these offer tantalizing glimpses of a reality which is neither "romantic" nor "idealist" in the denigratory senses of these terms, but which would allow the realistic expression of our natural endowment within a healthy ecological whole. The schizoid phenomenon, then, does not merely concern the human personality; for it also implies a repression of that vitality of the world which emerges from our relationship with it. To see it as a merely psychological problem is to accept the inevitability of a world in which this vitality *has already been repressed*, and to fail to recognize that the wholeness of the world in part depends on our healthy participation in it.

The suggestion that the schizoid personality is in some degree universal within industrial society is not, of course, a novel one: R. D. Laing, for example, wrote that "the normal person . . . is a shrivelled, desiccated fragment of what a person can be,"[73] a view which is consistent with Abraham Maslow's statement that "what we call normal . . . is really a psychopathology of the average."[74] More specifically, commenting on the personality structure of scientists, Evelyn Fox Keller has commented on the schizoid detachment necessary for scientific "objectivity."[75] However, until the recent appearance of an "ecopsychology" literature,[76] and with the notable exceptions of Harold Searles and Paul Shepard, few have noticed the relation between schizoid personality organization and our separation from the natural world. Many theorists have argued that pathological personality development is associated with inadequate parenting, which is one specific form of the environmental inadequacy suggested here. Some have discussed the cultural factors that may be associated with inadequate parenting. But the ideological fault line which divides the natural world from the constructed "human" realm seems to be so deep seated as to be an unnoticed, taken-for-granted aspect of the universe. Everyday psychological reality and the social sciences share a universe of meaning from which the natural world has been excluded except in the crudest material sense; and so there is an "ironic convergence," to paraphrase David Ingleby,[77] between an individual consciousness that is alienated from the biosphere, on the one hand, and an equally alienated social scientific praxis that myopically attempts to account for the problems caused by this alienation, on the other. In this situation, each legitimates the other within a hegemonic scheme from which it is difficult to escape. But escape we have to; and in the next chapter, we begin to explore the type of "consciousness" that might emerge if we were to succeed in realizing the subjective structures which would be consistent with the natural order.

Chapter 6

RESYMBOLIZING NATURE

Science comes to a stop at the frontiers of logic but nature does not. . . .

—C. G. Jung, "The Psychology of the Transference"

The Primacy of Relation

The object relational view that we have just discussed draws our attention to relationality as a fundamental characteristic of life, problematizing accepted definitions of the human animal as primarily autonomous and only secondarily related to context. This focus on relationality leads us toward a new understanding of us as naturally (and I use this term with no apology) integrated within larger structures; and so industrialism's fragmentation of these larger structures is easily seen as diminishing both selfhood and the rest of the world. True, much recent object relations theory permits itself to be drawn back into the orbit of individualistic assumptions through its preoccupation with those *inner* object relations that in culturally impoverished environments develop as substitutes for healthy external relationships; and, what is more, the revolutionary implications of object relations theory are seldom explicit and often only grudgingly acknowledged. Nevertheless, object relations theory implicitly problematizes the assumption of the self-contained individual, forcing us to admit that there are other, equally significant, structures in the world, and that healthy life is defined largely by the ways we interact with these other structures. This is not to say—in contrast to various postmodern stances—that the self is simply an artifact of social and linguistic forces: rather, the notion of relationality as I use it here carries the implication that any life form can be adequately defined only if it is recognized as existing at three levels: first, at the level of the functioning of those subsidiary organs that constitute it; second, as an individual entity with characteristic needs and purposes; and last but not least, as a constituent of larger systems. Any adequate theory of self must adequately incorporate the dialectic among *all* these levels of self.

The idea that relation to the world is an important constituent of identity is one that intrinsically raises the status of the world to that of subject. This is something which is intuitively recognized by environmentalists: after all, it is difficult to care passionately about a world that is dead or merely mechanical, and attempts by radical environmentalists to elevate the legal and moral status of other natural entities to levels previously enjoyed by (some) humans implicitly recognize the intelligence and purposiveness of these components. The discovery—or rather rediscovery—of the ontological status of nature is also a common theme of science fiction such as Fred Hoyle's *The Black Cloud* or Steven Spielberg's *ET*, in which something initially formless and lacking in discernible faculties is belatedly acknowledged to have not merely "structure" in the abstract, but some form of sentient intelligence. The opening of our awareness to other forms of structure, therefore, is not simply a reconsideration of the earth, but is also a complementary redefinition of ourselves, since it challenges the devastation of experience associated with the narrowing down of our relational possibilities to those involving a small number of fellow humans—family, close friends, lovers—and the exclusion of the entire nonhuman world. This devastation, whose prevalence is acknowledged in Guntrip's assessment that the schizoid personality is "more or less universal," is sedimented into the fashionable social scientific reductionisms that view meaning *either* as constructed by individual humans (as in most psychology) *or* as characterizing larger structures such as language (as in most cultural studies); both of which repress the vitality of life as an interaction between levels and structures. An extended object relational view therefore reintroduces us to the other members of our potential ecological community; and through this reintroduction, subjectivity extends beyond the constraints of human physicality, diffusing as well through a diversity of intelligently interacting natural systems.

The significance of relation for environmental theorizing is incorporated in Warwick Fox's "transpersonal ecology," which envisions a self that is sufficiently "wide, expansive, or field-like" to recognize "that we and all other entities are aspects of a single unfolding reality."[1] Extending Arne Naess' notion of the "relational field,"[2] Fox singles out identification as a key process in our healthy relating to other aspects of nature. He suggests that identification is based on "commonality," implying "a sense of similarity . . . even if this similarity is not of any obvious physical, emotional, or mental kind."[3] This in turn implies that we will be more likely to identify with those parts of the biosphere that are in some way "like" us, or that we have something in common with, than with those that are dissimilar to us. For example, we would presumably find it easier to identify with deer or bears than we would with, say, protozoa or scree slopes; and this would seem to raise questions about the wide-ranging type of identification which Fox, and Naess, have in mind. A relation to the world which rests on identification all too easily assimilates the other to self, since it is based on self's selection of those characteristics on which identification can be based: like Narcissus, we fail to recognize that the world we identify with may in fact be our own reflection. As Freud acknowledged, this "substitution of identification for object-love . . . represents a regression to original narcissism . . . the ego wants to incorporate [the] object into itself . . . by devouring it."[4] Identification, then, may often be a narcissistic ploy to avoid a more demanding, mature, and sophisticated form of relation that recognises otherness as well as similarity. In an attempt to overcome this criticism, Fox argues for a type of identification that he terms "cosmological," that "proceeds from a sense of the cosmos . . . and works inward to each particular individual's sense of commonality with other entities."[5] Here we are asked to identify with "all that is"—an experience that "can be brought about through empathic incorporation of mythological, religious, speculative philosophical, or scientific cosmologies."[6] However, this may be greatly to overestimate our capacity to recognise and relate to the consistencies, coalescences, and regularities which define ecological systems, since the behaviour of even simple systems frequently tends toward the sorts of "nondeterministic regularity" that consciousness struggles to understand. Over the past two decades, ecological theory has tended to move away from emphasizing equilibrium and homeostasis toward an awareness of instability and change; but as we become more aware of the significance of concepts such as "strange attractors" and the nonlinear character of much ecological process, so it becomes clear that relations that appear to us as disorderly and inexplicable may, in fact, embody forms of order that are beyond the reach of consciousness.[7] Thus while we may potentially understand or identify with particular natural forms—mainly creatures similar to ourselves—there is a huge range of other natural entities and structures that are likely to be consistently beyond our powers of empathy. An ethic that is based on our ability to identify with nature, while it may indeed be useful within the limited sphere of our capacity for identification, cannot incorporate that arguably much larger portion of nature whose orderliness is imperceptible to us. A better basis for an environmental ethic, then, may be the recognition that the order of nature is, often vastly, beyond our conceptual reach, and that therefore any attempt to understand, identify with, manage, or control nature unwittingly attempts to reduce the cosmos to the limits of the self, or even the Self. Recognising this, a more adequate ethic will abandon this attempt to extend our limited capacities for empathy and identification to encompass the cosmos; and will instead recognize the magnitude by which the latter exceeds the former, instilling in us a necessary humility and an ethic of noninterference, as we recognize that our best models of nature are merely "shadows of a vision yonder." Many indigenous mythological and spiritual systems incorporate exactly such a recognition of the limits of human intelligence.

A related problem which is common to most variants of deep ecology is that because of their preoccupation with commonality, continuity, and holism, they forget that *difference* is an equally essential basis of ecological structure. Jack Turner's insight that "to be absorbed in this life is to merge with larger patterns"[8] needs to be complemented by a recognition that ecological vitality depends on the *differentiation* that makes possible the structures and patterns of nature. Naess is clearly aware of this problem, arguing that although we "are part of the ecosphere just as intimately as we are a part of our own society . . . the expression 'drops in the stream of life' may be misleading if it implies that the individuality of the drops is lost in the stream."[9] For the same reason, Fox prefers the metaphor of the (individual) leaves of a (whole) tree in order to give "due recognition to the relative autonomy of different entities."[10] Such images are useful as metaphors; but are the images they summon up too static to convey the dynamic and structural qualities of life? "Delicate, elusive, quicker than fins in water," says Loren Eiseley, "is that mysterious principle known as organisation."[11] We betray this principle if we think of it in terms of a static image; for as the image of fins in water suggests, organization is temporal as well as spatial, an intricately moving dance that makes notions such as identification seem heavy-footed. Deep ecology needs to be positively grounded in this sort of elusive natural vitality rather than negatively in the rejection of industrialism—a stance that all too often smuggles in unrecognized some of industrialism's most destructive assumptions. The rejection of difference is an example of this tendency: if industrialism fragments the world into unrelated "bits," so the argument goes, then environmentalism should emphasize the continuity and overlap between these "bits." But difference is not a quality that was introduced by industrialism into a virgin world of seamless continuity, and nature is much more than a mirror-image of the industrialized world. Similarly, if the industrialist self can be understood as embodying a stagnant individualism, an ecological self is not simply one that "merges" with larger patterns by "softening" its boundaries. Rather, we need to restore a subjectivity that is dynamic, oscillating between the poles of separateness and relation, so defining and redefining temporal structures which today are almost inconceivable. As humans, for example, we are capable of behaving in ways that are entirely self-centered and egoistic; and we are also capable of immersing ourselves in some purpose or structure in ways that suggest a transcendence of self. We need a theory which incorporates this dialectic of separateness and integration, not through an uneasy compromise between them, or by replacing an individualistic approach by a "holistic" one, but as a dynamism that is central to the process of life. Such a theory would avoid the tendency to default into one of two opposite extremes: on the one hand, a holism that denies individuality and difference—a night in which, as Hegel put it, all cows are black; and, on the other, a disconnected, mechanistic universe in which all entities are defined only in terms of their intrinsic characteristics. While deep ecology, in its more recent incarnations, recognizes the necessity of avoiding these alternative reductions to holism or atomism, it has so far failed to fill the theoretical vacuum that it identifies. As Naess succinctly summarizes the central dilemma: "In unity diversity!, yes, but how?"[12]

The same difficulties also dog deep ecology's attempts to offer alternatives to the conventional, anthropocentric values of industrial society. For example, "biocentric egalitarianism," while it can be understood as usefully challenging the even more unsatisfactory anthropocentric scheme of values that is widely assumed within the industrialized world, does not itself offer a viable alternative to this scheme (as Naess agrees); and this is because beneath its apparent opposition to anthropocentrism, it embodies the same lack of integrative structure as we find within modern society. This lack of structure dupes us into accepting polarized views as the only possibilities. By analogy, in a society which lacks structures that constrain and articulate behaviors concerning food, we find eating disorders such as anorexia or obesity; just as a lack of structure relating the natural and the cultural realms predisposes academia toward the extremes of biologism and constructionism. Similarly, if the human-animal relation is poorly mediated by cultural structures, then we tend toward opposite extremes of seeing all nonhuman life forms simply as resources for us, on the one hand, or of claiming that all life forms are of equal value, on the other. To be fair to Naess, he agrees that the concepts of "rights" or "values" carry a misleading implication of quantifiability which poorly articulates the underlying intuition. But the question remains: how *do* we express this intuition in a way that carries practically realizable consequences?

Another symptom of this intangible structural absence is the way we conflate *difference* and *difference in value*, just as *equality of value* has been interpreted as meaning *identical with*. Thus, to digress into another realm by way of analogy, differences between people, racial groups, or men and women are often denied—for the entirely understandable reason that these differences have frequently been used to justify prejudice and right-wing political ideologies. But the denial of difference, while it may undermine racist or sexist prejudices, also undermines the respect for uniqueness, variation, and the cultural structures that they could support. As we noted earlier, the fact that many traditional cultural institutions in the modern world (and, for that matter, in many tribal societies) have ossified and become unfulfilling does not mean that cultural structure per se is oppressive. Take, for example, the specializations and roles that were previously accepted by men and women, and that have been seriously challenged by feminist theory and practice in recent decades. Most would agree that these changes have in many respects been beneficial, since institutions such as marriage have become infected by economic and power dimensions and so have little to do with the facilitation of individual potential or the integration of the individual into a larger world. Unfortunately, however, we have frequently rejected not merely the particular forms of social relation that have become oppressive, but also the awareness of *any* differences or tendencies around which cultural structures might be built. As Wallach and Wallach have argued, for example, the abandonment of traditional role models can be read, *not only* as freeing men and women from oppressive systems of categorization, but *also* as a move toward an androgynous self-sufficiency of the genders that denies the need for relation.[13] This stance, therefore, slides into the individualism which is so consistent with industrialism; and it is, therefore, subtly hostile toward the natural order. Rather than seeing our liberation from oppressive roles as the first step in the process of generating new, emancipatory ones, we have all too often seen it as an end in itself. Because those social roles that are based on natural biological distinctions have often become oppressive, we are inclined not merely to reject these particular social roles, but also to deny the biological distinctions that they are associated with. But what this rejection forgets is that other patterns of social relationship are possible in addition to those that directly and simplistically model the social world on the natural; and that the social order can be humane, flexible, and respectful of individuality while still being consistent with the natural order. Only within modern industrial society are these desirable properties usually viewed as deriving from our *emancipation from* nature.

There is, of course, much that the genders have in common; but there are also divergences. This is not to say that these divergences necessarily say anything about any particular individual: they are more often than not tendencies. Contrary to academic fashion, these differences are not always culturally constructed; some of them seem to be related, however indirectly, to basic biological differences between the genders. As Liam Hudson and Bernardine Jacot argue, the evidence suggests that

> While there are any number of economically primitive cultures in which it is the women, not the men, who carry heavy burdens, there is not one in which the women wage war while the men look after the home. More than that: the symbolically significant activity of fashioning weapons seems in every primitive culture known to anthropology to be largely or exclusively a male preserve. Far from the maleness or femaleness of an activity being biologically arbitrary . . . the ethnographic evidence suggests that the use of these categories is in fact biologically rooted very directly indeed.[14]

The influence of the natural, however, may also be very indirect, and this makes possible a great variety of cultural articulations, some of which may be experienced as oppressive and others as emancipatory. Our rejection of those articulations that are oppressive does not necessitate a further rejection of the natural structures that, in part, underlie them, and that could also be the starting point of a range of *emancipatory* social structures. A healthy relationality between biologically differing sexes would express in microcosm the broader relationality that could locate us within the natural world; and the denial that natural structures can provide any sort of basis for gender differentiation is symptomatic of the denial of natural structure as a whole. Unfortunately, however, while our liberation from ossified social institutions and roles that have come to reflect power relations is clearly desirable, our "liberation" from the natural is always illusory and invariably consistent with the aims of industrialism. The subtext of the stronger versions of cultural constructionism is just such a denial of the natural; and in thus divorcing the social forms within which we exist from our own naturally given predispositions, this is a denial that is deeply repressive.

But if we lack the sophisticated cultural structures that can align nature and social life, such simplistic oppositions and denials become the only alternatives to a genuinely primitive fusion, reducing the world either to "things" or to ourselves. If natural structure is obliterated by the pretence that the world derives from language, it is also denied by deep ecology's impulse to fuse with a nature so long distanced from us. Both these cases suggest a "colonialist" form of relation, as they impose *internal* structures on what is *outside* us. By analogy: a romantic relationship is not possible when the other is experienced as impossibly different and alien; but neither is it possible if I experience her as no different to, as an extension of, myself. This sort of too-simple fusion tends to occur when we *desire* a relationship but lack the skills and means to achieve it; and whether we are referring to the natural world or to an attractive human "other," these skills and means are essentially cultural. In contrast to the city dweller who tries to "love" nature in a simplistic way, the wilderness dweller who lives in intimate contact with nature knows that relationship with the other is more complex and elusive than this, involving mythological and spiritual aspects as well as down-to-earth practicalities. To collapse these sophisticated and diverse cultural articulations into the notion of "identification" is an epistemological short circuit that owes more to the fantasy of reunification than to its practical achievement.

That such an otherwise promising and fruitful approach as deep ecology tends toward the same denial of natural structure as social constructionism illustrates the pervasive absence, not only of specific cultural structure, but of the *idea* of cultural structure both in our lived experience and our theorizing. According to the constructionist viewpoint, all variation is "socially constructed" and so groundless: all structure is illusory, and ethical differences are merely rhetorical stances unrestrained by any necessary connection to a world that is real. In the case of deep ecology, the denial of structure is less obvious, existing more in the unfleshed-out character of the "identification" between the individual and the world. As was the case with the 1960s fashion for "peace" and "love" that it in some respects resembles, deep ecology's sugar-coating of well-meaning intention conceals a vacuum of necessary structure; and ultimately this stance drains the world of meaning, for a meaningful world is one defined by contrast, difference, and mystery as much as by harmony, similarity, and empathy. As Charles Taylor has argued,

> I can define my identity only against the background of things that matter. . . . Only if I exist in a world in which history, or the demands of nature, or the needs of my fellow human beings, or the duties of citizenship, or the call of God, or something else of this order *matters* crucially, can I define an identity for myself that is not trivial. Authenticity is not the enemy of demands that emanate from beyond the self; it supposes such demands.[15]

In other words, if our relation to the world is based solely on our identification with it, ignoring the differences and contrasts that also color it, both our own identities and those perceived to exist in the world itself become meaningless. And ultimately, any workable value system will have to be based on the realistic apprehension both of our kinship with and our distinctiveness from other entities, since bears, for example, are neither the humanesque creations of Disney nor the savage homicidal aliens of frontier lore, but rather demand altogether more sophisticated and multifaceted forms of understanding and relationship. The necessary complexity of such a value system precludes easy answers, but it is clear that these answers will not be forthcoming from any "human" sphere that is constructed so as to ignore the natural order. Such a sphere would be literally groundless, and as Taylor goes on to argue, "the critique of all 'values' as created cannot but exalt and entrench anthropocentrism."[16] However, as we have noted before, "anthropocentrism" is not quite the right term; for such a value system is centred not on human interests but rather on those of the industrialist order that has so extensively colonized us; so perhaps "technocentrism" would be more appropriate.

Either way, the essential role of *difference* in generating meaning and value is apparent. Just as we fetishize cultures that no longer threaten us, so we "love" a nature that is on the retreat; and this simplistic "love" is therefore a symptom of our alienation from nature. Just as the superficially liberating doctrine of seeing all difference as culturally constructed homogenises humanity into a bland cosmopolitan soup that is the human raw material of industrialism, so the deep ecologists' ambivalence about boundaries and distinctions all too easily falls into denying the structural characteristics of the world they claim to defend. Environmentalism, then, needs to develop a view of the world as containing entities that are both distinct and relationally immersed in the whole. While this may sound like a facile academic "whistling in the dark," I will attempt during the remainder of this book to show how this skeletal basis of a theory can be brought to life in a way that recognizes the existence and priority of the natural order as the basis of our existence. This is not, it should be noted, another form of reductionism, for "natural" in this context is not a synonym for "biological"; and the natural order is also simultaneously spiritual and much else besides that defies human categorization.

A relationality based on a recognition of difference as well as continuity is a precondition for adequate functioning, and an exuberant individuality can flow from an underlying confidence in the integrity of the social and natural worlds. A clear sense of one's boundaries is a necessary condition for effective functioning generally, and for healthy relationships in particular. Individuality, ideally, develops *in conjunction with* relatedness, not *in opposition to* it; and any approach that emphasizes one side of this dialectic while playing down the other will necessarily be unbalanced and unrealistic. As Guisinger and Blatt put it, "individuality . . . and the sense of relatedness to others develop in a transactional, interrelated, and dialectical manner, with higher levels of self-development making possible higher levels of interpersonal relatedness and vice-versa."[17] Overidentification with others, implying a lack of clarity regarding one's boundaries, is likely to be associated with personality disorders of a type which would make one's interactions with others chaotic and immature; while a cool and detached attitude is often symptomatic of schizoid personality disorders, suggesting an impoverished empathic ability and a tendency toward isolation. The need to "merge" with the other can be understood as a desperate attempt to regain contact with the other within a social context that makes real connection almost impossible; just as schizoid distancing and "objectivity" is the attempt to adapt to a world from which relationality has already been lost. But the essential point is that *both* these disorders originate in the lack of a structure that could mediate the relation between self and other.

But a qualification is necessary here. Emphasizing difference will not *by itself* assure the emergence of relationship, but will very likely lead us back into those prejudiced and totalitarian structures that so disastrously influenced the history of the twentieth century. The emphasis on difference needs to be balanced by an equal emphasis on complementarity; that is, the recognition of difference is a necessary but not sufficient first step in the evolution of larger structures that could beneficially integrate humanity and the rest of the natural world. And I have already, in earlier chapters, suggested that culture can provide this sort of "larger structure" that could orchestrate and integrate a community of individuals, both human and nonhuman. Difference and relation, then, together make up a sort of dance that is part of the vitality of the natural world. The need to choose between the isolation that has often been confused with a clear sense of identity, on the one hand, and the diffusion of the self into a cosmic whole, on the other, only exists within a metaphysic that is blind to the need for, and possibility of, integrative cultural structures that could allow both individuality and relation. An unbalanced emphasis on *either* difference or similarity is a denial of the structure that is embodied in the interplay between them.

In this regard, industrialism has fallen into the first of these traps; and as we saw in chapter 3, the emergence of consciousness has historically been associated with the developing sense of autonomy without an equivalent blossoming of integrative forms. As a result, this one-sided autonomy has expressed itself in the attempt to control and master the external world; and our potential for *relation to* others—particularly nonhuman others—has been repressed. If we are to retain a healthy balance between separateness and relation, the autonomy that is associated with consciousness needs to be balanced by and framed within a less conscious relatedness; and this, in turn, implies an expansion of subjectivity away from a narrow identification with existing forms of consciousness, so that we can accept a variety of subjective states as healthy. Only in a world whose basic integrity is endangered will individuality be renounced in favor of a desperate and regressive "merging" with the other. In the realm of nature as in personal life, a clear sense of one's own, separate, identity is a prerequisite for a mature relationship, one that recognizes the difference between oneself and the other while respecting the other. Paradoxically, it is also a prerequisite for letting oneself *temporarily* merge with the other. This paradox, however, is only superficial; since a vital relationship with the other is one that is dynamic, that recognizes the interplay between separation and relation. Like the butterfly and the flower, or the predator and the prey, each participant in a healthy relation will retain their distinctiveness within the dance of the natural world, even as their identities complement each other within a larger frame. A *static* theory, based in a literal language, cannot adequately model this dynamism, and so will inevitably incorporate a dualistic separation-versus-fusion viewpoint. Such a theory wavers, as Naess agrees, between "the ocean of organic and mystic views," and the "abyss of atomic individualism."[18] The assumption that the choice available to us is limited to our location on a continuum between "atomic individualism," on the one hand, and loss of identity in a relatively undifferentiated primal nature, on the other, betrays our ensnarement by a mechanical consciousness which is oblivious to the myriad shapes, forms, and fluctuations of the natural world. Neither the natural world nor the self can be fixed in formalin, or on a photographic plate, and their vitality is reflected by movement, flux, and indeterminacy—qualities that are difficult to model within the constraints of rational theorizing based on singular, static meanings. The necessity is not to abdicate our conscious separateness, but to recognize our grounding in the unconscious, relational, symbolic dimensions of being within which the world is unified;[19] and this requires the development of a subjectivity and a language which recognize that beneath the literalness of our concepts lies a symbolic realm that transcends the artificial boundaries and overused paths of consciousness. In James Hillman's terminology, "by treating the words we use as ambiguities, seeing them again as metaphors, we restore to them their original mystery."[20]

To express what under present social conditions is unconscious, therefore, we need revitalized forms of language in which the sap is still running strongly, in which metaphors of regrowth, diversity, resonance, and rediscovered relation transcend the petrified conventions of accepted expression. This, of course, is something that many environmental writers instinctively know and put into practice, using poetic forms or imagery to convey what is so difficult to express in prose. Warwick Fox, for example, considers, in the space of one page, our relation to the world in terms of "drops in the ocean," "leaves on a tree," "mandalas," "knots in a cosmological net," and "ripples on a tremendous ocean of energy."[21] These are images that have as much in common with dreams as with conventional, waking "reality," and they draw heavily on unconscious representation. Conscious, literal thought can express technical relations effectively—the world of unambiguous concepts, static, definable relations, and mutually exclusive alternatives, operating within a world that is reduced and simplified; but the recovery of a nontechnical relation to the world demands other styles of thought and communication—styles that we stumble upon in our efforts to express the currently inexpressible—multiple simultaneous possibilities, relations that alternate dynamically, imagery inaccessible to rational interpretation.

But the potential of these new styles of expression can only be realized if they give rise to new metaphors and new ways of experiencing that could suggest and promote alternative ways of behaving. The danger is that they will merely be fleeting glimpses whose character is so alien to everyday consciousness that they are blown away by the stream of everyday consciousness like smoke from a campfire. Probably the greatest barrier to new ways of experiencing is our unquestioned acceptance that conventional experience is in some sense "correct," and anything else is at best poetry and at worst pathology. So long as we "know" that the scientific or commonsense views of the world exhaust the possible explanations of the way the world is, so we will automatically reject other experiences of the world. The first step, then, is to open ourselves to the possibility that our understandings of the world and of ourselves are partial understandings, and that we need to take seriously metaphors other than those enshrined within scientific orthodoxy. What is more, we should allow these metaphors to gel and solidify so that we can explore their implications and develop them, letting them define new shapes and forms of relation. I am being deliberately vague here, of course, for one cannot specify in advance what these "shapes and forms" will be, only that they will exist.

This suggests a fundamental alteration in our psychological stance, reversing the trend that Theodore Roszak identifies as industrial society's "diminishing awareness of symbolic resonance."[22] But if it is to amount to anything, then this psychological change has to be complemented by cultural and political ones; otherwise we will be operating only in the rarefied spaces of the Cartesian *cogito*. To have any substance, psyche has to join with the physical world in defining an *integrated* cosmos, implying the need for cultural organization that connects the two. In the absence of these, "Self-realization" is akin to lifting oneself by one's (psychological) bootstraps; for, as Philip Cushman has put it, "experience uncontained in a constructed frame of reference is simply unliveable."[23] New awarenesses cannot survive unless they are supported by structures that can contain and articulate them, since direct contact with the unmediated unconscious is terrifying and overwhelming. Put in terms of an alternative metaphor: those psychological structures that transcend a narrowly egoic sense of self only become stable if they resonate dialectically with cultural structures—and, ultimately, with natural forms in the world "outside." For example, our intuitive recognition of the "intrinsic value" of, say, old-growth forest is quite poorly articulated at the moment, so that we tend to fall back on anthropocentric arguments such as the claim that so-called "trash" species may contain substances of use to humankind. As a result of this difficulty in expressing our intuitive insights, some are now arguing that the idea of intrinsic value should be abandoned—an argument which in effect retreats toward an environmentalism that accepts existing linguistic and philosophical conventions. This, of course, falls a long way short of what we need—which is a way of talking and thinking about the world that effectively expresses our *felt* connection to the world.

If this condition is met, however, then symbolic awareness can challenge the hegemony of the rational, allowing the patterned integration of a diversity of different life forms and affective structures, embodying that systemic character that ecology has at times tried to incorporate, and expressing a resonant interplay between similarity and difference. For example, if a tree is understood simply as a source of building material, then this limited meaning exhausts its significance for me. It is located within a solely rational world, and its meaning is a simple utilitarian one. But if, in addition to such instrumental meanings, I experience a tree, with its roots in the ground and its branches in the sky, *also* as a symbol of the integration of heaven and earth, or conscious and unconscious, or of intellect and feeling, then every time I see a tree I am reminded of the wholeness of the world, and the meaning of "tree" explodes outward from the prisonhouse of its literal meaning. In contrast to anthropocentric approaches that assimilate the natural to the human, such culturally defined meanings inhabit a symbolic space within which subjectivity, language, and biospheric diversity meet, contesting the often implicit dissociation between self and world. What is peculiarly destructive about instrumental views of nature is the freezing of this dynamic interplay into a frame of permanent difference; and there is a danger that environmental theory could become an equally frozen mirror-image of such views, asserting a homogenized vision of similarity, fusion, and equal value.

Such fixed models, even if they emphasize holism and biocentric egalitarianism, cannot accurately represent our participation in the natural world, since they exist primarily within the realm of conscious rationality that defines things uniquely and singularly. A more useful way of modelling this (potential) participation employs the conscious-unconscious distinction, especially that version of it developed within the Jungian tradition; and this allows us to see that the self is both—at a conscious level—largely separate from what is outside the self, as well as—at an unconscious level—rooted in, and connected to the rest of the natural world by a network of symbols and associations. Thus we are not fully defined by consciousness; rather, the consciousness within which we appear as separate is but a small part of a self that extends its roots deep into the world. It is not possible to encompass the wholeness of a self that is simultaneously conscious and separate, on the one hand, and unconscious and inhabiting a world of multiple symbolic relations, on the other, by means of a single definition that attempts uneasily to compromise between these two poles.

Both the psyche and the rest of the natural world are reduced by the conceptual and experiential stilling of the world that results from a narrow adherence to conscious rationality; and each of these reductions reinforces the other in a vicious dialectic. Marooned within a narrow consciousness, we can only relate to what is outside our consciously defended boundaries in simplistic, instrumental ways; for consciousness lacks the multiple resonances of the symbolic world. Deprived of this symbolic world, we turn to consumeristic substitutes, supporting a destructive economic system that assimilates nature to the destructive logic of capitalism. In turn, the economism that results from this assimilation continues to destroy what remains of the symbolic world, further impoverishing the psyche and completing the vicious circle. Our lifestyle can be understood both as a symptom of and as an attempt to compensate for our alienation from nature; and the experiential and physical destructions of the world can be seen to be profoundly enmeshed with each other. Merely technical solutions to environmental problems, then, will be superficial and ineffective, since they attempt to rectify at a *rational* level the destruction that has occurred at a *symbolic* level. It is easier, as Romanyshyn points out, to repair a broken pump than to heal a broken metaphor; and the former cannot substitute for the latter.

The Limits of Consciousness

Freud's view that the development of a technological civilization requires that "the intellect—the scientific spirit, reason . . . establish a dictatorship in the mental life of man"[24] has today become part of the bureaucratic and technological fabric of modern society. This "dictatorship" is explicitly based on the repression of the spiritual, intuitive, and affective faculties; and any movement toward a more complete relation to the natural world is likely to require that we "own" those intuitions, feelings, and awarenesses that are inconsistent with the technological rationality of our time, in a process which David Levin has beautifully referred to as the "body's recollection of being."[25]

However, the progressive eclipse of sympathetic cultural forms by technological and economic structures ensures that we have difficulty in expressing recovered aspects of self, so that they often remain voiceless, shadowy even to ourselves. Our embodied identities require complementary structures in the outside world if they are to be developed and realized: for example, as we argued in the last chapter, our eroticism can only be fully expressed in terms of relation to someone—or, more widely, something—beyond ourselves. And other behaviors such as those involved in growing, gathering, and eating food are, in a well-functioning society, interwoven with cultural structures in ways that make our relation to food much more than the simple satisfaction of a physiological need. The absence or dilapidation of cultural forms, therefore, far from freeing us to express ourselves fully, reduces our opportunities for self-expression to a minimal level. This leads to a restless searching for fulfilment—a predicament poignantly expressed by Edward Abbey:

> One begins to understand why Everett Reuss kept going deeper and deeper into the canyon country, until one day he lost the thread of the labyrinth; why the old-time prospectors, when they did find the common sort of gold, gambled, drank, and whored it away as quickly as possible and returned to the burnt hills and the search. The search for what? They could not have said; neither can I; and would have muttered something about silver, gold, copper—anything as a pretext. And how could they hope to find this treasure which has no name and has never been seen? Hard to say—and yet, when they found it, they could not fail to recognise it.[26]

Such primal experience, when unsupported by appropriate cultural forms, slips quickly through our fingers, leaving only a sense of loss and unrealized possibilities. By default, we tend to adopt whatever available concepts and language come closest to expressing what we feel, and we often lose sight of the discrepancy between such conscious devices and the experiential realities they struggle to convey. As Freud pointed out,

> We are not used to feeling strong affects without their having any ideational content, and therefore, if the content

> is missing, we seize as a substitute upon another content which is in some way or other suitable, much as our police, when they cannot catch the right murderer, arrest a wrong one instead.[27]

We tend to use ecology in this way, in spite of the limitations that I referred to earlier, largely because of its supposed recognition of the wholeness and integration of the natural world. However, if we acknowledge ecology's often mechanistic character, then—as we suggested in chapter 1—this adoption can be understood as reflecting a fantasy of wholeness projected onto a biological science, the term "ecological" becoming a code word for this wholeness.[28] The sort of wholeness actually suggested by ecology, however, is of a quite restricted kind that excludes, for example, the whole realm of subjectivity, and so falls far short of the integration of psyche and nature that we require. The fervor with which the term "ecology" is advocated, therefore, reflects its power as an unconscious pointer for the wider wholeness which we are searching for, so that we are unwittingly drawn toward the *symbolic* reality that is expressed metonymically by an *ecological* understanding. It is time, however, that we consciously recognized both the existence of this symbolic reality and its character as extending well beyond consciousness.

A similar psychological dynamic underlies the adoption of spokespeople for other cultures, whether apocryphal, such as Castaneda's Don Juan,[29] or historical, such as Chief Seattle,[30] who have been represented as speaking the truths that we know at some level, but cannot articulate within available—and academically acceptable—forms of discourse. If Castaneda's conversation with "Don Juan" did not actually take place in the context of anthropological fieldwork, as critics such as Richard De Mille have claimed,[31] this doesn't render it entirely illusory. Rather, the knowledge ascribed to Don Juan can be understood as a representation of the "unthought known," in Bollas' phrase:[32] that is, it expresses patterns and feelings that we know at some level to be valid, but that we are unable to consciously conceptualize or express in conventional language. It is in this spirit that James Hillman has described Castaneda's dialogue with Don Juan as a "creative interrogation of one's soul";[33] and the charges of fraud or dishonesty that have been levelled at Castaneda rest on a literal and psychologically naive interpretation. The widespread popularity of this type of writing demonstrates its resonance with repressed aspects of selfhood, suggesting a need for more politically potent and authentic ways of expressing such elusive truths. Modern physics, too, has been proposed as a candidate for environmental beatification, but usually without any recognition that its implications are symbolic rather than literal. In this respect environmental theory sometimes follows the unfortunate tendency of the humanities, famously identified by Alan Sokal and Jean Bricmont, to adopt the concepts of physical science in a crassly literal way rather than as metaphors.[34]

The difficulties of expressing arational knowledge has practical consequences for environmental debate. Symbolism, while subjectively meaningful, carries little weight in modern political argument, leaving us in the frustrating situation of appearing to be powerless to fight for what we deeply feel to be of value. For example, it is possible that, from a narrowly rational perspective, the loss of a particular species may have few, if any, deleterious consequences, and may even, in some respects, be beneficial.[35] Nevertheless, the right of this species to exist would probably be passionately defended by many environmentalists—a passion which arises from the unconscious recognition that this species' continued existence represents metonymically the quality of wholeness that we struggle to express within a reductive discursive universe. Thus, as Robert Pogue Harrison has argued, "ecological concern goes beyond just the forests insofar as forests have now become metonymies for the earth as a whole."[36] But this is not merely a lyrical way of expressing a flight of the imagination; for the relation of forests to the earth as a whole possesses more, and more material, dimensions than the psychological. It is, in other words, a real relation with practical consequences, even if these consequences are often elusive to conventional scientific methodologies. The separation between subjectivity and materiality is one that consciousness imposes; and their reintegration can only occur in the symbolic realm. The fact that many of the awarenesses recognized by environmentalists are difficult to express in conscious terms suggests not the invalidity of these awarenesses, but the limitations of consciousness.

Recognizing this point enables us to avoid such fruitless recurring questions as whether we should be devoting our attention to species or individuals, with the consequent charge that the ecocentric emphasis on species, for example, implies a covert fascism. This debate is only meaningful within a conscious realm in which language has lost its symbolic overtones and resonances; for if we understand the individual metonymically, it acquires symbolic value as a representative of, a part of, a manifestation of, larger entities such as species and ecosystems which are relatively inaccessible to consciousness. The fact that a species may not be endangered does not mean that individual members of this species are of no value; for, according to the framework developed here, an individual is not *only* an individual, but is also part of larger eco-symbolic structures. This understanding of value does not lead to any simple principles of action which are easy to apply—"I should kill this animal" or "I should not kill this animal": rather, it throws us back on a complex process of valuing that can only generate solutions according to the particular circumstances prevailing in any specific situation. This conclusion, however, needs qualifying: consciousness, and the rational principles that consciousness can articulate, are of some use as "first approximations," as general guides to action. We should be ready, however, to modify the conclusions they generate by an awareness that transcends rationality; and in this respect cultural structures are essential in maintaining continuity between conscious and unconscious, as Erich Neumann lucidly argues:

> The world of symbols forms a bridge between a consciousness struggling to emancipate and systematise itself, and the collective unconscious with its transpersonal contents. So long as the world exists and continues to operate through the various rituals, cults, myths, religions, and art, it prevents these two realms from falling apart, because, owing to the effects of the symbol, one side of the psychic system continually influences the other and sets up a dialectical relationship between them.[37]

For example, the meaning of the term "water" varies according to whether we experience it in a constricted, rational way, or whether we allow ourselves to be aware of its symbolic overtones. Such symbolic alternatives to the literal understanding of water, whether as rain, river, floods, tears, sea, fountain, spring, storm, or well, are unconsciously present in all of us whether or not they are articulated by the cultural traditions available to us. The Christian baptism, for example, originally involved immersion in the flowing, living water of a river or a spring, the power of the flowing current symbolizing the spiritual flow of life within the person's own body.[38] Today, we are more likely to experience a dab of water from a font—an apt if unintentional expression of the stilling of the waters of life within much contemporary religion as well as in the modern world as a whole. Environmentalists' traditional dislike of dams may in part reflect our awareness of their contribution to the stagnation of life in ways that are symbolic as well as literal, even if we find it difficult to articulate this symbolic awareness within any existing cultural form. It would be a mistake to understand such symbolic experience as referring only to an ethereal psychological realm; for the narrowing of awareness toward an industrialist rationality is often, as in this case, continuous with an ecological narrowing that has a clear material dimension, and conversely, the natural world is incomplete in the absence of its symbolic dimensions. As Paul Shepard expresses it:

> The history of mythology is rich in signs of plants' affinity for evoking aspects of one's inner life. One need only remember the radiant mandala effect of the rose, the lotus, or the cross-section of a tree trunk; King Solomon's "garden enclosed"; the visionary, gemlike, preternatural luminescence of flowers as doorways to "another world"; the syllogisms of symbolic fruits and seeds; and the trees of life and of the knowledge of good and evil. We are inclined by modern culture to see these references as literary or artistic devices. This huge body of symbolic allusion has been tainted by . . . the assumption that natural appearances are merely the raw matrix of creative analogies in art. Poetic reference to rootedness, flowering, and fruitfulness are misunderstood as arbitrary rather than essential processes.[39]

This continuity between conscious and unconscious, between individuality and embeddedness in the world, that culture can embody can be seen within a range of societies, both past and present. Among Athenians in the fifth century BC, for example, any attempt to enlarge one's worldly powers and boundaries was seen as a act of hubris, inevitably arousing the envy of the gods and inescapable nemesis;[40] and so "environmental ethics" resided in the balance between a conscious knowing experienced as a nascent individuality and an unconscious awareness which was given form through mythology. In the modern world, however, our one-sidedly conscious orientation diminishes our ability to articulate the truths that such myths embodied; and so we are left merely with residual feelings of unease over the asset-stripping of the natural world. Unable to give form to such awareness, we can only see it as *outside* ourselves, and so we project it elsewhere. Under current conditions, as Luigi Zoja puts it: " The notion of limits belongs to unconscious myth, and unconscious myths make themselves manifest through a process of projection: limits thus come to be projected onto the scientific data."[41]

Of course, science *does* eventually discover the limits, often by belatedly assessing the effects of their transgression, as in the case of atmospheric ozone depletion or the Bovine Spongiform Encephalitis crisis that plagued British beef farming in the 1990s. To take the latter case as an example, problems such as BSE are unlikely to arise in peoples such as the Bella Coola, who sing a song to call the wolf to one of their kills—a bear. They would take the bear's hide, but believed that bears did not wish to be eaten by humans.[42] Such culturally based taboos, which may appear to the "rational" mind as mere superstitions, maintain a symbolic order which is continuous with the ecological order, even if their pragmatic effects are untraceable by consciousness. A sound environmental ethic, grounded in workable cultural structures, is preventative rather than palliative, giving us a positive model of healthy living rather than the sort of trial-and-error learning that stumbles from one crisis to the next. As Zoja argues, science lends us precision; but also has a way of "imperceptibly cutting us off from what we know, . . . In addition to suppressing myth, [scientific] knowledge also did away with the psychological glue that holds the whole of human experience together."[43] If science is often *reductionist*, culture is potentially *integrative*; and these two modes of knowing need to exist in balance, so that conscious intelligence and symbolic awareness exist within an integrated cosmos in which the particular constantly reminds us of the whole, and the whole finds expression through the particular.

Within the humanities and the social sciences, as we saw earlier, our preoccupation with human consciousness manifests itself in the currently fashionable "turn to language," which holds that the realities we inhabit are primarily *discursive* ones—most extremely, that there is "nothing outside the text."[44] However, loss of biodiversity, climatic changes, and the spread of pollutants possess a reality which is, unfortunately, not merely rhetorical; and our difficulty in articulating environmental awareness indicates not the invalidity of this awareness, but rather the hegemony of a technological/discursive system that represses by omission, while claiming completeness, and that blinds us to the need for those cultural forms through which we could articulate our felt loss. Environmental problems can be understood as symptoms that attempt to remind us of what *is* "outside" the texts of consciousness and language. However, as Roszak has noted, an absence is more difficult to recognize than a presence; so while consciousness can record the symptoms of environmental degradation, it has difficulty in recognizing that forgetfulness of the symbolic realm which gives rise to a range of environmental and psychological problems. We need to acknowledge that long-term solutions cannot be found within the technological/discursive system whose basis is the alienation of self from environment, and that the restoration of a healthy natural world is impossible if we fail to restore the symbolic foundations of this world.

Reinhabiting the Symbolic World

We will also need to recognize that at a symbolic level, self and culture are continuous; although this is not, as we have already seen, to deny the separateness of self from context that exists at a conscious level. As Geertz explains this point, "becoming human is becoming individual, and we become individual under the guidance of cultural patterns . . . in terms of which we give form, order, point, and direction to our lives."[45] Only in the industrialized world do we assess our individuality in terms of *non-conformity to* cultural mythologies and traditions, insisting, as Robert Bellah and his colleagues have emphasized, "on finding [our] true selves independent of any cultural or social influence, being responsible to that self alone, and making its fulfilment the very meaning of [our] lives"[46]—an insistence that is central in weakening the self, making it vulnerable to commercial pressures and narcissistic fashions, and assuring that our relation to the rest of the natural world, by default, will be one of uncomprehending exploitation. The form taken by modern individualism involves the withdrawal of subjectivity from culture; and Stanley Diamond has contrasted this with the "primitive individuation" found in aboriginal cultures, which he defines as "the full and manifold participation of individuals in nature and society."[47] Ideally, myth, ritual, and religious belief can express subjectivity so that it resonates with rather than opposes natural structure, contradicting the repressive assumption, prevalent in the industrialized world, that what cannot be expressed within consciousness and articulated through rational discourse is necessarily invalid and unreal.

Until we develop such structures, we will be prone to interpreting environmental "correctness" in a *negative* way, with the emphasis on avoiding activities that are destructive, rather than interacting with the natural world in a full and aware manner—a guilty withdrawal that tends toward an asymptote of not living. This retreat from participation in the world is understandable given that current social, demographic, and economic conditions often shape our archetypal human inclinations toward forms of expression such as consumerism that are destructive. This situation is exacerbated—crucially—by the population pressures which cause potentially benign activities such as hiking, felling trees to build or heat our homes, and most forms of transport to appear less and less acceptable. It also accords with the spirit of an age that encourages a privatistic withdrawal into the self and a bland emotional equanimity, in which maturity is often equated with adjustment to conditions that are experienced as unchangeable. We are, as Christopher Lasch puts it, "like animals whose instincts have withered in captivity . . . [who] long precisely for a more vigorous instinctual existence."[48] For such caged animals, almost any behavior except withdrawn passivity and a resigned acceptance of the invisible bars that constrain our natural predispositions will appear as, and often will actually be, antisocial or destructive. Lasch's description of the individual in modern industrial society is precisely accurate: "Outwardly bland, submissive, and sociable, they seethe with an inner anger for which a dense, overpopulated, bureaucratic society can devise few legitimate outlets."[49] The long-term answer to this situation, however, in addition to involving a frank recognition of the need for population reduction, is not a quietistic withdrawal from engagement with the world, but rather the envisioning of frameworks that would allow more positive and constructive expressions of our nature.

In a healthy culture, the wildness within us resonates with parallel aspects of the rest of the natural world. But the development of a conscious realm which perceives itself as actively *mastering* nature—that is, both internal and external nature—sets consciousness in opposition to the unconscious, although our identification with consciousness usually ensures that this split is misleadingly posed as one between "humanity" and "nature." Under these circumstances, consciousness is no longer a flexible means of focusing momentarily on whatever circumstances demand, but instead becomes a rigid structure that attempts to dominate and defend itself against the natural order. The war between these two systems—the techno-economic rationality that has colonized consciousness, and the symbolic rationality of the natural order—is the defining tension of modernity; and the terrain over which it is fought encompasses both the human personality and the nonhuman world. In this situation, whatever characteristics and relations are denied by conscious rationality are repressed, and those that remain become frozen into what we understand as "reality." When this occurs, the world lapses into apparent lifelessness and amorphousness, approximating the mechanisms posited by post-Cartesian science; and the alliance between conscious and unconscious is severed. The destruction of the natural world is thus underpinned and paralleled by a covert, psychocultural destruction in which the multiple metaphors of symbolic process are replaced by the reified categories of the language of rationality. The "middle ground" of culture, which could align us with the natural world, has been allowed to dissolve through our identification with consciousness. Consciousness is powerless to rectify this situation, since it is, by definition, unaware of what is unconscious; and so environmental theory finds itself attempting to comprehend environmental destruction, and to suggest solutions, through the terminology and concepts that themselves arise out of our alienation from the earth.

However, if theory grows in the direction of acknowledging and incorporating those symbolic, metaphoric dimensions of experience that inhabit a more meditative, trancelike state of mind, then we may be able to move toward a more accurate empathy with the natural world. In other words, if we embody viscerally as well as acknowledge intellectually the psychoanalytic metaphor of depth, so that we conceptualize and experience the self in broadly Jungian terms as a continuum from a separate and flexibly selective consciousness through levels that become progressively less conscious, toward a realm in which individual identity gives way to the integrity of the natural world, then we can both retain a sense of personal identity and be aware of our rootedness in the earth. In these terms, our loss of relation to nature is not a necessary consequence of the separateness of the conscious self, but is rather the result of our forgetting the other layers of selfhood—those that are part of a symbolic natural community, and in which the idea of "separateness" tends toward meaninglessness. Instrumental attitudes and practices, to the extent that they have become hegemonic, rely on the reification and freezing of our subjectivity within current forms of consciousness, and on the corresponding loss and pathologization of other, more integrative styles of subjectivity; and from this perspective, technological and rational capability become problematic only if they are elevated above, and so displace, those spiritual and cultural ways of knowing which embody a relatedness that is symbolic rather than rational. But this displacement has already happened in the industrialized world, according to Jung:

> Today, for instance, we talk of "matter." We describe its physical properties. We conduct laboratory experiments to demonstrate some of its aspects. But the word "matter" remains a dry, inhuman, and purely intellectual concept, without any psychic significance for us. How different was the former image of matter—the Great Mother—that could encompass and express the profound emotional meaning of mother earth. In the same way, what was the spirit is now identified with the intellect and thus ceases to be the Father of All. It has degenerated to the limited ego-thoughts of man. . . .
>
> As scientific understanding has grown, so our world has become dehumanised. Man feels himself isolated in the cosmos, because he is no longer involved in nature and has lost his emotional "unconscious identity" with natural phenomena. These have slowly lost their symbolic implications. Thunder is no longer the voice of an angry god, nor is lightning his avenging missile. No river contains a spirit, no tree is the life principle of a man, no snake the embodiment of wisdom, no mountain cave the home of a great demon. No voices now speak to man from stones, plants, and animals, nor does he speak to them believing they can hear. His contact with nature has gone, and with it has gone the profound emotional energy that this symbolic connection supplied.[50]

In the absence of this "symbolic connection," our exclusive identification with a rationality which is abstracted from the world rather than expressive of it causes order outside egoic boundaries to seem incoherent, so that—for example—the natural world appears chaotic, lacking in inherent order or intelligence; and those aspects of self that are not conscious appear "primitive" and "uncivilized," as in the Freudian id. In either case, the "wild" region appears to require ordering and controlling, so seeming to justify the interventionist and controlling approach characteristic of technological society. If, however, we are prepared to recognise, with Eugene Gendlin, that nonegoic experience may be "*too complex* to fit the forms of reason and experience," and to question "the assumption that order is always something imposed,"[51] then we realize that what is not understandable in rational terms may nevertheless contain an orderliness that makes *symbolic* sense. As Jung argues:

> What we call complicated or even wonderful is not at all wonderful for nature, but quite ordinary. We always tend to project onto things our own difficulties of understanding and to call them complicated, when in reality they are very simple and know nothing of our intellectual problems.[52]

Thus, ecosystemic functioning may seem complex when viewed through the lenses of conventional logic and discourse; but it is the everyday stuff of life within natural systems whose intelligence we deny. Cultural form can clarify these otherwise inaccessible aspects of nature, as in the case of the Lakota, whose "very thoughts were drawn from the land he called native, and the winds that blew over its soil, the rivers that ran through it, and the mountain peaks that drew his gaze upward [and] coloured his consciousness with their subtle influence."[53] The sort of "environmental ethic" which this implies is not one that has to be imposed as a sort of moral code on a reluctant personality, but rather is one that arises out of a basic alignment between the self and the natural world.

Theory which fails to recognize and embody this unconscious, culturally expressed alignment with the natural world, even if well-intentioned, will inevitably further the domestication of nature. As John Rodman points out, measures such as "saving" a species by breeding the few survivors in captivity, or "saving" the San Bernadino National Forest by replacing trees killed by smog with a smog-resistant variety; or issuing wilderness permits specifying where the backpacker intends to spend each night may be "environmentally sound" in some respects; but they are also powerful expressions of an impulse that is anything but wild.[54] A reliance on consciousness ensures that dissociations within the psyche come to be incorporated into our environmental "solutions." Even "wilderness preservation," while unquestionably necessary—and I emphasize this—as a stopgap measure, may not be above suspicion if it is adopted as a longer-term end in itself. As the film *Sophie's Choice* illustrated, there are occasions when the act of choice itself enslaves us within a totalitarian rationality, and the refusal of the choice may be the only possibility open to us that does not betray a wholeness to which rationality is blind. The partitioning of the world into "wilderness" and "developed areas" may result in a world that contains "wilderness"; but it will not be a "wild" world. We must hold on to the long-term aim of reintegrating humanity fully into nature, even when current social, political, and demographic conditions make the realization entirely impractical at the moment. If we lose sight of this ultimate aim, then we are accepting some of the most basic assumptions of industrialism—for example, that humanity is *necessarily* destructive to nature, and that nature is something that is *outside* ourselves. In contrast, a theory which insists on and expresses the membership of humanity within the symbolic community of nature, so furthering the resonance between our own wildness and that of the rest of the world, will reinforce the wholeness of nature, and so will avoid extending those splits between what is "wild" and what is "civilized" that unwittingly express a technocratic system of categorization.

Jung recognised the archetypes of the collective unconscious as "the hidden foundations of the conscious mind . . . the roots which the psyche has sunk . . . in the earth."[55] Furthermore, "[a]ll the mythologised processes of nature, such as summer and winter, the phases of the moon, the rainy season, and so forth, are in no sense mere allegories of these objective occurrences; rather they are symbolic expressions of the inner, unconscious drama of the psyche."[56] Such insights recall our commonality with "wildness" wherever it may occur, recognising that a separation between the "wild" and the "not-wild" assimilates both to a world that is anthropocentrically ordered and literally conceptualized. The fundamental tenet of any radical environmentalism must be that *the natural frames and includes the human*, not vice-versa. An environmentally healthy world cannot be one in which wildness becomes merely a human category; but rather must be one in which we are fully integrated *within* wildness, understood as the "forest of symbols"[57] which is the fundamental integrative matrix of the world. It is the clear-cutting of this symbolic "forest," invisible to consciousness though it may be, which underlies the destruction both of the natural world and of our own integrity; and superficially distinct psychological, social, and environmental problems can all be seen to be rooted in a common ontological predicament.

The destruction of the physical reality of nature is thus the final, and most visible, act of a long historical drama, the psychological and cultural foundations of which were laid over many centuries. The search for an environmental ethic based on the conscious intention to evaluate, order, control—even, perhaps, to "save"—may therefore be misguided, since it already *assumes* the experiential order which emerges from this drama, along with the social controls that maintain it. In other words, whatever rational consciousness may be "saving" about a world that is in many ways beyond its comprehension is likely to lack precisely those qualities that make it "wild," and the natural world thus "saved" will end up looking like a cross between Kew Gardens and a Safari Park, a simulacrum superficially similar to, but profoundly different from, original nature. In this situation, environmental theory will have converged with social constructionism in identifying nature with our conscious representations of it. At the same time, the wild aspects of the self will fall silent, so that there will be a specious harmony between a reduced self and a reduced nature; and the vision of a world fully colonized by technique will have been realized.

Reuniting the Symbolic and the Material

As should by now be clear, our potential interwovenness with the natural world cannot be realized within the confines of a consciousness made sick by its colonization by industrialist rationality. Terms such as "biodiversity" or "ecosystem" are probably as close as consciousness can get to a recognition of the qualities of the natural world; and notions such as "wildness" test the limits of what consciousness is capable of, pointing to a realm beyond its boundaries without really specifying the character of this realm. Just as a concerto is not equivalent to the theme picked out one note at a time on a piano, so the "marker species" and crude measurements we make of ecological health give only a superficial picture of the full spectrum of resonances and interrelations which make up the natural world. As Jack Turner puts it:

> If I have an interest in preservation, it is in preserving the power of presence—of landscape, art, flora, and fauna. It is more complicated than merely preserving habitat and species, and one might suppose that it is something that could be added on later, after we successfully preserve biodiversity, say. But no, it's the other way around: the loss of aura and presence is the main reason we are losing so much of the natural world.[58]

Among the most significant harmonics that are lost are those that resonate between the human psyche and the rest of the world; and without these, the world becomes dissociated into "human" and "nonhuman" fragments. Consequently, myths such as Sophocles' Oedipus story, which Freud, in keeping with the literalizing spirit of his, and our, times saw as expressing individual psychological structure, can also be seen as conveying analogous patterns in the world outside the individual psyche. The marriage of Oedipus and Jocasta resonates with the perverted relation between humanity and the natural world, a marriage that in reappropriating the symbolic realm to conscious intentions demonstrates a forgetfulness of the necessary ontological boundaries between the literal and the symbolic. The perverted understanding of the natural order *primarily* as "raw material" drags it into the literal realm of instrumentalism, forgetting the maternal, nurturant, and symbolic qualities which are implied in the etymology of the word "material."[59] We were begotten not merely by our biological mothers, but also by that maternal world that is the revered *matrix* of life, and not simply the sum of its chemical constituents. Just as a mother is not merely a machine for providing milk, so the world is more than a source of "raw materials." The act of appropriation of the symbolic to the literal therefore redefines our relation to the natural in a violently reductive way, and this redefinition provides the underlying pattern for such apparently diverse situations as a colonial conquest, a scientific experiment, the damming of a river, or a conventional psychoanalytic interpretation. It is also a patricide that represses the sort of deep moral awareness that the world embodies a spiritual code—an awareness that is clearly expressed, for example, in the Ashanti myth that "a human being is formed from the blood of the mother and the spirit of the father," so that "man is both a biological and a spiritual being."[60] We blind ourselves to what we have done, reducing awareness so that it is consistent with the world we are producing in an effort to escape the depression, horror, and guilt that would be the natural consequences of this historical forgetfulness. The denial of the symbolic realm, in transforming the world into a loose assortment of "things," is therefore an essential first step for industrialism, which can only relate to the world in terms of fragmented pieces ready to be commodified and assimilated. Freud fell prey to the literalism which is the signature of our betrayal of the symbolic realm, accepting the Oedipal imagery his patients produced as if it were *simply* the literal expression of repressed sexual desires. Freud's story of patricide and incest is, therefore, itself a literalization of a more profound symbolic truth; and this literalization reinforces the manufactured dissociation between subjectivity and our natural context, trimming symbolism to fit an autonomous "human" realm, and cutting us off from the natural roots of our own experiencing. The recovery of these roots, conversely, requires us to allow myths such as the Oedipus story to speak to us, once again, symbolically—a change that is both psychological and epistemological.

Freud's tendency to literalize mythological symbolism protects rationalism's monopoly on meaning, aligning psychoanalysis with science's preference for unequivocal meanings and definitions—a "single vision" whose limitations it is essential to recognize if we are not to descend into scientism. Just as Blake saw, not a sunrise, but "an Innumerable Company of the Heavenly Host crying, 'Holy, Holy, Holy is the Lord God Almighty,' " so we must learn to see the landscape in more than geographical terms and to feel its spiritual and emotional resonances. Science is pragmatically useful; but if its meanings are taken to be the only meanings there are, then the question of what is hidden by these meanings becomes inadmissible. Science's environmental culpability is one of omission rather than commission, since the environmental crisis arises precisely in our forgetting of the natural structures which science occludes. The scientistic vision—and the likely end-point of humanity—is an existence in which only technical meanings are allowed; and the world falls into a kind of coma, animated only by the life-support machinery of industrialism.[61]

But if science develops a sense of its own limitations, its own partiality, then it can learn to live in peace with other structures, recognizing that scientific truths need to be framed within more general truths which are of a more than merely rational character, and which possess moral, spiritual, and intuitive dimensions. Within this larger frame, the more profound insights of science begin to overlap with moral and spiritual insights, so that, as Gregory Bateson put it toward the end of his life, sin can be understood as involving "certain kinds of epistemological error."[62] Science reaches out at its edges toward those fundamental ontological truths that should frame it; and conversely, recognizing our past denial of these truths, and the de-struction of the world implied by this denial, may be as close as we can come to identifying our fundamental moral error. This coalescence of epistemology and morality characterizes such larger, integrative frameworks as the Australian aboriginal, Navaho, and Andean cultural systems that we discussed in chapter 4; and within such frameworks, subjectivity spills over into the rest of the natural world.

Our inability to sense this "wild" order is a direct result of our fetish of rationality, which denigrates wildness as threatening to "civilized" consciousness. If the structure of nature is *symbolic* as well as logical, then rationality is revealed as blind to much of the natural order. "To the scientific mind," Jung argues, "such phenomena as symbolic ideas are a nuisance, because they cannot be formulated in a way that is satisfactory to intellect and logic."[63] It follows that if we are to recognize the intelligence of the natural world—as opposed to imposing a conscious rationality on it—then we have to become familiar with the subversive languages of the unconscious, and the subjective states which this requires. If we are prepared to recognize, with Gregory Bateson, that "[scientific] language depends on nouns, which seem to refer to things, while biological communication concerns pattern and relationship," and that the relation between species is more a metaphorical than a taxonomic one,[64] then this recognition suggests an alteration in consciousness and an awareness of the disjunction between language and the natural order. For example, Roderick Nash notes the difficulties involved in using the noun "wilderness" to point to something more elusive to language:

> "Wilderness" has a deceptive concreteness at first glance. The difficulty is that while the word is a noun it acts like an adjective. There is no specific material object that is wilderness. The term designates a quality (as the "-ness" suggests) that produces a certain mood or feeling in a given individual.[65]

Nature's structure is, in a healthy world, that of the unconscious, invoking multiple meanings, ambiguity, and symbolism. As Gary Snyder modifies Thoreau's famous dictum: "wildness is not just the preservation of the world, it *is* the world."[66] The technical order is a reduced version—a special case—of the natural order; just as consciousness is a special case of the unconscious. As Bateson says, "metaphor is not just pretty poetry, [it] is the logic upon which the biological world has been built."[67] Animals, thus, communicate metaphorically, as is illustrated by one of Bateson's examples:

> [W]olves . . . go out hunting and then come home and regurgitate their food to share with the puppies who weren't along on the hunt. And the puppies can signal the adults to regurgitate. But eventually the adult wolves wean the babies from the regurgitated food by pressing down with their jaws on the backs of the babies' necks. . . . [T]he previous year one of the junior males had succeeded in mounting a female. Up rushed the lead male—the alpha animal—but instead of mayhem all that happened was that the leader pressed the head of the junior male down to the ground in the same way once, twice, four times, and then walked off. The communication that occurred was metaphoric: "You puppy, you!"[68]

Bateson argues, more generally, that natural processes do not follow the laws of logic so much as symbolic relations such as syllogism. Take, for example, the syllogism:

Grass dies;
Men die;
Therefore, men are grass.

From a logical point of view, the conclusion that "men are grass" is clearly "incorrect"; indeed, it has been taken as diagnostic of schizophrenic thought disorder. Within a "logical" framework, men and grass are entirely distinct; humans are 'separate' from the natural order; and the metaphoric relations which knit the world together are denied. But, as Bateson points out, to completely deny the validity of such syllogisms "would be silly because these syllogisms are the very stuff of which natural history is made."[69] Furthermore, they are, as Chapman and Chapman have pointed out, "reality oriented and adaptive."[70] To say that "men are grass" is not just meaningless nonsense, because it expresses something important about our mortality and our place within the natural community. Given this, it is hardly surprising that syllogistic reasoning has survival value. Take, for example, the syllogism

Some fruit are berries
Some fruit are poisonous
Therefore, some berries are poisonous

The conclusion "some berries are poisonous" is logically invalid, but is nevertheless quite likely to be correct. Denying the ecological awareness embodied in such syllogisms may ensure one's survival in a mathematics department, but heaven help the mathematician should he get lost in a wilderness area. Industrialism functions by fragmenting the world into separate categories; and these categories are linked through a logical system which selects certain specific properties of these categories, and so expresses only partially the possible forms of relation that exist in the natural world. For example, an area of forest may be conceptualized as a mixture of "timber" and "scrub"; but this is a conceptualization which denies validity to a myriad of alternative understandings and structures that are the context of life for a diversity of native flora and fauna. Symbolic relationships such as that of syllogism express relationships which are metaphorical and probabilistic, and which *suggest* possible conclusions rather than positing them with certainty. This lack of certainty does not render symbolism useless, but merely difficult for formal systems of logic to handle; and this difficulty is used extensively in industrial society to delegitimate it. Symbolic relationships such as syllogism are the connective tissue of the world, the patterns which define our membership in the natural community. As Bateson puts it, "[t]he whole of animal behaviour, the whole of repetitive anatomy, and the whole of biological evolution—each of these realms is within itself linked together by syllogisms . . . whether the logicians like it or not."[71] Small wonder, then, that the denial by industrialism of the natural order rests upon its denial of the "logic" which expresses natural patterns and rhythms; and that the cultures which take these patterns seriously, integrating them into their own psychological, spiritual, and social realities experience industrialism not only as pragmatically objectionable, but also as evil.

To state this is only to express more-or-less formally what many environmentalists already know "in their bones." Although we are constrained to speak in conscious, rational terms, much of the motivation of environmentalists comes from a preconscious recognition that species such as the Northern Spotted Owl are symbols of an elusive reality whose disappearance we sense but have difficulty in articulating. Similarly, the term "ecology," as we noted earlier, has become an indicator of a wholeness and integration that is difficult to articulate consciously. Environmentalists are not merely fighting for the survival of wilderness in the world "outside"; we are also fighting for the survival of patterns and relations which suffuse and connect psyche and nature. The danger is that we limit our understanding of terms such as "biodiversity" or "ecology" to their literal, biological definitions, forgetting that such terms are passports to the forbidden terrain of the symbolic, keys that can open doors to what lies beyond industrialism, the visible tips of unconscious conceptual icebergs. For at a more-than-conscious level, men (and women, too) are not only grass: we are also the spotted owls, the grizzly bears, and all those as yet unrecorded species which we unknowingly trample underfoot; and the habitats of these syllogistic kinfolk are our homes in a more profound sense than are the square brick buildings we actually live in.

To the extent that we embody this symbolic way of being, our identities are more than merely egoic; they resonate with other living and nonliving aspects of the world, in ways that rationality finds impossible to explain. We are talking here of more than the psychoanalytic concept of "identification," for it is not just a matter of possessing some properties in common with something "outside" ourselves. For example, I have never felt particularly "at home" in the English Midlands, where I have lived for much of my life. I experience it as sometimes bleak and always overcrowded, and often find myself recoiling from its inhospitable character (or so I perceive it). In contrast, I often feel most completely myself when I am in one of several areas in the American Southwest, a part of the world that was unknown to me until my early twenties and where I lived for less than two years. I have no idea why this should be; and so far as I am aware, there is no rational explanation for it. Clearly, I am not "identifying" with this area of the natural world which is not, in the usual sense of the word, a living creature like myself; and obviously it does not physically resemble me. And yet there is a bodily resonance which was immediately apparent to me, and which has remained ever since. It is one of the most difficult things in the world to accept that this type of relation cannot be rationally "explained" or articulated, but can nevertheless be part of our wider bodily awareness. It is a matter of *complementarity* rather than *similarity*, just as an animal can be said to "fit" its environment. The animal is not *like* its habitat, neither does it *resemble* it, nor does it *identify* with it: rather, they fit together, defining a wholeness which rationality finds difficult to handle, but which is one of those elusive but crucial properties that environmentalists instinctively defend. This complementarity does not imply an aggrandizement of the ego, or an enlargement of its boundaries, and it suggests a reinhabiting of the world rather than a use of it to feed one's egoic structure.

Consider the following scenario. I move to a new area, and am immediately excited by its possibilities. My world of desires and hopes spills outward onto the landscape, reconfiguring it according to my ego's vision. I *assimilate* the landscape to my egoic desires in much the same way as Columbus "claimed" the New World for his European monarchy.

However, after a time, I become more familiar with this place: I become more attuned to its rhythms, noticing the coming and going of its nonhuman inhabitants and recognizing the way it changes with the seasons and the time of day or night. Then I begin to realize that I am actually enjoying these characteristics, feeling less need to impose my own structures on them, or to change them. I become receptive to an order and a structure which I perceive as outside me, but recognize at the same time that it affects me, awakens something in me, complements me, and so is not *just* outside me. Rather, I feel myself extending into it—not as conqueror, but through my acceptance of its character. This second attitude does not demand an assertive "intelligence" which looks for ways to assimilate the landscape; rather, it can only occur if I relax my usual conscious controls, letting subjectivity be structured by what is outside myself.

Occasionally, we find our colonialist assumptions upset in a more active manner, reminding us that the world is not equivalent to our cognitive representations of it. On a large scale, environmental problems such as climate change contradict our aspirations to "control" nature and remind us of the incompleteness of a rational explanation of the world;

but such contradictions can also occur in more personal and immediate ways, as the wildlife photographer Joel Sartore realized when photographing wolves in Yellowstone National Park. Sartore found that the wolves seldom allowed him to get close enough to photograph them, and that they would get up and walk away however silently he approached them. One evening, however, Sartore hiked toward a pack near an elk carcass, expecting that the animals would walk away as they had previously done. As he relates, however,

> One of the wolves looked at me and got up. The others got up and barked and growled at me. One started to gallop towards me . . . and that's when I looked back and saw that my rental vehicle was a dot in the distance. . . .
>
> It was a pack that had been released without an adult. They knew they didn't like people, but didn't have an adult around to teach them to avoid people. . . . The pack circled me for a few minutes, about 20 yards away, just thinking about it, I guess . . . I started walking for the road. It was pitch dark by then, and although I couldn't see them, I could hear them panting and whimpering behind me. They followed me all the way back. I was so shaken that I could barely get my key in the car door's lock.[72]

But this example understates the extent of our predicament; for in the industrialized world, our assumptions are often widely accepted and underly global practices, and may therefore ultimately be fatal to civilizations rather than to individuals. Furthermore, although such individual experiences may bring us face to face with a natural order that undeniably transcends our cognitive structures, our awareness of this order will need to be embodied in cultural forms such as mythologies, spiritual practices, or rituals which give them permanence and consensual recognition. The task we face, therefore, amounts to a regeneration of culture.

Culture and Symbolism

In a culture that favors the individualistic explanation and experience of subjectivity, it is not surprising that the poverty of our relationship to what is "outside" us has often come to be felt and theorized as an *inner* emptiness. Christopher Lasch, for example, refers to "the void within"; and Paul Zweig admits to the "suspicion of personal emptiness."[73] This metaphor of "emptiness" is not only an expression of our lack of relation to the world, but also an ideological camouflage for it, exemplifying the way we are encouraged to experience the loss of social and natural form as *personal* inadequacy. As Philip Cushman succinctly describes it, "our terrain has shaped a self that experiences a significant absence of community, tradition, and shared meaning. It experiences these social absences and their consequences 'interiorly' as a lack of personal conviction and worth, and it embodies the absences as a chronic, undifferentiated emotional hunger."[74] This fateful deception, which substitutes an inner emptiness for an outer absence, captures us within the spiral of alienation from the world, since the dominant metaphor of "emptiness" ensures that we seek "the experience of being continually filled up by consuming goods, calories, experiences, . . . romantic partners, and empathic therapists in an attempt to combat the growing alienation and fragmentation."[75] Attempting to compensate for our alienation from the world, therefore, we actually increase that alienation by chaining ourselves more firmly into an economistic order.

But to imply that inner emptiness is simply a disguised representation of an outer bleakness is too simple, for this maintains the separateness of inner and outer as alternatives, and disguises the extent to which inner and outer are dialectically constructed. If the outer world is bleak and unreflective of our needs, then our relationality will have nowhere to go; and this will cause the turning inward we have described as narcissistic, so reinforcing the dissociation between inner and outer. In the short term, then, the structure of self will adapt itself to outer reality; while in the longer term, we are physically constructing an outer reality which accords with our inner vision of the world. There is, therefore, a dialectic between a self which is narcissistic and a world which is in the process of becoming mechanical: each complements and reinforces the other, so that our empirical experience of the world increasingly appears to confirm its "reality." Under these circumstances, it is important to hang on to the idea that not only is the world not necessarily the sort of carpentered, constructed, mechanical world which accords so comfortably with consciousness, but also that our apparent separation from the world is less inevitable and "natural" than it seems.

We find an illustration of this dialectic between historical change and psychological reality in Romanyshyn's account of the emergence of our current understanding of the human heart. Earlier meanings attached to the heart are preserved in the words we still use to describe emotional experience—*heartfelt, warm-hearted, big-hearted, heartrending, heartbroken, downhearted,* and so on. It is also connected etymologically to the notions of *cordial, concord, discord,* and *courage,* and, through the latin term *credere,* to believe, to ideas of *belief.*[76] These terms suggest the strong emotional connotations of the word *heart,* as well as its intrinsic relationality; but although they remain as metaphorical figures of speech in our everyday conversation, they have lost their rootedness in a believable reality, becoming mere flowery adornments to an accepted scientific understanding, the shards of an extinct belief-system. Emotionality has become dis-embodied, poorly articulated, and lacking the argumentative clout accorded to rational argument.

Several years ago I was listening to a Canadian government official defending the annual seal cull then taking place in the Arctic areas of that country. What struck me particularly was one sentence: "The problem is that people get so *emotional* when they think about a man taking a seal pup." Here we have the power of language to conceal, to peddle ideology, and to deny those aspects of selfhood which are inconsistent with the comfortable, business-as-usual assumptions of most citizens of the developed world. Feelings, lacking a discourse that could effectively articulate them, seem inferior to rational calculations (how many fish a seal eats during the course of its life; how the population has grown; etc). Laundered language ("taking") is used to deprive the situation of its physical and emotional reality, to reduce it to pure quantity, calculation. We are not in an integrated world of mind and feeling, complete with feelings, intuition, and heartfelt reactions; rather we are in Descartes' realm of the pure intellect. Small wonder that existential psychiatrists such as Laing have referred to "an unbelievable devastation of our experience."

The fading of the passionate, emotional heart occurred over many centuries. A significant milestone, however, was the publication in 1628 of Harvey's *An Anatomical Disquisition on the Motion of the Heart and Blood in Animals*, which described the heart's role in the circulation of the blood. After 1628, the heart-as-pump was seen not merely as one metaphor among many, but as a reality, or, rather, *the* reality—a view which, as medical science has demonstrated since, is extremely powerful. As a result of this change, the literal understanding of the heart-as-pump does not merely supplement alternative understandings—for example, of the heart as the organ of courage—but rather drives them out. "[T]he pumping heart [is] regarded as real, factual, empirical, and/or literal," says Romanyshyn, "while the courageous heart [is] ignored as unreal, fictional, psychological, and/or metaphorical."[77] And this mechanical conception of the heart-as-pump exemplifies in microcosm a more general shift: the world, including our own bodies, has become a predominantly *mechanical* world, explicable according to the sciences of biology, chemistry, physics, medicine, and so on. The heart was a pivotal notion in relating inner and outer, connecting the world of feeling to the "external" world beyond our physical boundaries; and conversely, the mechanization of the heart is an important step in the isolation of a largely rational self from a largely mechanical world.

But even a powerful model is not necessarily a complete model; and we need to question the fate of those feelings and qualities which the heart-as-pump cannot contain or express. Harvey suggests that the natural state of the heart is *empty* rather than *full*, an emptiness, Romanyshyn argues, that parallels a corresponding and more general emptiness in the human relation to the world. It is no coincidence that Harvey was writing in the same era in which the "method of doubt" became the dominant force in philosophy—a method which rejected feeling, intuition, and sensory experience as bases of knowing, while emphasizing the primacy of the intellect. A form of selfhood emerges, then, that complements an emerging world—a primarily intellectual, rational self that relates only through understanding to a world conceived as mechanism.

However, to accept the self as *entirely* constituted by the epistemological fashions of the day would be to repeat the constructionist error of accepting the disembodied self as an inevitable reality, and it is important not to forget what is omitted by this dialectic which co-constructs the thinking self and the mechanical world. What has happened to the *emotional* reality which was previously expressed through ideas related to the heart? And is it surprising, if this emotional reality has become more difficult to express within a world understood as mechanical, that we experience a certain emptiness? The shift from a feeling heart to a mechanical one is also a shift from an integrated world to one which is fragmented, which fails to resonate with our relationality; for a heart which can feel is only possible within a world which is alive, and an unfelt world is one we cannot relate to passionately. These conditions provoke an increasingly desperate search for sufficient sense of relation to the world to enable us to feel real; for, in Romanyshyn's terms, "the exuberance which characterises an outward, vigorous movement toward the world betrays a quiet inner despair, the loneliness of an empty heart."[78] It is symptomatic of the current lack of integrative cultural structures that this search itself is fragmented, expressing itself, for example, in a proliferation of new religious cults and in diverse forms of sexual expression. Symptoms such as this express both the human yearning for an integration that is spiritual and sensual as well as intellectual, and the inability of industrialist "culture" to meet this need. Despite this inability, the patterns available to us are often those of industrialism; and as a result, "new" spiritual, cultural, or environmental movements are often infected with narcissism and spiritual materialism, becoming vehicles for aggrandizing the ego[79] and reproducing industrialist assumptions. Thus a drive toward wholeness, realised within a cultural context which is fundamentally at odds with it, results in repeated, and repeatedly perverted, attempts at cultural regeneration; and we experience a procession of "radical" social, religious, and personal styles, each of which is quickly assimilated by the commercial system before it can establish a serious challenge to that system.

But if we can resist the temptation to fill the emptiness with the too-easy solutions of industrialism, then we may find that it is not so much an emptiness, but a space from which new sources of meaning can emerge. For a hint as to how this might occur, we need to return to Romanyshyn's tale of the human heart, which, you will recall, was defined by Harvey as "empty." Romanyshyn relates how a vision of the Sacred Heart of Christ appeared in 1673 to a sister Margaret as she knelt before the altar of the convent chapel:

> One day, when I was before the Blessed Sacrament, and having at the time more leisure than usual, I felt myself wholly invested with the presence of God. Thus I lost all thought of myself, and the place where I was . . . surrendering my heart to the power of His love. My sovereign Master granted me repose for a long time upon His divine breast, where he uncovered to me the marvels of His love, and the inexplicable secrets of His Sacred Heart, which He had hitherto concealed from me. He opened for me for the first time His divine Heart.

Romanyshyn comments:

> These words describe the opening of another heart. But this heart which opens in 1673 is not the same heart which opened in 1628. It is, on the contrary, the heart which has been forgotten with the appearance of the pumping heart. . . . It is a heart which speaks with the fullness of love. The other heart, which appeared in 1628, is an empty heart . . . in the seventeenth century the fullness of the heart begins to matter when the emptiness of the human heart becomes a theme. The appearance of this heart which is filled with love is an acknowledgement that one cannot live with an empty heart, with a heart without belief.[80]

In an age rooted in the maturity of Descartes' method of radical doubt, it is not surprising that we express what Rilke referred to as our "unlived lines" precisely through those embodied senses which Descartes, and modern "rationality," reject; nor is it coincidental that such moments of dawning awareness as that described by Romanyshyn often occur in situations where denial is most extreme, so that what is omitted clamors for attention. Thus, in the well-known passage in which Aldo Leopold describes watching the death of a she-wolf he has just shot, precipitating the beginnings of his conversion from a view of forest management inspired by Gifford Pinchot toward what later became the "land ethic," his awareness is not so much of positive learning but rather the awareness of an *absence*. "I realised then," he writes, "and have known ever since, that there was something new to me in those eyes, something known only to her and to the mountain."[81] In seeking to fill this absence, to become aware of what previously was known only to the wolf and the mountain, we first have to recognize the empty space within us, and the ignorance that lies at the heart of our ways of knowing. "Those unable to decipher the hidden meaning know nevertheless that it is there," states Leopold, hinting that an environmental sensibility may be found not so much in rational thought, but in a recognition of the limits of rationality, and an awareness of what it discards.

My argument here converges with Romand Coles' discussion of the writings of Adorno and Lopez. In *Of Wolves and Men*,[82] Lopez details at some length what we know about the wolf; but equally, he emphasises the significance of what we *cannot* understand, as Coles illustrates in his use of quotations from Lopez:

> "The truth is we know little about the wolf." "No one . . . knows why the wolves do what they do." "In a word, not enough is known." "Wolves are wolves, not men." "We know painfully little about wolves. We can only ask questions and guess." Even more profound than his explicit gestures toward nonidentity are the constellations of questions he raises about the wolf for which he has no definitive answers. What happens when the eyes of wolves meet the eyes of their potential prey? Lopez depicts this event not primarily as an answer to the question—though he toys with suggestions—but rather as an event that poses a question that probably exceeds any answer we can ever give. He presents the bits of the world that he can grasp in such a way that they not only unfold a rich knowledge of identities, but—in an *interrogative mode of appearance*—gesture infinitely beyond their identities for us to an inexhaustible surplus of nonidentity.[83]

This "inexhaustible surplus of nonidentity" is what I have referred to as the "otherness" of the natural world—that unconceptualized wild space within which the wolf mainly exists. Restricting ourselves to rationality will simply expand the realm of rationality, driving into deeper obscurity what it cannot encompass. What is obscured therefore becomes increasingly inexpressible except in the form of inarticulate feelings and mute symptoms whose sources are unclear. And this is where the work of Lopez diverges from that of Adorno; for while Lopez recognizes that the natural order can be a basis, however indirect, for healthy living, Adorno perceives beyond the boundaries of the rational only a bleak landscape of disorder, superstition and regression to the primitive. Cut off from the natural, and denying validity to cultural forms that could articulate it, Adorno's critique of modernity has nowhere to go.

But while the feelings and symptoms that are inexpressible by rationality are potential sources of relation to the world, as Roszak has emphasized,[84] there is a danger that they are allowed to remain simply as feelings and symptoms. No doubt a new breed of "ecological psychoanalyst" would make a comfortable living interpreting them in an "environmentally aware" manner. Freudian psychoanalysis should be taken as a cautionary tale in this respect; for as we have seen, psychoanalysis has largely remained a procedure for assimilating unconscious material to consciousness, rather as an angler drags a fish out of the water. While consciousness and rationality have useful and necessary roles to play, let us also recognise that the fish which lies on the merchant's slab with a piece of parsley in its mouth is not the same fish which swims in the shallows of the river, and that the latter is not reducible to the former. Nor, for that matter, is the river the same river when the fish has gone; for the unconscious, too, is distorted by the polarization of conscious and unconscious. The symbolic world represents the essential matrix out of which the rationality we value so highly is abstracted; and attempting to "understand" the symbolic world, with its vitality, diversity, and wildness, in rational terms is grossly to misconstrue its fundamental character. What we *can* do, however, is to open ourselves to this symbolic realm, and to allow our embodied selves to resonate to its rhythms, a possibility explored in the final chapter.

Chapter 7

HEALING THE WORLD OF WOUNDS: RESTORING AN ECOLOGICAL SUBJECTIVITY

> There is a silence needed here before a person enters the bordered world birds inhabit, so we stop and compose ourselves before entering their doors. . . . The most difficult task . . . is that we learn to be equal to them, to feel our way into an intelligence that is different to our own.
>
> —Linda Hogan, *Dwellings: A Spiritual History of the Living World*

Integrating Visions

Cultural structures can resonate with and support the less permanent shapes of individual experience; and as we learn to live within culture, so subjectivity transcends its purely individual expression, integrating us within a world which outlives us and which extends ecologically beyond individual consciousness. But culture does not only shape and articulate experience: it is also, over many generations, shaped *by* it, providing a constructive interplay between the spontaneity of moment-to-moment involvement in life and the sedimented knowledge of previous generations, so locating our individual lives within less fleeting time-structures. An analogy here would be the river bed, which is shaped by water over centuries, but also channels water in the present in a way that often maximizes the diversity of life, reflecting the apparently intrinsic tendency of life to evolve toward more sophisticated, varied, and differentiated forms. The evolution of difference is an essential part of this evolutionary process, and as an ecologically simpler landscape evolves toward the more complex systems allowed by dry land, water, and riparian zones, possibilities arise as a result of these differences. Just as the *Alcedininae* family of kingfishers, for example, depend on all three of these zones, so naturally occurring differences and continuities such as day, twilight, and night are the central metaphors that structure subjectivity, ensuring its consistency with the world outside. This is, of course, a somewhat idealistic view of culture which ignores its possible, and frequent, perversion, ossification, and colonization by principles such as those of industrialism. But even if cultural practice inevitably falls short of its highest ideals, it is important to strive toward and maintain the vision embodied in these ideals if culture is not to degenerate into merely pragmatic relations that forget the origins of morality in an order beyond the human. A healthy culture, then, is one which, although guiding us toward some articulations of experience rather than others, embodies a diversity of ecological and subjective possibilities. It is an indication of the integrative power of a healthy culture that difference becomes the basis of creative diversity rather than conflict, something which is celebrated rather than feared.

If culture can integrate difference, it is also true that a lack of difference stills the cultural imagination. As Octavio Paz expresses it, "[w]hat sets worlds in motion is the interplay of differences, their attractions and repulsions. Life is plurality, death is uniformity. By suppressing differences and peculiarities, by eliminating different civilisations and cultures, progress weakens life. . . . The ideal of a single civilisation for everyone, implicit in the cult of progress and technique, impoverishes and mutilates us. Every view of the world that becomes extinct, every culture that disappears, diminishes a possibility of life."[1] In this book I have argued that in a healthy world, this cultural dialectic will also be a natural dialectic. Consequently, just as the "monocultures of the mind" referred to by Vandana Shiva allow the destruction of nature, so an impoverished nature stills our cultural imagination—a problem that is particularly apparent in "overdeveloped" countries such as Britain, where most native large mammals are already extinct and where wilderness has virtually disappeared. In this situation, environmental activism necessarily expresses itself through opposition to the building of more motorways, or to the destruction of remaining woodlands, hedges, wetlands, and so on; but having little experience of a truly natural environment to refer to, there is often an acute sense of puzzlement among environmentalists about the ultimate aims of such activism. In this sort of situation, it is harder to envision a healthy world than would be the case, for example, were one defending an area of ancient forest in the western United States, since the coalescence of psychological and "external" realities to form an integrated world occurs more readily in the latter case. Here, the continuity between personal emancipation and ecological health emerges clearly, since the relational extension of self into the world is essential for both.

A positive vision of a future natural world, therefore, also contributes to our own psychological survival. Joanna Macy[2] has written of the despair frequently experienced by environmental activists—a despair that is entirely authentic and understandable given the speed with which the natural order is being demolished, and our apparent powerlessness when faced with overwhelmingly powerful interests. But while protesting against the destructiveness of affluence may be an essential *part* of environmental activism, it needs to be complemented by a clear vision of our long-term goals. As Erik Erikson pointed out, one cannot adequately define one's sense of identity by what one is *not*.[3] While promoting a positive vision *integrates* and *heals* one, devoting one's life purely to opposing what one is *against* is an act of nihilism in which the culmination of one's own success is self-destruction. The greatest environmentalists, however, from Muir to Watson, have also been visionaries; and what I say here is in no way intended to demean the actions of those courageous enough to take direct action against environmental vandalism, risking—and sometimes suffering—injury, imprisonment, or death. They are the saints and heroes of the natural world, and will one day be recognized as such. The power of their vision, however, has not been widely matched within an environmental movement noted for its fragmentation and its suspicion of structure; and the need for an integrating vision is overwhelming.

Recovering the Ecology of Experience

The recovery of an "ecological" integrity, as I argued earlier, involves the recollection of a resonance between self and world that redefines both and that is more-than-conscious; and this process is reminiscent of some of the more radical variants of psychotherapy. Most psychotherapeutic practice, like Freud's psychoanalysis, can be described as exegetic: that is, its purpose is to assimilate regions which are beyond the boundaries of consciousness to conscious understanding. Such approaches are of no use to radical environmental theory, since they reproduce key characteristics of colonialism.

There are psychotherapeutic traditions, however, that implicitly challenge the exegetic principle, in that they place the significance of experience above that of any particular interpretative scheme. Foremost among these is the client-centered tradition of Carl Rogers, which we looked at briefly in chapter 2. This may not be a tradition that normally lays any great claim to radicalism, since its theoretical framework is deceptively conventional; but its potential lies in its openness to the client's experience and the way it tries to avoid imposing any selective or interpretative structure on this experience. If "organismic experience," as Rogers terms the sense of *bodily* knowing which seems to come from outside intellectual understanding, embodies the radical potential that I have claimed for it, then a method which remains true to this experience will itself possess radical potential. The "fully functioning" person, according to Rogers, is one who is "congruent"—that is, one in whom conscious awareness is harmoniously integrated with this organismic experiencing. The task of therapy, then, is the recovery of congruence through the realignment of consciousness with those deeper layers of the self that are present as bodily awareness, and that embody "the pattern, the underlying order, which exists in the ceaseless flow of . . . experience." The real self, according to Rogers, "is something which is comfortably discovered in one's experiences, not something imposed upon it."[4] What exists outside the imposed order of rationality, then, is not *dis*ordered, but ordered in an alternative way, implying that we need to be able to refocus awareness if we are to recover our resonance with natural structures.

As we saw in chapter 2, Rogers' approach fails to realize its real potential because the experience that is recovered is viewed as essentially *individual* and unconnected with any order outside the self. Although fully functioning persons, according to Rogers, "feel a closeness to, and a caring for, elemental nature . . . and . . . get their pleasure from an alliance with the forces of nature, rather than in the conquest of nature,"[5] and are also open to "the sensory and visceral experiencing which is characteristic of the whole animal kingdom,"[6] there is no recognition in Rogers' work that these experiences may not come simply from within the individual, but might transcend the boundary between individuality and the world outside. In this respect, Rogers is true to such existential predecessors as Kierkegaard, implicitly rejecting any structure outside the self as a source of order.

As a result of this sort of assumption, the individual is separated from any potentially coherent and evolving cultural context, and each generation repeats the same painful processes of developing incongruence in childhood, and—perhaps—attempting to recover congruence by means of therapy or other forms of "personal growth." Existentialism therefore repressively incorporates one of industrialism's most important dissociations: that which alienates the person from natural and cultural forms, and so makes us easy prey for the industrialist structures that appear to us as the only recognizable sources of coherent order. Like a scratched gramophone record, modern humanity keeps making the same mistakes, repeating the same experiences with each generation, while a near-hegemonic industrialism develops its power and reach with each year that passes. So long as congruence has to be recovered anew through individual effort by every child, unassisted by an evolving cultural framework, our ability to challenge industrialism will be slight. Our task, then, is clear: to articulate the insights, recovered relations, and reestablished congruence that we have individually realized in a form that is transferable to the next generation, developing a cultural ecology that is capable of effectively challenging industrialism. This cultural ecology, in turn, can provide the basis for the unifying vision which environmentalism so badly needs.

Others have developed the potential that remains latent in Rogers' work. In particular, Eugene Gendlin has explored the extension of subjectivity into previously uncharted territory in his process of "focusing"—a process that arises from the awareness of an unclear "edge," a "sense" of more than one says or knows. Change, says Gendlin, comes from "an unclear, fuzzy, murky 'something there,' an odd sort of direct datum of awareness."[7] This process is one which relies heavily on bodily sensing, and it tends to emerge, according to Gendlin, within the middle of the body—"throat, chest, stomach, or abdomen." While change can arise from "the felt sense of reliving the past," "what matters most for [change] is precisely the new implicit complexity of bodily living."

This style of sensing is inherently holistic, and challenges the experiential taxonomies of rationality. "Whether one attends to a whole situation or to some tiny aspect of it, the bodily felt sense of it will be a whole. This sounds contradictory, I know. But the bodily sensing of the smallest aspect of anything is an implicitly complex whole." Here Gendlin is pointing to the metonymic qualities of symbolic experiencing, the way something can be simultaneously a specific *part* of the whole and the whole itself. This wholeness, he says, "is a characteristic of the felt sense." What "the whole" is remains undefined, and consciously undefinable; but it is nevertheless something we can sense and react to. This is a point which is highly significant for environmental theory, for it suggests that even as consciousness focuses on specifics, at another level we can be aware of the holistic qualities of a situation. And most importantly, Gendlin argues, we allow experience to assert its own form, rather than imposing preexisting, "rational" structures:

> The inward process we are specifying involves keeping quiet and sensing the unease in the body directly, wholly,

> as it comes, without putting one's maps, cuts, distinctions on it . . . Then let that *unease* make the map, let *that* sort itself into whatever parts or pieces it falls into on its own.[8]

Applying this to the environmental problematic, the "unease" we sense may involve the "depression," "anxiety," and so on which stem from our awareness of what is happening to that potential part of us which also exists in the natural world. It "comes between the conscious person and the deep, universal reaches of human nature where we are no longer ourselves." There is a sense here of discovering an additional dimension of selfhood that needs to be allowed the space to speak. "It turns out that the deliberately speaking client to whom we relate is not the one to whom our responses are chiefly addressed! Rather, we hope the speaking one will take our responses down to consult that other one, the felt sense. We hope the client will let that one speak, will wait for what comes from it, will work to find words that 'resonate' with it, rather than interrupting, lecturing, or interpreting it."

Here, we are discovering an unsuspected coalescence of subjectivity: one for which the distinction between "inner" and "outer" is put to one side, which is somehow larger and more inclusive than the conscious self, and which supplements "normal" conscious awareness. Furthermore, in contrast to the defensive assumption that "letting go" of the egoic self is to open ourselves to disorder and chaos, we may discover another, unsuspected order: "Our felt sense may at first seem less sophisticated than our reasoning. If we receive and resonate that initial contact, however, soon what comes is more intricate and more correct than anything we could *think*. . . . We learn that what comes from the felt sense has its own logic and its own good reasons, even if these are not immediately apparent." Nature, says Gendlin, "is vastly more organised than [conceptual thought]."

The "felt sense," then, is not "corrected" by more conscious layers of experience, but is instead allowed to determine its own evolving form. "For example, some little thing went wrong today. We tell ourselves 'It's all right. . . . It doesn't matter. . . . Soon I will have forgotten it. . . . Mature people don't get all upset about such trivia. . . ' and so on. Each of these responses is contradicted by the discomfort that 'talks back' and vividly corrects our attempts to think it away." Here, Gendlin is adopting a very different approach to that of many therapists who induce a sense of order by locating subjectivity more firmly within the currently dominant system of rational *thought*. In this respect, he is reversing the usual order of things whereby we take human concepts of intellect, rationality, and civilization as our starting point. Instead, he begins with nature itself as we experience it within us. There is a parallel here with Edward Casey's geographical metaphor: "Is the natural world really something we *edge towards*? The very idea of edging *out* from built places into the wild world beyond presumes the primacy of a humanocentric starting point . . . [But] what if nature is the true a priori; that which was there first, that from which we come, that which sustains us even as we cultivate and construct?"[9] This is a truly challenging idea for the modern person: that instead of applying an order produced by thought *on to* the world "outside," we open ourselves to the natural order, allowing resonances to develop which define our commonality with that order.

Consistently with the spirit of Casey's remarks, the process of "focusing" uses words when and if they seem to "fit" the felt sense, but rejects them when they don't. "In focusing, when a felt sense arises, one concentrates on its *quality*, and tries to find a *handle-word* for that quality. Just trying for a word helps one stay with a felt sense as a bodily sensation, rather than going into the familiar feelings and thoughts of the problem. Is it 'jumpy' or more like 'heavy'? Is it 'flat' or perhaps 'crowded' or 'pushed back' . . . ? If nothing fits, call it *that quality*." The client in this process needs to check the therapist's words "not against what they said or thought, but against some inner being, place, datum . . . 'the felt sense'; we have no ordinary word for *that*." *Experience*, in other words, takes precedence over the *language* through which we struggle to articulate experience.

Gendlin's focusing approach is an important bridgehead into those repressed realms beyond everyday consciousness. Environmental theory needs to reach out in a similar way, recognizing that the words and concepts that are available to us distance us from the natural world. Like Gendlin, we need to be ready to reject words when they compromise our experience of the natural order, and to remain with the ambiguity when this ambiguity is closer to our felt sense of the world than a more literal "correctness." Here, the work of David Grove in arguing for the use of "clean language" is important. Language is "clean" when it does not carry ideological baggage—at least, not more than is unavoidable. For example, if a person has just watched while their favorite wood has been bulldozed to make way for a bypass, the questions we might ask them are important in determining whether that person maps their experience onto a conventional epistemological frame, or whether it can be articulated in alternative ways. "What went through your mind while it was happening?" implies that the person "has" a "mind," and that what occurred was in this "mind" rather than in, say, their abdomen, or in a space which includes both their abdomen and the wood. Given that the split between intellectual structures and bodily experience is, as we have seen, of fundamental significance for industrialist epistemology, such a question strongly orients the person away from the natural order and their bodily awareness of it. "Were you upset?" implies that any reaction was in the past, dissociating them from what they are experiencing in a present and continuing way, and prioritizing a conventional temporal scheme in which event A follows event B. Clean language has a minimum of compromised metaphors that might draw the person into the orbit of a particular ideology. Rather, feeling has "its own epistemology, a something else-ness, possibly in the stomach."[10]

A cleaner question, in the situation of the person viewing the bulldozed wood, might be: "What's happening?" This question does not locate the "happening" within the individual or within their mind or, indeed, in the wood; but allows the location to remain indeterminate, perhaps involving the *relation* between these elements. The question does not assume the separateness of the person from the wood, but allows their response to reflect whatever degree of relation might exist. For example, the person might answer: "It felt as though part of me was being chopped down." Clean language therefore allows alternative epistemological structures to emerge out of bodily awarenesses—battered, bruised, intimidated though these awarenesses might be. Of course, in an era which lacks the rituals for mourning such losses, the internality of the feeling tends to be dissociated from the physicality of the act of destruction; but by avoiding the linguistic conventions that cement such dissociations, we open the doors to other ways of construing such events.

Experience has for so long been assumed to be a purely individual matter that rediscovering its relational possibilities requires a major effort of imagination. However, if we allow experience to find its own form, then the "wholeness" to which Gendlin refers need not relate only to the felt wholeness of a congruent self, but also to other possible "wholes," such as the wholeness of a self-in-context, a self-which-is-part-of-the-world. Words, of course, fail us here, as we attempt to communicate what modern Eurocentric language and consciousness implicitly deny. Part of our task is the development of a new language: one which, as R. D. Laing playfully put it, will be as "unambiguously paradoxical as is feasible so as not to be disastrously understandable";[11] for only in such ways can we avoid the seductive and illusory clarity of purely conscious formulations, and so point to what lies beyond language.

Now, if nature "within" is continuous with nature "outside," then the "felt sense" of wholeness reported by Gendlin's clients should also be stimulated through direct contact with aspects of the natural world which resonate with our own deep structures—something which a number of early social scientists such as William James noted before psychology retreated into the experimental laboratory. This, of course, many of us know from our own experience, although articulating this knowledge through conventional communicational means is far from easy. As Frederick Turner describes this sort of experiential awakening:

> In some truly unimproved natural settings—one well removed from the reach, the sights, and maybe especially the sounds of our wonted culture—surrounded by the immemorial phenomenal world, whether trees, ocean, or the waves of prairie grasses, a change may overtake us, precisely to the extent that we are willing to remain where we are and resist what will be a gathering temptation to return to more certain comforts. It will not quite be fear, but it will be next to this: a kind of existential humility born of a sense of all the life that surrounds and includes us and that will go on without us. And this is the ground of myth—fear or humility and submission to the still unfathomed mystery of life.[12]

The change to which Turner refers can be understood as involving the shift from an exclusively egoic stance in which self is separate from context, toward a subjectivity which is also part of nature as a whole rather than being the exclusive property of the human brain. If, as Gregory Bateson argued, "*any* ongoing ensemble of events and objects which has the appropriate complexity of causal circuits and the appropriate energy relations will . . . show mental characteristics,"[13] then complex ecological systems are likely to possess a degree of subjectivity; and if humans are incorporated within such systems through the mediation of cultural organization, then this subjectivity will necessarily be more sophisticated than that of the isolated human individual. As this normally dormant style of subjectivity begins to stir, experiences that previously would have seemed incomprehensible and fleeting can begin to take on a more definite and positive significance. Fragments begin to integrate into recognizable wholes; dreams and fantasies assume a certain significance; and political and social issues relocate themselves within a broader, natural frame, as subjectivity expands to include the natural order. For example, Paul Watson, founder of the Sea Shepherd Society, was a volunteer medic at the occupation of Wounded Knee in 1973 by members of the Oglala Sioux; and during an initiation ceremony in which he was honored as a warrior of the tribe, he had a vision in which

> I saw a buffalo struck by an arrow which had a string attached to it, and I was a wolf, and I ran down the hunters of the buffalo. The whale is to the ocean what the buffalo is to the plains, and the arrow was significant

> to the harpoon. What I took from that vision was that I should do everything in my power to protect the whales and other wild life.[14]

Epistemological dead-ends may generate this sort of visionary leap if we resist the temptation to cover them up with rationalistic "solutions," instead opening ourselves to creative and radical alternatives. There is a considerable literature, in fact, which suggests that such dead-ends, together with the despair that often accompanies them, are *essential* prerequisites to any radical paradigm shift.[15] Periods of personal or societal turmoil, in other words, may be the epistemological rapids we necessarily travel through before we can reach the uncharted water beyond, allowing us to engage with understandings previously out of reach. A year after the Wounded Knee episode, Watson was taking part in the first Greenpeace campaign against whaling. While the protesters were in a small inflatable between a Russian whaler and its prey, the harpooner fired over their heads, striking a female sperm whale. Watson describes what happened:

> The bull came full out of the water and dived, and we'd been told that the whale would attack us, and we were waiting with a lot of anxiety for the whale to do just that, when I turned because the ocean erupted behind me, and the sperm whale had thrown himself up out of the water straight at the harpooner in the bows of the Russian vessel. And they fired a harpoon that was not attached, and it exploded and the whale fell back and died, and as it was screaming and rolling in the water, blood everywhere, I looked up . . . into an eye the size of my fist, and what I saw there was understanding—the whale understood what we were trying to do, because he could very easily have come forward and crushed us or seized us in his jaws and killed us. Instead what he did was slowly slide back into the water—eye-to-eye contact all the way—and went beneath the waves and died. What I also saw in that eye was pity—pity not for himself or his kind, but pity for us that we would be able to commit such an abominable act.[16]

This is a type of experience drastically different from the conventional one in which whaling is viewed with the alienated eye of one who is immersed in and colonized by industrialism, protected from the bloody reality of the situation by words such as "harvesting," assessments of "sustainable yields," and other abstractions. On the contrary, this sort of direct experience recognises a world that is fully personalized—full of spirit, feeling, intelligence, relation; a world that in its diversity and fascination shames the tendency of the human ego to categorize or explain.

But the insights and awarenesses which arise out of our contact with the wild world—including that part of it which is within us—will only be able to challenge the industrial order and to provide an alternative to it if they are collectively integrated within a coherent mythology/epistemology/psychology. Approaches such as Gendlin's are an excellent start in this direction; but if they produce only "individual" insights and "personal" growth, then their potential to initiate radical political and environmental change will have been wasted. The repression of the wholeness of the world is not simply a personal tragedy, to be dealt with by each of us privately; rather, its very universality betrays its character as a lasting feature of the industrialist landscape, including us but also reverberating beyond us. In this respect, environmental theory can take the lead by illuminating industrialism's denial of our relationality, and by pointing toward the cultural forms which could fulfil our relationality, insisting on it as a defining element in the wholeness of the world.

Resonance and the Natural World

An adequate model of our relation to the rest of the natural world will avoid the sort of reductionism which can focus only on one level of coherence, such as that which imprisons subjectivity within the individual human mind. Rather, we need to release subjectivity into the world in a way that suggests what Geertz calls an "epistemological empathy," not merely as academic stance but also as ontological reality. I have pointed, in a deliberately undefined way, to this involvement with nature by means of the term "resonance." The notion of resonance—I shy away from referring to it as a "concept," with all the ideological baggage which that implies—is not one which exists simplistically within the conscious world of rational concepts, laws, and debate; but neither is it a purely symbolic reference, for in this case it would have little conscious meaning. In contrast to either of these alternatives, I have converged on the term "resonance" as a notion that exasperatingly but, I hope, fruitfully, has a foot in both camps. On the one hand, it begins to give shape to those shadowy forms that exist in unconscious nature, making them more accessible to consciousness and to conventional forms of language; while on the other hand, it draws us down into those depths beneath the surface level of rational consciousness, problematizing its assumptions, and allowing subjectivity to diffuse into realms wherein the world is neither separate from nor opposed to the individual. Its purpose, then, like that of any metaphor, is that linkage of conscious and unconscious which is an essential first step in the project of reestablishing an integrated reality. At the same time, an integrated reality need not be one in which the parts are permanently fused together so that they lose their distinctiveness; and the notion of resonance is designed to give form to the relation between the individual and the external world while recognizing that they are, at some levels, distinct entities.

There will be those who argue that I should define the term unambiguously and specify its relations to other ideas with which it seems to share something in common, such as complementarity, understanding, empathy, and so on. To do so, however, would be to sever connections with what is beyond consciousness, and with all those aspects of the natural world that cannot be contained within a rational perspective. If the reader has got this far, then probably he or she will recognize that I am resisting this approach in order to avoid anchoring my argument within the established domain of conventional debate that achieves intellectual clarity by deadening our relation to the natural world. Defining a term unambiguously, through selected connections to other established concepts, is to ground it within an ecologically and ontologically impoverished reality—a process analogous to pinning a butterfly into one's collection. In both cases, we capture something—but the original reality we hoped to grasp has somehow disappeared, converting the flitting, the unpredictable, the dynamic, into something measurable but without vitality. Or, to borrow another metaphor, the phenomenon that emerges from a particular interaction of light, rain, and vision is referred to as a "rainbow," the implication being that it is a "thing" with particular, and presumably measurable, properties. However, when we try to isolate this "thing" from its context (which includes us), it disappears! Of course, there are aspects of the natural world that don't evaporate when isolated from context; but to equate nature with the assorted debris which results from this process of fragmentation and reduction is rather like describing a symphony as a collection of musical notes.

Perhaps the most important characteristic of "resonance" is that it *recognizes and respects the structure of the other*, so that the wholeness which evolves out of the joint resonance of self and other incorporates the individual structures of both or all such components. In auditory terms, two things—say, a cello string and the cello sound box—can resonate to the same note, or harmonically related notes, only if they complement each other, often in a way that is not obvious. Therein lies their power of relation. The cello, the score, and the cellist, too, are very different in many obvious ways; but when they interact in a mutually respectful manner, the resulting music is unique and not predictable by analyzing any one of them separately. I glimpse a similar possibility when I experience a beautiful landscape: there is a subjective resonance which includes myself and the landscape, and will not occur in the absence of either. This contrasts with a type of relation that I have referred to as "colonialist," in which one, egoic, structure overwhelms and subdues another; and in which I impose my own vision on it, responding only minimally to its present form. Resonance, then, implies an openness to and complementarity with the other, a willingness to let go, at least temporarily, of individual defenses and presuppositions, implying a sort of confidence in the world, an acceptance that the world is basically "good." These are characteristics that define self and land in a mutual, interactive way, expressing "a moral universe . . . in which the structures of kinship reach out to all living men, to all his fellow creatures, and to the rivers, the rocks and the trees."[17]

The notion of resonance offers us an alternative way of envisioning our superficially paradoxical separateness-from and relatedness-to nature. It is not difficult to conceive a range of resonances, some of which indicate an intra-individual integration, reflecting the dynamic interaction between parts of the person; and other larger-scale ones that could occur between the person and structures larger than the individual, or simply among these larger structures. Resonance, in other words, can be understood as expressing that coordination, joint functioning, dynamic interrelation which defines the systemic character of life—a quality which is not well communicated by our understanding of the world as made up of individual entities, connected only by physical "laws."

Furthermore, a resonant world is one in which ecology is continuous with subjectivity, drawing together the conventionally separated realms of materiality and spirit; and this is a world which many environmental writers reach out toward. Gary Nabhan, for example, referring to the dwindling population of bighorn sheep in the Sonoran Desert, worries that "as wild sheep slip out of sight, then out of mind, then out of dreams, a vacuum is created not only among desert people but among all people." Later, lamenting the absence of pollinators which could maintain a healthy population of Kearney's Blue Stars, a rare desert flower, he remarks that "[i]t was as if the Blue Stars' bodies were there, but their spirits had flown away."[18] Such remarks imply a world suffused by subjectivity, in which spirit and matter are fully integrated.

Resonance also has something in common with systems theoretical notions. Systems theory, however, while it usefully articulates the holistic character of sets of interacting units, is less successful when it comes to articulating the ways in which the structure and character of individual elements influence the extent and manner of interaction. In other words, its focus tends to be fixated at the level of the system, and it articulates only poorly the dynamic relations between this level and other levels, such as that of the individual. What is it, for example, that makes the Californian condor so well fitted to its habitat? What are the particular characteristics of "wildness" that strike such a powerful chord with experience? And returning to Arthur Kleinman's concern, noted in chapter 1, for "the defining human element in individuals—their moral, aesthetic, and religious experience," how is this "defining element" related to our cultural and natural environment? Conventional science, at best, offers partial answers to each of these questions; and the notion of resonance can begin to fill the gap between such scientific explanations and our felt experience, a gap that involves a subjective dimension unknown to science. Our answers to this sort of question need to recognize that the entities we refer to as condors, individuals, "wildness," or "cultural environment" possess certain characteristics which constrain the types of relation they may have with other entities; but also that these relations are not entirely reducible to these characteristics. To put this another way, the notion of resonance expresses the subjectively obvious fact that while we are partly what we are even when we don't interact with anything beyond our boundaries, we can also be much more than isolated individuals in those situations which involve interaction with an other—whether this other is another person, a landscape, an idea or whatever. We therefore realize ourselves most fully in relation to the other; and when significant others are absent from our lives, we adopt the "default" position of the isolated individual—a position which, as we saw in chapter 2, mainstream psychology generally accepts as the only possible one. This psychological definition of individuality, then, is not so much *wrong* as grossly incomplete in denying that the self-structure can transcend that of the isolated individual as it resonates with structures beyond its physical boundaries.

To illustrate these characteristics by means of an analogy, consider my daughter's relation to her cello. The music that results from this relation is not a property *either* of her *or* her cello or, indeed, of their sum: it emerges, rather, out of an interaction between cello, cellist, and score which is more than material, and cannot be fully understood in material terms. The combination daughter + cello + score allows emotional structures that are otherwise latent to achieve some sort of physical, material reality in the form of sound waves; and these, in turn, can arouse similarly latent structures in those who listen to her music. Subjectivity, then, has become articulate, integrated within cultural and physical structures; and my daughter has extended herself into these structures in a way that makes her more than herself-as-individual. Music expresses characteristics of both daughter and cello—characteristics which are otherwise unlikely to be expressed. The cello is merely a "thing" when separated from the cellist, just as the cellist is "merely" an individual; and each requires the other to bring out what otherwise remains latent. The attitude of the cellist is therefore reminiscent of that of the Eskimo wood carver, and of Heidegger's cabinet maker, mentioned in chapter 2. The cello, the player, and the score act together systemically to produce a result which is impossible in the absence of any one of these, and which is not reducible to the observable characteristics of any or all of them. The result of this joint interaction cannot adequately be quantified except in a trivial sense: we are talking about something which, while it has material, measurable components, is also more than these. If daughter and cello were separated, they would remain, to all appearances, the same as if they were together; but the music would be missing. Similarly, our separation from the natural world *appears* to leave us intact, at least as far as modern consciousness or science can tell; but nevertheless, something even more elusive and insubstantial than music is lost. Resonance, then, is the music of ecosystems, the voices of its parts expressing their complementarity. Unlike music, however, those resonances that could define our healthy relation to the natural world are largely unrealized in modern industrial society, and so are as difficult to imagine as a Beethoven sonata would be in a land without music.

The notion of resonance problematizes the idea that any particular entity can adequately be defined in terms of a static, fixed, structure, since its resonance with a range of other entities may change from moment to moment, implying the spontaneous emergence and disappearance of alternative structures. Just as the cello appears to be no more than an odd-shaped wooden object with particular physical properties and dimensions until the cellist reveals its hidden potentialities, so the cougar or the cottonwood appear to be at least potentially explicable in terms of biological mechanisms or physical reality. And yet, just as cello and cellist transcend their material understandings when united, the isolated cougar in a cage is not the same cougar which stalks among the canyons and mesas. As we interact with parts of the natural world, so we become more than our individualist definitions; and so, too, do those parts of the world we interact with. Science can often recognize those structures that are permanent and static, and defines entities according to such reliably reproducible structures; but it is much less capable when it comes to recognizing structures which are fleeting or which alternate with other structures. As it is these latter structures, and their moment-to-moment variation, that define the vitality of nature, it is not surprising that conventional definitions of natural entities usually ignore them, focusing instead on more-or-less fixed, replicable qualities such as chemical composition, density, or biological makeup. Resonance is not a quality which can be isolated, but part of what Gregory Bateson called "the pattern which connects"; and this pattern is not just material, but also temporal, symbolic, subjective, and ecological. Our sense of a place's history and evolution, for example, is part of a pattern that makes sense of the world, just as our ontological security can be rediscovered by our making sense of our personal history. Such subjective patterns are not merely ephemeral, transitory phenomena, even if the resonances that constitute them are; for like the flash floods which carve the plateau into islands and canyons, they leave permanent traces; and these traces can influence our behavior so as to allow the world to ecologically rebuild itself in a way that incorporates them.

They also carry the kind of ethical implication which is contained in Leopold's dictum that a "thing is right when it tends to preserve the integrity, stability, and beauty of the biotic community." Leopold implies by his use of words such as "beauty" and "community" that "rightness" has something to do with the subjective, and not just the biological, dimensions of nature. Restoring subjectivity to nature, however, does not mean that we have to ascribe any particular thinking or communicative powers to, say, trees; or to empathize with rabbits as if they were Disneyesque "bunnies." This would be to reproduce the individualism (and the anthropocentrism) that underlies our own apartness from the world, in effect populating the world with autonomous beings just like ourselves and perceiving it through the filter of our colonialist fantasies. A healthy subjectivity, in contrast, will be one which is capable of alternating between a focus on the wholeness and autonomy of the individual, on the one hand, and a resonant sense of participation in the whole, on the other. The first of these is (sometimes) well articulated within current conceptions of selfhood; but the latter is virtually ignored. Some other cultures—and I have already referred to the Navaho in this respect—already have a culturally articulated sense of relation to the world; and it is significant that the *primary* unity expressed by this relation is not the individual, but the "Great Self" of which the individual is a part. This unity is not amenable to scientific detection or measurement, but it is nevertheless intensely real for the Navaho; and it is this sense of subjectivity as transcending the boundaries of the person that we have lost as it has become squeezed into the confines of the individual. The problem, then, is not that we experience ourselves as individuals, but rather that we *only* experience ourselves as individuals. A healthy subjectivity, in contrast, is one which moves freely and undefinably between the two poles of pure individuality and complete immersion in other wholes as a diversity of resonances waxes and wanes. During this dynamic process, the self will change, sometimes approximating the "individual" assumed by psychological theory, sometimes becoming part of some larger entity. This is a vital process which industrialism freezes, so that the world suffers an ontological collapse in which entities retreat within their physical boundaries, precipitating a stability that is a sort of deadened inertness rather than a vital tranquillity.

If, in addition, we understand resonance as a complementarity that integrates the poles of subjectivity and physicality, then we can glimpse its power to relate realms and entities that are kept separate within reductionist understandings. Of course, we would not expect that current forms of consciousness would be able to make much sense of this idea, since the *integrative* qualities of the world are precisely those which they omit; and, indeed, the apparent coherence of consciousness depends in part on its wilful denial of these integrative qualities. Consequently, the difficulty of formulating and defining resonance within existing languages is not just inevitable, but a *necessary* characteristic of any notion sufficiently radical to challenge conventional knowledge structures. Let us not be in too much of a hurry to drag it back into the court of existing understanding.

The particular resonances that will occur in any particular set of circumstances will be closely linked to the character of the physical environment. For example, if I live in a city, then my relations to my physical surroundings would probably be most accurately described as a sort of instrumental complementarity: I will use the subway, the stairs, the supermarket, the bus, and so on—but I would hardly describe my relation with any of these as "empathic." If it makes any sense to talk of resonance in such situations, then these resonances will be of a very low order. That is, they will not possess the complexity or the multifaceted character of the moment-to-moment complementarities I experience when I am in a place which is sacred to me, or with a close friend, or with the plum tree in my garden that I have nurtured since it was a sapling. Put colloquially, when I am in the city, the situations I meet do not draw me out of myself or extend myself beyond egoic boundaries in the way that these other situations can do, so that I usually remain "stuck" within my individualistic definition. Complementarily, if our relations to the world are already devoid of empathy, then we are likely to construct a physical world that is incapable of resonating with an empathic subjectivity. Subjective and physical thus coalesce, hiding from view possibilities which become increasingly elusive and unreal, and seeming to confirm that the world conforms to our reductive understanding of it. Conversely, if I live in an "environment" whose aliveness reflects and complements my own at a more than superficial level, then my "individual" subjectivity will constantly overflow into the world, discovering resonances and complementarities within it and creating a sense that "I" am much more than my egoic self, and that there is part of me which owes its existence to what, conventionally, is "outside" me. My behavior is therefore likely to reflect this felt sense of involvement in the world, becoming simultaneously self-enhancing and responsible. As an illustration of this, let us return once again to the Eskimo wood carver, who

> [Holding] the unworked ivory lightly in his hand, turning it this way and that, . . . whispers: "Who are you? Who hides there?" And then: "Ah, seal!." He rarely sets out to carve, say, a seal, but picks up the ivory, examines it to find its hidden form and . . . carves aimlessly until he sees it, humming and chanting as he works. Then he brings it out: Seal, hidden, emerges. It was always there: he did not create it, he released it; he helped it step forth.[19]

This is not a process of imposing structure on a passive "raw material," but rather the discovery of what is already there—a very different sense of "discovery" to that in which Columbus is said to have "discovered" the New World. Proust's suggestion that "the real voyage of discovery consists not in seeking new lands, but in seeing with new eyes" expresses a noncolonialist openness, allowing a space within which resonances could occur. The healthy self, then, is not simply one that can masterfully manipulate the "external" world by imposing its own structure; like the sound box of the cello, it also contains an emptiness, a pregnant space, that invites relation with the world. The carver digs deep within himself to find a shape that resonates with what the world offers, allowing their complementarity to emerge, defining a subjectivity which includes both carver and seal. This sensitive attunement of self to world is the basis of the traditional Eskimo's ability to survive; for in the absence of a technology that can rebuild the world according to the form one has in mind, insensitivity to natural structures will tend to have dire consequences. It is in this sense that subjectivity, for the Eskimo, includes the world, as indicated by the term *sila,* which, as we saw in chapter 2, "refers to the world outside man, especially weather, elements, the natural order . . . [but also] to the state of the inner mind." Intelligence, according to this perspective, is not a property of an individual person, but rather expresses the degree to which the relation between person and world is sophisticated, flexible, and mutually responsive, defining a larger realm that includes both.

This argument may lead to accusations of anthropomorphism. However, what we loosely refer to as anthropomorphism conflates two quite different—and, from a scientific perspective, equally taboo—practices, depending on whether these reflect a genuine openness to the other, or simply the projection of a merely human consciousness onto some nonhuman entity. The latter case is epitomized by the "Disneyfication" of cuddly animals, so that we see them as "cute," "savage," "cunning," and so on—a fundamentally colonialist style of experience that extends the reach of a narrowly human subjectivity over an essentially amorphous world. The other sense of "anthropomorphism," however, acknowledges the subjective qualities of the natural world, and embodies a willingness to allow resonances to develop between self and world.

But, one might argue, if "minds" exist in humans but less in wolves and not at all in canyons, how can "resonance" encompass both human and wolf, or human and canyon? In answering this point, let us return to the analogy of the cello and the cellist. Music, we pointed out, cannot exist purely in the brain of the cellist, for it needs both cello and cellist, as well as score, for its realization. While the "individual" cellist would be a perfectly "normal" individual if examined in a psychological laboratory, she would transcend the reduction assumed in this situation when extended into the musical structures and traditions and practices that she expresses; and our relations with wilderness can achieve a similar transcendence. A musically gifted individual born into a hypothetical society without music of any sort would be a sorry creature indeed, probably being dogged by a vague sense of unfulfilment, although unable to point to the reason for this feeling. Similarly, without features of the natural world to which we humans could subjectively relate, we live—or, perhaps, survive—in a spiritual and relational vacuum quite incapable of offering us a basis for the meaning-making so essential to a healthy life. The mind, in a sense, *needs* the canyon, the clouds, the wolf, to function effectively, just as, in Geertz's view, it needs a culture, and just as the cellist needs a cello; and so the mind engages with what is outside itself to form a system, albeit one that is temporary, dynamic, and only one of many possible such systems. There is a parallel here with Fairbairn's recognition that it "is impossible to gain any adequate conception of the nature of an individual organism if it is considered apart from its relationships to its natural objects, for it is only in its relationships to these objects that its true nature is displayed";[20] and as Fairbairn recognised, growing "up" is also a process of growing *into* the world, aligning our psychological and physical capacities with those of the world. This is what makes too great a reliance on such manufactured experiences as those provided by computer games and theme parks so pernicious: for these experiences are essentially extensions of our own fantasies and exist largely within the industrialist sphere, so offering no intelligent external world with which we can learn to relate. One of the necessities for a healthy subjectivity, in these terms, is that it is not confined within a single "mind" or other entity, but can psychologically and physically engage with what is, initially, "outside" itself, so reaching out to define larger entities than the individual self. Thus our hypothetical musically gifted individual, according to this view, will be "healthier" if she is born into a social world in which musical expression is facilitated than if she inhabited one in which music was unknown. Equally, each of us will be healthier to the extent that we manage to find structures outside ourselves which we can engage with so as to define that elusive wholeness that is the Holy Grail of much environmental writing. As part of the natural order, we bring a particular quality of awareness to it; but to locate this awareness solely within the mind is perhaps the fundamental epistemological mistake of industrialism. How many people in the industrialized world, one wonders, are aware at some level of this "vague sense of unfulfilment" that is so difficult to explain?

For example, the traditional hunter, if he is to be effective, needs an awareness that is part of the natural system he inhabits. He must be attuned to, and resonant with, the character of the landscape, the psyche of the prey and of other animals, and the relations among them. On the other hand, the "sport" hunter *may* rely for his "success" more upon a high-powered rifle and expensive binoculars than on such intimate knowledge of the natural world. Sport hunters may, although not necessarily, have little resonance with the landscape; and in this case, the prey becomes a "thing" rather than a nexus of relation. We can see this scenario at its most grotesque in the behavior of those affluent individuals who "buy" the "right" to shoot a lion or an elephant in certain African countries, approaching the animals in their 4x4's and shooting them at close range. For some traditional hunters, the relation embodied in the eating of meat is only one part of a more general relationality; while at the other extreme, a literalization occurs in which the only form of relationality becomes that of killer/killed—perhaps the most extreme expression of a colonialist mentality. Whether hunting can be regarded as healthy then, depends as much on the "eco-spiritual" condition of the specific hunter-in-context as on overt behavior or conventional ecological criteria; and to assess the ethical status of any specific form of hunting *only* by means of the analysis of population dynamics is to adopt a sort of ecological behaviorism that ignores the subjective and structural characteristics of nature. "Eco-spiritual" concerns, obviously, must ultimately refer to the same world as the material and biological processes more familiar to science, although the conventional dissociation between them makes this unity hard to perceive.[21] And conversely, it is possible to demonstrate that the sort of "wilderness experience" investigated by the Kaplans, and discussed in chapter 2, will have certain measurable psychological effects; although it follows from the argument presented here that an adequate understanding of these effects will require that we go beyond a purely scientific frame of reference.

Such a transhuman subjectivity, of course, will not be simply intellectual. The expulsion of passion from academia and science has played a prime role in distancing us from the natural world, and so we will need to restore these emotional links if we are to recover our capacity to participate in the world. But this is not simply a matter of adding a measure of emotion to our intellectual activities: rather, a whole natural world would be one in which affect and intellect are functionally intertwined, combined into a passionate intelligence or, if you prefer, an intelligent emotionality. Such a suggestion may seem bizarre within the industrial world, but is often unremarkable outside it. Geertz's description of Javanese subjectivity offers us an example:

> For the Javanese . . . the flow of subjective experience, taken in all its phenomenological immediacy, presents a microcosm of the universe generally; in the depths of the fluid interior world of thought-and-emotion they see reflected ultimate reality itself. This . . . sort of world view is best expressed [in the concept of *rasa*]. *Rasa* has two primary meanings: "feeling" and "meaning." As "feeling" it is one of the traditional Javanese five senses—seeing, hearing, talking, smelling, and feeling, and it includes within itself three aspects of "feeling" that our view of the five senses separates: taste on the tongue, touch on the body, and emotional "feeling" within the "heart," like sadness and happiness. The taste of a banana is its *rasa*; a hunch is a *rasa*; a pain is a *rasa*; and so is a passion. As "meaning," *rasa* is applied to the words in a letter, in a poem, or even in common speech to indicate the between-the-lines type of indirection and allusive suggestion that is so important in Javanese communication and social intercourse. And it is given the same application to behavioral acts generally: to indicate the implicit import, the connotative "feeling" of dance movements, polite gestures, and so forth. But in this second, semantic sense, it also means "ultimate significance"—the deepest meaning at which one arrives by dint of mystical effort and whose clarification resolves all the ambiguities of mundane existence. *Rasa*, said one of my most articulate informants, is the same as life; whatever lives has *rasa* and whatever has *rasa* lives. To translate such a sentence one could only render it twice: whatever lives feels and whatever feels lives; or: whatever lives has meaning and whatever has meaning lives.[22]

The term *rasa*, then, like the notion of resonance, presents itself to the industrialized mind as vaguely formulated and unsatisfactorily defined; but both terms point to something central to life. For the Javanese, feeling and meaning are closely integrated. And also, presumably, whatever does not feel does not live; and so the nonfeeling world described by science is necessarily one that is dead, the corpse of another, more vital, world. Resonance, however, implies interaction and integration: just as the cello and cellist, when isolated from each other, cannot produce music, so emotion and intellect, in isolation, remain mere dead fragments of a dead world. Only when they dance together can they revitalize the Cartesian world-as-corpse. And herein lies a common pitfall of our emotional life, originating in one of the less profound fashions of the 1960s: learning that the emotions are repressed by the intellect, we have been encouraged to put aside intellect so as to allow free emotional expression. Such emotional flatus gets us nowhere; for it is in the *articulation* of emotion through narrative, myth, relationship, action—and music—that we achieve fulfilment. Catharsis offers temporary relief, amounting to a sort of emotional masturbation that expresses our isolation and inarticulacy rather than healing it. The notion of resonance, then, challenges the dissociation of intellect and emotion, obstinately defining an integrated world in which human practices synthesize feeling and rationality.

There is a convergence here between an environmental theory that embodies this integration and object relations theory's recognition that "instinct" reaches out into the world, striving for relationality and meaning, although this recognition is a rare exception to psychology's otherwise relentless individualism, and is generally restricted to mother-infant relations. An *extended* object relations theory, however, implies that we are not whole unless we relate in a fully dialectical way with a world that is outside us; and for our purposes, the "world" extends far beyond the boundaries of the immediate family environment. This is an implication which ultimately problematizes the Western concept of the "individual" as a self-contained, static, and largely autonomous entity, as well as the character of the world as something devoid of subjectivity.

A World That Is Heard

Resonance, as an *auditorily* derived metaphor, emphasizes the harmonies and discords which govern interactions between systems. This is an important feature, since, as we noted earlier, the technological world's bias toward *visual* representations predisposes us toward fragmentary understanding. It is not easy to imagine the intertwining of visual images in ways that define a whole; but auditorily, this is an everyday experience, as when we listen to the interwoven harmonies of a song or a symphony. Intentionally, then, resonance is an auditorily-derived notion, since an overreliance on visual perception is an initial step in the process of destruction, fragmenting the world in a way that makes its subsequent physical fragmentation seem quite unremarkable, and leading us toward a world populated only by "things." In contrast, the metaphor of resonance is a fundamentally integrative notion, since resonances occur *between* things, expressing relation rather than independence, interaction rather than autonomy, and dynamism rather than stagnation, as in Michael Taussig's depiction of Aymara culture: "the enchantment of nature and the alliance of its spirits with mankind form an organic resonance of orchestrated social representation. The organisation of kith and kin, political organisation, use of the ecosphere, healing, the rhythm of production and reproduction—all echo each other within one living structure that is the language of the magical landscape."[23]

If auditory metaphors have the potentially integrative abilities that I am claiming, then we might expect that societies which articulate predominantly auditory understandings might be more profoundly integrated into the world than more visually oriented societies. Although no formal attempt to test this rather speculative hypothesis has been carried out, at least to my knowledge, there are some relevant hints in the anthropological literature. For example, James Cowan points out that Australian aboriginal music is lower in pitch than European music, in keeping with the natural sounds they hear about them. "Nature, it seems, is the tuning fork that Aborigines listen to."[24] Furthermore, aboriginals' representation of their world in terms of "songlines" is a particularly profound integration of culture and nature.[25] Music and nature, then, are closely linked in this most environmentally aware culture. In contrast, Cowan argues, modern secular music offers a much more overtly emotional and intensely personal form of expression—the release of emotion trapped within a self which is dissociated from an emotionless world.

A particularly suggestive piece of work is Steven Feld's *Sound and Sentiment: Birds, Weeping, Poetics, and Song in Kaluli Expression.*[26] The Kaluli, who inhabit a forest region of Papua New Guinea, possess a highly elaborate culture based on sound, and particularly on birdcalls; and the relation between the Kaluli and their forest world is intensively mediated by their cultural articulation of sound. In some respects, the sounds of the forest are considered to be directly expressive of human feeling: for example, the three- or four-note melody of the *muni* bird (the Beautiful Fruitdove—*Ptilinopus pulchellus*) is used as a sound metaphor for sadness. Forest sounds also pattern Kaluli songs: when Feld began writing songs in the Kaluli style, the feedback from his mentors included sentences such as "Your waterfall ledge is too long before the water drops"; "The water stays in the pool too long," and "There is much splashing."[27]

But the place of sound in Kaluli culture is not simply a matter of direct translation between natural sounds and their cultural equivalents, and the coherence of the Kaluli social world is derived from the natural world in complex ways. "Ecology and environment," says Feld, "as a model of balance and a mediator of social identity, as expressed in themes of journeying and bird metaphor, focus the desired states of identification with place and geographical history."[28] This ensures a consistency between the social and the natural orders which is essential for the survival of the tribe, for "myth can accomplish the work of establishing homologies and analogies between the social order and the external world, thus causing nature and culture to mirror each other."[29] Furthermore, "sound . . . is not only an alternative method for organising bird categories but is a dominant cultural means for making sense out of the Kaluli world. . . . Actively listening to birds on a day-to-day basis is a way of reckoning time, space, season, and weather. Living with birds is an extension of living by myths, maintaining the coherence of bird and human analogies that make domains like gender, beauty, colour, and [language] logically patterned."[30] Sound therefore provides the vehicle whereby the Kaluli inhabit their forest, not merely in the instrumental sense of using it, but more profoundly, as an extension of their sense of self.

What emerges most clearly from Feld's ethnography is the extraordinary extent to which integration characterizes the Kaluli world. But this integration does not imply any diffuseness or loss of boundaries: on the contrary, Kaluli make the most precise distinctions between different sounds, individual roles, and other aspects of life. However, as in any healthy culture, these clear distinctions and boundaries are complemented by integrative metaphors. Among the Kaluli, a clear sense of individuality is one component of a subjectivity that also extends far beyond the boundaries of the individual; and individual experience is but one focus for a common fund of myths and songs that also includes the world. I am not arguing here that a predominantly auditory experience of the world is essential for a healthy integration between the "human" realm and the rest of the natural world; or that such integration is impossible within a mainly visual society. But I am suggesting that a society in which visual terms greatly outnumber those deriving from other senses will be predisposed toward interpretations that emphasize static, tangible, distinct, "things," and that the members of such a society would have difficulty in articulating the integrative properties of the world. Consequently, an environmental ethic would, in such a society, necessarily be reduced to the value and preservation of "things."

In contrast, consider the parallels between a natural ecosystem and an orchestral symphony. In both cases, particular possibilities—rhythms, tones, practices—can either fit into the whole, or will be discordant. In both cases, their "value" transcends the sum of their recognizable parts. In the case of the orchestra, the "value" of the symphony is hard to measure operationally; but has something to do with its form and holistic character: the particular sequences of notes, the particular combinations of instrument and pitch. It reflects the joint functioning of many instruments, since the melody played on one instrument would have less "value" than the ensemble playing of the whole orchestra. Nevertheless, the "value" of the symphony is not just a function of the number of instruments playing at any one time: an exquisitely beautiful passage may involve just the wind section, for example. "Value," although elusively, has something to do with organization; and simplistic attempts to express it in terms of complexity, diversity, or the number of instruments involved are doomed to failure. Of course, the economic value of the orchestra or the symphony as "raw material" would be derisory—perhaps having to do with the calorific value of the assembled instruments and scores—in other words, how big a bonfire they'd make. And equally, a scientific analysis of the symphony—perhaps involving the identification of particular patterns, phrases, harmonies, and sequences—is likely to be unenlightening. We might say that the instruments, the players, and the musical score have a certain amount of "intrinsic value"; but overwhelmingly, their value derives from a form of *interaction* that is imperceptible to science. What is it about all those membranes, strings, and pipes, whose patterned reverberation is so closely linked with subjectivity? And what is it that makes a zoo, even if it contains an enormous diversity of species in their separate compounds or cages, so much less than a wilderness area? We cannot say, any more than we can distil the essence of natural scenery or its ability to extend subjectivity. The "value" of an ecosystem lies not only in the presence or absence of particular species, nor even in their diversity; but more elusively, in the ecological, subjective, and symbolic patterns of interaction that define it. And just as music is as much felt as thought, so this is also true of the patterns of nature. It is hard to "explain" the meaning of a piece of music in the same way as it is hard to explain the value of an ecosystem. Although the economic world has encroached on the aesthetic world of music, music itself remains impossible to articulate in terms of technical or economic criteria. It has been packaged and commodified; but like the mountain lion in a cage, it still retains an essence which resists the ideological implications of this commodification. Perhaps the lesson here is that much of what is most valuable about the natural order escapes the grasp of consciousness, and so is likely to elude our attempts to preserve it. It is a lesson in humility: while our rational procedures—measuring, tracking, captive breeding—may preserve a few parts of the world, the most significant form of action we can take, at least until we develop an authentic resonance with the natural order, is to leave things as they are, respecting what we glimpse but cannot define.

Implications of a "Resonance" Perspective

The term "resonance" is deliberately vague, and I have allowed it to flutter around in an elusive way which, I hope, manages to retain some of the vitality, indefinability, and variability of the natural world. Experimentalists will, of course, complain about these qualities, and would feel happier in the static, predictable world of laboratory testing and unambiguous concepts. However, as Gina Abeles put it, referring to another slippery notion, the "double bind," "[t]he experimental method is an excellent approach for studying many things: it is not however appropriate for *everything*. Some phenomena do not survive its application; it is not fair to say that things which die on the dissecting table never existed."[31] Part of the vitality of the natural world lies precisely in its variability, its moment-to-moment dynamism, and its immeasurability, which the experimental method finds so difficult to encompass. Any of the established methods of analysis face the same difficulty to some extent.

Let us return for a moment to our debate about "value," and my suggestion that the value of something may vary according to the extent of its interaction with other aspects of the natural world. Here, we come up against the limits of words that have evolved within a *fragmented* world: since "value" is usually understood as a relatively static, definable measure, relating to a "thing" understood in isolation, we are probably stretching it too far in applying it to an envisioned world that emphasizes *integration*. In order to make this discussion of value less abstract, consider Wallace Stegner's account of Dinosaur National Monument in Colorado:

> A place is nothing in itself. It has no meaning, it can hardly be said to exist, except in terms of human perception, use, and response. The wealth and resources and usefulness of any region are only inert potential until man's hands and brain have gone to work; and natural beauty is nothing until it comes to the eye of the beholder. The natural world, actually, is the test by which each man proves himself: I see, I feel, I love, I use, I alter, I appropriate, therefore I am. Or the natural world is a screen onto which we project our own images; without our images there, it is as blank as the cold screen of an empty movie house. We cannot even describe a place except in terms of its human uses.
>
> And as the essential history of Dinosaur is its human history, the only possible destruction of dinosaur will be a human destruction. Admittedly it would be idiotic to

> preach conservation of such a wilderness in perpetuity, just to keep it safe from all human use. It is only for human use that it has any meaning, or is worth preserving.[32]

The anthropocentrism of this passage, by one who has done much to preserve the natural world, says a good deal about how difficult it is to escape from the conventional assumptions of our time. Here, value is located firmly in human judgment which *ascribes* meaning rather than recognizes it. But value cannot be ascribed anthropocentrically by humans to place; and neither is value something intrinsic to place-in-isolation. Rather value can be understood as having to do with the resonances which become possible through the interaction of a place and the diversity of life forms that inhabit or visit it. On this basis, human presence might add some additional resonance, and so increase its "value" somewhat. On the other hand, if it is visited by too many humans, other species and relations would be driven out, so reducing its overall resonance and therefore its "value." An understanding based around the notion of resonance, then, understands value as neither "extrinsic," as ascribed by humans to place; nor as entirely "intrinsic," since its value will vary according to the extent of its resonances with other natural forms.

Value, if the term has any place at all within an adequate environmental understanding, can be understood as reflecting the strength, multiplicity, diversity, and patterning of the resonances which characterize an area of the world. This is why attempts to reconstitute landscapes, as may occur after strip-mining, are rather like putting the components of a washing machine in a box together, and expecting them to function. What is missing is invisible to us, although because of nature's self-regenerative powers, it may slowly return. The notion of resonance, then, attempts to articulate more adequately our intuited sense of the importance of ideas such as biodiversity and ecosystemic interaction which have been used to point to those qualities of the natural world that are imperceptible to conscious rationality. It also recognizes the particular importance of certain species such as the grizzly bear, which is capable of more, and more complex interactions than, say, the woodlouse. Such factors as these can be conceptualised as closely related to the extent of resonance, and therefore the vitality, of the whole. Their absence or impoverishment will reduce the resonance, and vitality, of the whole: a reduction in the number of species inhabiting an area, or the restriction of species to particular portions of the area, or the substitution of the grizzly by the woodlouse will all reduce the resonance of the area. But equally, there may be occasions on which *introducing* exotic species into a well-functioning ecosystem will reduce resonance, destroying existing patterns of interaction. Again, our orchestral analogy may be useful here; and by substituting musical instruments for species, the parallel impoverishment becomes clear.

Resonance, then, may be understood as reflecting the integrated functioning of all the components of an ecosystem. The "value" of the Grand Canyon, for example, reflects not only its sheer scale, but also the many resonances which derive from the variety of species that have adapted to and so learned to complement its particular characteristics. The essentially *relational* basis of resonance may be clarified if we envisage an identically sized canyon on the moon which would, of course, be devoid of life. Such a canyon would conform quite closely to its definition as merely a "thing" of a particular size, of particular geological interest, and so on. Dimensionally, it would be identical to the Grand Canyon; but it would be of immeasurably less "value" for reasons which are harder to measure, but which I have tried to express above. The lunar canyon would not structure the lives of any living inhabitants, human or otherwise; it would be the scene of no history, and would form the basis of no stories or mythologies. It would be a skeletal, corpselike canyon, whose physical form would be at least potentially measurable, but which lacks the resonances that would bring it to life, giving it membership within the teeming, changing, multiply-relating symbolic world. And this recurring image of the corpse is also, perhaps, an apt expression of the character of the person reduced to the individual, for we cannot be fully alive within a world that is dead. Ultimately, however, the term "value," with its implications of separateness, fragmentation, and static, isolable qualities, may be redundant within a more relational subjectivity, an anachronistic hangover from a barbaric past world; although the difficulty of finding an alternative term may require us to use it a little longer.

Economic value is a quite different concept to resonance, being based on the "raw material" value of a part of nature individually torn from its context—that "lowest common denominator" that identifies value not with the most exalted potentialities possessed by an entity, but rather with its untranscended materiality. The animal-as-edible-flesh, for example, usually represents the total value of the animal within the economic system; but its value as part of the natural world, while this will include its value as flesh, extends well beyond this material value into realms involving types of value that are harder to quantify. In shooting an animal, we reduce to near zero its ecological value, since it loses all resonance with its context except the low level ones associated with its chemical nutrient value, or value as flesh to another life form. However, it gains in economic value as "raw material" for pet food, or whatever. The latter is a relatively static value determined by market forces; whereas resonance will vary from moment to moment as the creature interacts with its context in different ways, and as the condition of that context varies according to many factors, not least human encroachment. As its "environment" becomes degraded, so the resonances that are possible for the animal will narrow, and it will become increasingly isolated within an environment which less adequately complements its qualities. Just as industrialism reduces us to a conventional, static individuality, so it reduces nonhumans from their wild forms, in continuously changing resonance with the rest of their natural context, to relatively isolated individuals; and in both cases, there is a reduction of the world to our conscious image of it. What is lost through degradation of the environment cannot be located separately within species, or individuals, or in their contexts: but rather in the dynamic interaction of all of these things. Resonance, and "ecological value," are not located in any particular place or places, and may shift dynamically as particular resonances are realized or forgotten.

Just as the notion of resonance offers us a potential basis for the qualitative evaluation of the effects of human action in the world, so we can similarly evaluate the influence of other species. One of the criticisms that has been made of "ecocentric" approaches is that they accord no more value to humans than to any other species. Now, this criticism may in part be traceable to an anthropocentric overvaluation of humanity; but even the most diehard biocentrist would find it difficult to justify the evaluation of the AIDS virus as equal to humans. From a resonance-based perspective, the AIDS virus—along with invasive exotic species or, regrettably, humans—can actually reduce the strength of resonance through their "success," depending on how they interact with other species. AIDS, in other words, can be conceptualized as assimilating and "colonizing" other structures rather than resonating with them, in contrast to other microscopic organisms that enhance the structures of their hosts. Particular species may promote or diminish resonance within the ecosystem as a whole, depending on their behavior; and the more highly evolved the species concerned, the greater their power to determine whether their behavior will increase or decrease resonance. It is hard to see how the AIDS virus, however, can increase the overall resonance of an ecosystem, unless we take an extraordinarily pessimistic view of humanity; and so the notion of resonance suggests that species—to the limited degree that we can assess them in isolation from their ecosystemic context—vary in "value." It should be emphasized, however, that this "value" will be highly dependent on the many other factors and relations that complexly determine the overall level of resonance.

What is "valuable" about a species, therefore, is the way it has the potential to enlarge, express, and develop vitalities which would otherwise remain latent; and in doing so, the species will articulate and develop its own potentialities. The notion of resonance therefore neither sacrifices the individual to the species or the system, nor takes an exclusively individual-level viewpoint that is blind to species- or ecosystem-level considerations. In doing so, the resonance perspective recognizes the diminution in value which flows from the isolation of a creature from its habitat, holding on to the awareness of that value that it *potentially* embodies. As we look at the mountain lion in its cage, we poignantly recognize this reduction. Unlike conventional assessments in which the value of a creature is independent of its context, the notion of resonance forces us to recognize not only *what is*, but also *what could be*; and the difference between these two embodies both a reproach and an ethical demand.

Similarly, exotic species such as the non-native grasses and knapweed which, aided by overgrazing by cattle, have so changed the ecology of the American West have clearly simplified ecosystemic functioning by driving out many native plants, so reducing overall resonance. What this experience teaches us is that a healthy biosphere is not simply an ecological free-for-all any more than a healthy global community is the result of an economic free-for-all: *boundaries* and bioregional *structure* are necessary in order to maximize diversity, vitality, and, in our terms, resonance. Put crudely, simply mixing the ingredients of the biosphere in a sort of grand ecological stew achieves the degree of damage in the ecological realm which economism's near-hegemony has achieved culturally. A healthy natural world, like the healthy cultural world that is closely associated with it, is one that contains a diversity of structures, ensembles, alternatives. These are difficult qualities for rationality to model; and much social science, consistently with industrialism, denies their existence. The metaphor of resonance, however, simultaneously emphasizes individual structure, contextual structure, and the resonances between them; and so it does not have to retreat into an exclusive focus on any one of these. By analogy, if we listen to a cello concerto, our attention moves fluidly between the soloist and the orchestra, and we enjoy both the individual performances and the overall shape of the piece.

It is not, I emphasize, that we need to replace egoic consciousness by other forms; but rather that a healthy individuality is one that can move flexibly within a wide spectrum of states of consciousness. As we move between states, so the integration of life is expressed by our awareness of alternative forms of organization, including those which map our own egoic structures on to the world, and those in which we open ourselves to other structures outside ourselves. The "colonialist" mentality I have criticized is defined by its rigid prioritization of those structures which are consistent with technological rationality, and which are recognizable only within the narrow form of *individual* subjectivity which we recognize as consciousness, and within which only a limited range of structures appear "real," all others being "illusory" and insubstantial.

Environmental philosophy, in its more traditional forms, makes use of terms such as "values," "instrumentality," and "rights" that presuppose this limited form of subjectivity. Like the notion of "value," the concept of "rights" fits easily into individualistic analyses; but this concept, and, for that matter, that of altruism, fade within a resonance paradigm, to be replaced by that sublimated impulse to survive which grounds individual welfare in that of the natural order as a whole. In this case, concepts such as "rights" and "responsibility" which seem to oppose each other at a conscious level begin to converge at another as the boundary between the welfare of the individual and that of the natural world becomes more diffuse. When Aldo Leopold realized that protecting deer from predators resulted in a defoliated landscape and ultimately mass starvation of deer, he stumbled on the insight that competition, if it exists within a larger, ecological, paradigm, mutates into something like cooperation.

We need to keep in mind that the notion of resonance is, like any other description of the world, a *metaphor*. I am not suggesting that there "are" resonances in some narrowly material sense, or that these are in some potentially quantifiable manner related to established concepts such as "value." Such terms exist only within human models of the world, and should be treated with the caution and disdain they deserve, lest they suffer the same fate of reification suffered by most conventional concepts. Some readers, nevertheless, will be impatiently insisting that I show how resonance can be measured, demonstrated, quantified; and how it connects with, or is different from, existing concepts such as biodiversity. But to accede to this demand would be to drag it, like the rainbow, into the preexisting, humanly constructed world of single definition and mathematical explanation. Like the butterfly trapped in a spider's web, resonance would die within such a context, deconstructed into its components, its vitality stilled. Geertz's critique of analyses of culture which rely too much on intellectual constructions applies with equal force to our attempts to comprehend nature:

> One cannot run symbolic forms through some sort of cultural assay to discover their harmony content, their stability ratio, or their index of incongruity; one can only look and see if the forms in question are in fact coexisting, changing, or interfering with one another in some way or other.[33]

The notion of resonance, then, will live or die depending on its ability to reach out from the limited sphere of human awareness, enabling us to recognize and relate to patterns and processes that are currently invisible to us, and ultimately enabling human life to be a process of participation in rather than control of the world. Instead of trying to define it in the terms of the industrialist order through its inquisition by that order, we should allow it to flit around the boundaries, engaging in an ideological guerrilla war against the destructive certainties and definitions of technologism. Resonance needs to be a reminder of uncertainty, of alternatives, bringing back into focus structures that are as real as those of industrialism, although denied by it. It is a metaphor for that felt experience which can locate us within the world; and like any metaphor, its power lies in its ability to bridge the gap between conscious, rational thought and the symbolic world we are attempting to articulate. It is, in Sam Gill's terminology, a "shadow of a vision yonder." Like the Northern Spotted Owl, it signifies something much larger than itself, which otherwise might slip into oblivion unnoticed. The uncaptured butterfly may be exasperating in its elusiveness as well as impossible to measure; but it is, nevertheless, alive!

The notion of resonance does not itself constitute a theory, but is better considered a metatheory. It is a scaffolding upon which a number of theories or ways of experiencing can grow, a seed-bed for a range of actions and more detailed ideas, a template which offers a new direction and shape for environmental theory; and such templates can be persuasive long before the entities they suggest are given any more definite or quantifiable form. The atomic theory of early Greek philosophy, for example, was influential for two millennia before anybody observed or measured an atom; but its fateful influence in populating the world with discrete entities, and in confirming the assumption that there is a material basis to all reality, has been altogether broader and more profound than its strictly physical interpretation. In inheriting this influence, we forget that this theory once coexisted with one based on the gods, and that the modern world has seen the almost complete victory of the former over the latter. As a result, the dominant form of explanation is a material reductionism, and ideas of structure that are, at best, distantly anchored in the material level—and here I am thinking of spiritual, cultural, and even ecological structures—are often seen as mystical and insubstantial. But a complete environmentalism is one which recognizes that it is the interaction between both these epistemological poles which brings the natural world to life, and that the attempt to reduce nature to one pole is to deny it either substance or form. Since our particular cultural prejudice is to emphasize materiality and to deny pattern, the environmentalist is left struggling to explain a vitality that originates in the confluence of these two poles. The time for an alternative, relational understanding of the world is long overdue.

Culture and Resonance

The individual who is isolated from culture is, as we saw in chapter 4, a poor thing indeed, an "unworkable monstrosity" in Geertz's phrase. And yet even Geertz (like any of us!) does not entirely escape the industrialist imperative to fragment the cosmos and to isolate the human from the natural; for while humanity and culture are profoundly interwoven within his scheme, the natural world is often represented through such abstractions as "fundamental reality"[34] or "the way the world is."[35] But perhaps we can extend Geertz's remarkable insights so that nature is not merely the amorphous context of human symbolic activity, but becomes the larger symbolic whole within which human life takes place. After all, if the evolution of the human central nervous system was associated, as Geertz argues, with a co-evolving cultural context, then we can hardly deny that it also took place within a *natural* context. We are, as he suggests, information-seeking creatures who seek out and respond to structure outside our individual bodily boundaries; and there are too many traditional cultures that are dialectically interwoven with the natural world for us to pretend that they have evolved only through a sort of self-sufficient, ingrown maturational process unconnected with the natural order.

As Geertz himself suggests, "in sacred rituals and myths values are portrayed not as subjective human preferences but as the imposed conditions for life implicit in a world with a particular structure."[36] Such a world, then, embodies its own forms and imperatives and values, demanding that culture accord with nature, even if only indirectly, if it is to be the basis of a sustainable lifestyle. Here we should note that the idea that morality, for example, should have nothing to do with the way the world *is*, as enshrined within the "naturalistic fallacy," is really a plea for a view of morality based entirely on disembodied abstract principles which exist only on the "human" side of the manufactured human-nature divide. Ideas such as that of the "sacredness of life" have often become literalized, so that morality appears in terms of rigid, simplistic injunctions such as "thou shalt not kill." Life, however, does not reside only in those individual entities that are perceptible to consciousness, but also in what happens between them, in that elusive vitality of the natural world that the notion of resonance tries to convey. A less fragmented morality, then, would imply a more sophisticated basis for action than merely the avoidance of killing, although it may often invoke this principle, rather as Einsteinian theory is for many practical purposes reducible to its Newtonian predecessor. Such a morality will seek to maximize the vitality of the world in ways which may not always be simply predictable, since they will be filtered through the complexities of specific cultural systems. The attempt to develop straightforward ethical principles through disembodied rational thought bypasses the complexities of this process, implying a relation to the world that is imposed by rationality rather than negotiated with natural realities—a situation which is inevitable given the dereliction of cultural structures that could align human awareness with the natural world.

If cultural structures are as essential to the articulation of primal experience as I have suggested, then a central question is how we might foster the emergence of new cultural structures capable of such articulation. But to put the question this way implies some sort of intention realized in action, which in turn implies an involvement with consciousness; and consciousness, as we have seen, is suspect as a messenger of bodily awarenesses. Culture, however, is essential to human functioning precisely because it operates, at least to a significant extent, outside of consciousness; and attempts to make culture understandable in terms of conscious rationality may also smuggle in the same tired old impulse to assimilate all experience to consciousness. If we are to reach out into the world beyond our solipsistic intellectual islands, we need to open ourselves to forms of relation and experience that cannot be consciously formulated; and this applies both to our everyday experience of the natural and cultural worlds and to our attempts to model them.

Psychologists, as we have seen, have been particularly prone to retreat to the laboratory when faced with the messy realities of life; but anthropologists, too, have sometimes behaved similarly. Geertz tells how Lévi-Strauss, attempting to communicate with his Tupi-Kawahib informants, was utterly unable to understand their language; and so he withdrew, disappointed, into building theories about them, "trying in vain to repiece together the idea of the exotic with the help of a particle here and a fragment of debris there." Tellingly, Lévi-Strauss concludes that "to reach reality, we must first repudiate experience,"[37] a decision that aligned his project unambiguously within the mainstream of modernism. A similarly cerebral approach has been widely accepted among those who seek to "understand" nature, as they attempt to cement together the dislocated fragments left over from its dismemberment—the species, typologies, physical characteristics that are accessible to consciousness—using whatever logical connections seem to be available. To recognize that a scientific understanding of nature results in a sort of paradoxical distancing from it[38] is not a denial of the power and insights of science, only an acknowledgment of its necessary limitations.

In academia, as we saw earlier, a parallel distancing that commonly occurs among nonscientists takes the form of narcissistic symptoms such as the loss of belief in the meaningfulness and reality of the world, and its replacement by a linguistic "reality."[39] Underlying both these forms of alienation from the world, which have strongly infected environmental writings, is an oppositional stance which can critique and reject the assumptions of both science and nonscience, but which is incapable of envisioning any academic discourse that could *positively* articulate natural form. Nevertheless—to quote Geertz again—"To abandon the hope of finding the 'logic' of cultural organization in some Pythagorean 'realm of meaning' is not to abandon the hope of finding it at all";[40] and a similar point can be made in relation to *natural* organization. In both realms, the "logic" may not be of a kind to which consciousness is normally receptive; and a refocusing of subjectivity may be necessary—perhaps, for example, in a way which is inspired by audition rather than vision, and which is as sensitive to those invisible harmonies, silences, sequences, and rhythms that exist in the spaces *between* "things" as to the "things" themselves. Again, however, we must beware of interpreting this task as one of *replacing* one scheme which is visual, or scientific, or reductionist (or, for that matter, semiotic) by another which is auditorily based, or arational, or holistic. Both schemes should be regarded as metaphors, rather than as mutually exclusive alternatives, one of which is right and the other wrong. In this respect, the framing of scientific research as one of "conjecture and refutation" has been profoundly misleading; and the notion that one system—whether of understanding, or ecology, or culture—should competitively triumph over others has significantly influenced modern society away from consistency with the natural order.

If consciousness as currently constituted is poorly fitted to articulate these multifaceted and constantly shifting realities, it can nevertheless hint, at least indirectly, at its own inadequacies. Environmental and social problems, psychological stresses, and our own sense of unease are messages that suggest to consciousness its own incompleteness, thereby pointing toward a subjective space within which alternative structures might coalesce. A conscious recognition that we do not, and in principle *cannot*, know all the answers is a better basis for an environmental ethic than the presumption that we *might* know, given enough scientific sophistication. An admission of ignorance is an implicit recognition of the existence of what is unknown; and this recognition leads us toward a fertile humility that restrains itself from imposing structure, but rather is ready to open itself to those structures which exist outside of consciousness. Environmental theory, then, will be most valuable when it legitimates what it cannot itself articulate, clearing a space within which resonances between ourselves and the rest of the natural world can emerge. This space will be both internal and external; for such resonances depend on the existence of what consciousness would experience as empty, quiet places both in the world and in ourselves. Eventually, as these initially timid resonances become more established within supportive cultural forms, it will dawn on us that these apparently separate spaces are in fact the same place: and in that moment, our long-envisioned reintegration with the world will have become a reality.

In the absence of such a reintegration, the dream of human independence from the world, crystallized most explicitly within Cartesian philosophy, is today reflected in the tarmac and steel of our physical environment as well as in our own experiences and relationships, and in the theories by which academics attempt to comprehend these. But most of all it is materialised in the *absences* around which our lives are constructed: the lack of emotional articulation, the paucity of vital spiritual frameworks, the poverty of cultural forms, and the absence of a meaning-laden relation to the earth.

It is time for a new dream. Given the positivist, materialist emphasis of modern society, any genuinely alternative vision will be materially unrealized and is bound to seem insubstantial; and it is important to realise that this insubstantiality is a *necessary* characteristic of any approach that escapes from the widespread materialist reductionism of our age. But the historical reverberations of Descartes' philosophy teach us that a dream can be the first stage of a powerful reconfiguration which has both material and subjective components; and if alternatives to industrialism are to emerge, we need the courage to dream what will at first be insubstantial. Ideas such as resonance can potentially be substantiated in ecological, cultural, and psychological structures as readily as those divisions between thought and feeling, human and animal, or cultural and natural which we assume today. The prejudice that only a materiality shorn of subjectivity is real should be challenged by materialising subjectivity in cultural forms, inscribing on our legal, moral, and social structures a resonant understanding of the world that will eventually become a physical and ecological reality. Just as a freeway network is the material realization of one particular idea of relation, so an ecological community which includes the human is another; and the notion that the former is somehow more real than the latter is an epistemological prejudice the world can no longer afford.

The institutionalized severance of felt experience from the outside world begins in our schools, wherein emotionality, if it is not directly prohibited, is expressed through forms which are relatively independent of any of the issues that children feel passionately about, including the state of the natural world. The experience of the anthropologist Dorothy Lee is not untypical here. Asked to construct a frieze of horses, she recounts, her son was discouraged from portraying the "skinny, elongated beasts, full of straining movement and savage life" that he had previously painted, the teacher insisting that he copy "pinkish, sleek, placid, fat, lifeless" animals from a history textbook. Lee was appalled; but also relieved to find in the corner of her son's painting "the tiny figure of a bird, of no known genus, scraggy, leering, menacing." Her son's experience, she recounts with relief, "had not been entirely mowed down in the drive for uniformity."[41] Such stories are usually portrayed as involving the repression of individual creativity rather than any potential relation between child and world, so *assuming* the prior emotional separation of the child from the world even when, as in this case, the world is explicitly depicted. In such situations, a groping attempt to articulate a passionate relation to nature is replaced by a more detached, socially constructed representation; and an emotional sense which is struggling to find its object in the external world is replaced, under duress, by a detached representation which abandons this struggle, embodying an acceptance that art exists in an autonomous world that has little use for passionate relation to anything outside itself. Clearly, such styles of socialization involve a psychological loss to the child; but to see this as the *only* loss is implicitly to accept an individualistic frame within which those cultural forms that could integrate the child within an ecological and subjective whole have *already* been lost. While the growing acceptance of emotional expression in schools is to be welcomed, this should be seen as only the first step along a lengthy path which leads toward the reintegration of emotion and intellect, and to their joint engagement with the world beyond human consciousness. Feeling, in other words, seeks the structure that could articulate it; and the way love seeks the other can be seen as the prototype of a much more general way of relating to what is outside ourselves. Love, as the seer suggests, "moves and spurs the intellect to go before it, like a lantern, to the forests . . . ", implying that while feeling and intellect may maintain a partial separateness, they can also form parts of a transcendent integration that includes the physical world. The structures with which we resonate may sometimes be ideational; but they are also, in the world we must move toward, mythological, spiritual, ecological, and physical, drawing us out of ourselves and confirming our participation in a world that is alive. Within such a world, we experience nature multiply—as simultaneously frightening, awesome, exhilarating, symbolically meaningful, as home and "other"—as something that we engage with both emotionally and thoughtfully, not just *use* for our (material or emotional) needs.

It is not so much a question, then, of *teaching* environmental awareness as one of *allowing* it to develop. It is, in other words, a natural developmental process from which we have been diverted only by the most strenuous efforts of socialization into a world permeated by dualisms. As we saw earlier, children *look for* relation; and only the overpowering focus on individual personality and dispassionate manipulation prevents relationality from developing. It is therefore more a question of providing a context wherein such relationality could develop than of prescribing some socially approved form of relation; and such a context must incorporate access to enough healthy wilderness that direct experience of the natural world can become a fundamental part of the curriculum. Some answers may be found through introspection or in books; but others can only be found by walking in the forest or sharing a campfire. Complementarily, education needs explicitly to recognize the limits of knowledge, and to communicate an awareness of these limits to children. We should be ready to answer questions with responses such as: "We don't know the answer to that. Some people think that x is the case; others that the answer is y. What do you think?" Some questions have definite, clear answers. With others, the answers are more tentative. And some cannot really be answered at all. We need to communicate to children not only what we know, but also what we don't, or can't, know; and this, in my experience, encourages children to seek answers in other realms which we conventionally assume to be inside us, but which in a healthy culture also extend into the world outside.

I am not suggesting that we should retreat into what Eugene Hargrove has referred to as "environmental therapeutic nihilism"—the doctrine that because our understanding of ecological systems is so partial and inadequate, "doing nothing is better than doing something, because any action will most likely have bad consequences whether it succeeds in solving the initial problem or not."[42] Both the uncontextualized technological hubris that so disastrously pervades the post-Enlightenment world and the "therapeutic nihilism" that is often a reaction against it reflect a lack of balance and coordination between the known and the unknown, the conscious and the unconscious, the wild and the domesticated, and the rational and the arational. It is not a question of choosing between each of these apparent "alternatives," but of regaining the sense of integration between our technological power and our participation in the natural order; and this in turn implies a realistic awareness of both human abilities and human limitations.

Recognizing that we do *not* always know, then, is the first step toward recognizing that something out there, while it may sometimes be felt at some level, is *not known*; and this is an environmentally crucial realization. But while much of nature may not be *known*, its resonances with us may nevertheless be in some way familiar; and cultural form may sometimes enable us to express this glimpsed commonality between our own nature and wildness in the outside world. When this happens, as John Dewey wrote in describing Wordsworth's poetry of nature, "we do not find ourselves in a strange, unfamiliar land," but rather in one about which we "already had some dumb feeling;"[43] and a rejuvenated language, rather than sealing our dissociation from the world, can enable our extension into it by expressing symbolic as well as rational awarenesses. Such a momentous change will also require the development of additional expressive forms less anthropocentric than human language, which, as we have seen, has all too often become both a means of communicating with other humans and a means of avoiding communication with the nonhuman. Although I do not have the space to argue for it here, a "critical realist" perspective[44] allows us to view language (in the most general sense of the term) as potentially reaching out toward the natural world "in the light of an imaginative generosity that seeks to enter the other's voice into the dialogue."[45] Outside the industrialized world, as Barry Lopez notes, "language is not something man imposes on the land. . . . The very order of the landscape, the ecology of its sounds and thoughts, derives from the mind's intercourse with the landscape."[46] This intercourse between the human and the nonhuman can become a dance which defines both our separateness and our interdependence; and as we recover our place in this dance, so the world regains its vitality and integrity.

Notes

Preface

1. Regarding this point, I consider Colin Turnbull's work in chapter 4. Another important step in this direction is Don Kulick and Margaret Willson, *Taboo: Sex, Identity, and Erotic Subjectivity in Anthropological Fieldwork* (London: Routledge, 1995).

Chapter 1. Introduction: In Search of the "Natural"

1. The classic text is: Max Horkheimer and Theodor Adorno, *Dialectic of Enlightenment* (New York: Continuum, 1991). For an accessible introduction, see: David Held, *Introduction to Critical Theory: From Horkheimer to Habermas* (London: Hutchinson, 1980).

2. Christopher Lasch, *The Culture of Narcissism: American Life in an Age of Diminishing Expectations* (New York: Norton, 1979).

3. Robert D. Romanyshyn, *Technology as Symptom and Dream* (London: Routledge, 1989), p. 228.

4. Harry Guntrip, *Schizoid Phenomena, Object Relations, and the Self* (London: Hogarth, 1980).

5. See Edward Goldsmith, "Gaia: Some Implications for Theoretical Ecology," in P. Bunyard and Edward Goldsmith (eds), *Gaia: The Thesis, the Mechanisms, and the Implications* (Wadebridge, U.K.: Wadebridge Ecological Centre, 1989); and Neil Evernden, *The Natural Alien* (Toronto: University of Toronto Press, 1985).

6. Robert P. McIntosh *The Background of Ecology* (Cambridge: Cambridge University Press, 1985). An example which McIntosh discusses is the seldom challenged emphasis on competition within niche theory.

7. Robert H. Peters, *A Critique for Ecology* (Cambridge: Cambridge University Press, 1991), p. 10.

8. For an excellent introduction to ecopsychology, see Theodore Roszak, Mary E. Gomes, and Allen Kanner (eds.), *Ecopsychology: Restoring the Earth, Healing the Mind* (San Francisco: Sierra Club, 1995).

9. Ibid.

10. John Rodman, "The Liberation of Nature?" *Inquiry* 20 (1977): 104.

11. Val Plumwood, *Feminism and the Mastery of Nature* (London: Routledge, 1993).

12. R. D. Laing, *The Divided Self: An Existential Study in Sanity and Madness* (Harmondsworth, U.K.: Penguin, 1965), p. 39.

13. Edwin A. Burtt, *The Metaphysical Foundations of Modern Physical Science* (London: Routledge and Kegan Paul, 1959 [1924]), pp. 236–37.

14. Ibid., p. 302.

15. See, for example, Irving Velody and Robin Williams (eds.), *The Politics of Constructionism* (London: Sage, 1998).

16. Michael Buerk interviews Philip Angier, BBC Radio 4, June 2, 1998.

17. Rodman, "The Liberation of Nature?" p. 104.

18. Jim Cheney, "Ecofeminism and Deep Ecology," *Environmental Ethics* 9 (1987): 115–45.

19. Plumwood, *Feminism and the Mastery of Nature*, p. 42.

20. John Shotter, *Images of Man in Psychological Research* (London: Methuen, 1975), p. 13.

21. Timothy Reiss, *The Discourse of Modernism* (Ithaca: Cornell University Press, 1982).

22. Fredric Jameson, *The Prison-House of Language: A Critical Account of Structuralism and Russian Formalism* (Princeton: Princeton University Press, 1972), pp. viii–ix.

23. Stanley Aronowitz, *Science as Power: Discourse and Ideology in Modern Society* (Minneapolis: University of Minnesota Press, 1988), p. 22.

24. Phil Macnaghten and John Urry, *Contested Natures* (London; Sage, 1998), pp. 19, 3–4, 15.

25. Macnaghten and Urry, *Contested Natures*, p. 3.

26. Sandra D. Mitchell et. al., "The Whys and Hows of Interdisciplinarity," In P. Weingart et al. (eds.), *Human by Nature: Between Biology and the Social Sciences* (Mahwah, N.J.: Lawrence Erlbaum, 1997), p. 135.

27. Macnaghten and Urry, *Contested Natures*, pp. 25, 35.

28. Ibid., pp. 21, 22.

29. Roy Bhaskar, *Reclaiming Reality* (London: Verso, 1989); *The Possibility of Naturalism*, 2nd ed., (Hemel Hempstead, U.K.: Harvester Wheatsheaf, 1989).

30. Frederick Turner, *Beyond Geography: The Western Spirit against the Wilderness* (New York: Viking Press, 1980), p. 89.

31. See, for example, R. Edward Grumbine, "Wildness, Wise Use, and Sustainable Development," *Environmental Ethics* 16 (1994): 227–49.

32. Romanyshyn, *Technology as Symptom and Dream.*

33. See for example, Don Ihde, *Technology and the Lifeworld* (Bloomington: Indiana University Press, 1990).

34. Kenneth McLeish, *London Sunday Times*, August 5, 1990.

35. Quoted by Christopher Manes, *Green Rage: Radical Environmentalism and the Unmaking of Civilisation* (Boston: Little, Brown, 1990), p. 43.

36. Romanyshyn, *Technology as Symptom and Dream*, p. 14.

37. Bill Devall, *Simple in Means, Rich in Ends: Practicing Deep Ecology* (Salt Lake City: Peregrine Smith Books, 1988), p. 146.

38. Walter J. Ong, *Interfaces of the Word: Studies in the Evolution of Consciousness and Culture* (Ithaca: Cornell University Press, 1977), p. 72.

39. David M. Levin, *The Opening of Vision: Nihilism and the Postmodern Situation* (New York: Routledge, 1988).

40. Philip Cushman, "Why the Self Is Empty: Toward a Historically Situated Psychology," *American Psychologist* 45 (1990): 599–611.

41. Arthur Kleinman, *Rethinking Psychiatry: From Cultural Category to Personal Experience* (New York: Free Press, 1988), p. 51.

42. Arthur Kleinman, " 'Everything That Really Matters': Social Suffering, Subjectivity, and the Remaking of Human Experience in a Disordered World," *Harvard Theological Review* 90 (1997): 333.

43. Donna Haraway, "A Cyborg Manifesto," In *Simians, Cyborgs, and Women: The Reinvention of Nature* (London: Free Association, 1991).

44. Arthur Kleinman and Joan Kleinman, "The Appeal of Experience, the Dismay of Images: Cultural Appropriations of Suffering in Our Times," *Daedalus* 125 (1996): part 1, 1–23, p. 2.

45. Ros Coward, "Wild Shots," *The Observer*, December 6, 1997.

46. *The Guardian*, January 16, 1999.

47. Coward, "Wild Shots."

48. Tim Luke, *Ecocritique: Contesting the Politics of Nature, Economy, and Culture* (Minneapolis: University of Minnesota Press, 1997), p. 61.

49. Susan G. Davis, "Touch the Magic," in William Cronon (ed.), *Uncommon Ground: Rethinking the Human Place in Nature* (New York: Norton, 1996), p. 211.

50. Anthony Giddens, *The Consequences of Modernity* (Stanford: Stanford University Press, 1990), pp. 127, 133.

51. Fram Abrams and John Davison, "Apocalypse Now Turns Us Green with Worry," *Sunday Times*, June 10, 1990.

Chapter 2. Psychology's Betrayal of the Natural World

1. Quoted by Tzvetan Todorov, *The Conquest of America: The Question of the Other* (New York: Harper & Row, 1984), p. 30.

2. "Oath Sworn Regarding Cuba," June 1494. Quoted by Todorov, *The Conquest of America*, p. 22.

3. Todorov, *The Conquest of America*, p. 29.

4. Ibid.

5. Ibid., p. 22.

6. Ibid., p. 97.

7. Ibid., p. 69.

8. Ibid., p. 127.

9. Thomas Moore (ed.) *The Essential James Hillman: A Blue Fire* (London: Routledge, 1990), p. 30.

10. James Hillman, *Re-Visioning Psychology* (New York: Harper & Row, 1975), p. 25.

11. David Kidner, "Why Psychology Is Mute about the Environmental Crisis," *Environmental Ethics* 16.4 (Winter 1994): 359–76.

12. Todorov, *The Conquest of America*, p. 69.

13. Rita L. Atkinson, Richard C. Atkinson, Edward E. Smith, and Daryl J. Bem, *Introduction to Psychology*, 10th ed. (San Diego: Harcourt Brace Jovanovich, 1990), pp. 170–72.

14. Bernard McCrane, *Beyond Anthropology: Society and the Other* (New York: Columbia University Press, 1989), p. 67.

15. Frederick Turner, *Beyond Geography: The Western Spirit against the Wilderness*, p. 73.

16. Don Bannister, "Psychology as an Exercise in Paradox," *Bulletin of the British Psychological Society* 19.63 (1966): 21–26.

17. Alfred Korzybski, *Science and Sanity* (New York: Science Press, 1941).

18. It might be objected that statistical techniques such as the ANOVA *do*, in fact, recognize interactions between factors. However, the sensitivity of such techniques to interactions is much less than their sensitivity to main effects; and the types of "interaction" that are reflected in the results fail to do justice to the complexity of the term "interaction" in its ecological sense. See Douglas Wahlsten, "Insensitivity of the Analysis of Variance to Heredity-Environment Interaction," *Behavioural and Brain Sciences* 13 (1990): 109–61.

19. Sigmund Koch, "The Nature and Limits of Psychological Knowledge: Lessons of a Century *qua* 'Science.' " In: Sigmund Koch and David E. Leary (eds.), *A Century of Psychology as a Science* (New York: McGraw-Hill, 1985), p. 89.

20. Kurt Danziger, *Constructing the Subject: Historical Origins of Psychological Research* (Cambridge: Cambridge University Press, 1990), p. 2.

21. Seymour Sarason, *Psychology Misdirected* (New York: Free Press, 1981), p. 183.

22. Michel Foucault, *The Birth of the Clinic: An Archeology of Medical Perception* (New York: Vintage Books, 1975), p. 197.

23. Lev S. Vygotsky, *Mind in Society* (Cambridge, Mass.: Harvard University Press, 1978), p. 57 (my emphasis).

24. Frederick Turner, *Beyond Geography: The Western Spirit against the Wilderness*, p. 15.

25. Kurt Danziger, *Constructing the Subject: Historical Origins of Psychological Research.*

26. Kenneth Gergen, "Social Psychology as History," *Journal of Personality and Social Psychology* 26 (1973), 309–20.

27. Rachel and Stephen Kaplan, *The Experience of Nature: A Psychological Perspective* (Cambridge: Cambridge University Press, 1989).

28. Ibid., p. 16.

29. Ibid.

30. Ibid.

31. Ibid., p. 17.

32. Ibid., p. 176.

33. David Ingleby, "Ideology and the Human Sciences: Some Comments on the Role of Reification in Psychology and Psychiatry," *The Human Context* 11 (1970): 159–87.

34. K. Gergen. "Experimentation in Social Psychology: A Reappraisal," *European Journal of Social Psychology* 8 (1978): 507–27.

35. See, for example, Kenneth Gergen, "The Social Constructionist Movement in Modern Psychology," *American Psychologist* 40 (1985): 266–75; Ian Parker and John Shotter (eds.), *Deconstructing Social Psychology* (London: Routledge, 1990).

36. Quoted by Edward S. Casey, *Getting Back into Place* (Bloomington; Indiana University Press, 1993), p. 229.

37. Terry Eagleton, *Literary Theory: An Introduction* (Minneapolis: University of Minnesota Press, 1983), p. 60.

38. Michael Billig, *Arguing and Thinking: A Rhetorical Approach to Social Psychology* (Cambridge: Cambridge University Press, 1987), pp. 41, 44. Karl Popper, *The Open Society and Its Enemies*, Vol. 1, (London: Routledge and Kegan Paul, 1966), p. 57.

39. Billig, *Arguing and Thinking*, p. 134.

40. Ibid., p. 111.

41. André Gorz, *Critique of Economic Reason* (London: Verso, 1989), p. 176.

42. Vivien Burr, *An Introduction to Social Constructionism* (London: Routledge, 1995), p. 4.

43. Peter Mason, *Deconstructing America: Representations of the Other* (London: Routledge, 1990), p. 15.

44. Elizabeth A. R. Bird, "The Social Construction of Nature: Theoretical Approaches to the History of Environmental Problems," *Environmental Review* 11 (1987): 255–64.

45. Burr, *Introduction to Social Constructionism*, pp. 19–20.

46. Jane Bennett and William Chaloupka (eds.), *In the Nature of Things: Language, Politics, and the Environment* (Minneapolis: University of Minnesota Press, 1993), p. xii.

47. Ibid., p. 5.

48. Ibid.

49. William Cronon, "Introduction," in William Cronon (ed.), *Uncommon Ground: Rethinking the Human Place in Nature* (New York: Norton, 1996), p. 36.

50. Ibid., p. 69.

51. Bruno Latour and Steven Woolgar, *Laboratory Life* (London: Sage, 1979), p. 128.

52. Holmes Rolston III, "Nature for Real: Is Nature a Social Construct?" In Timothy D. J. Chappell (ed.), *The Philosophy of the Environment* (Edinburgh: Edinburgh University Press, 1997), p. 53.

53. Timothy Reiss, *The Discourse of Modernism*, p. 35.

54. Gergen, "Social Psychology as History," pp. 309–20.

55. Billig, *Arguing and Thinking*, p. 24 (my emphasis).

56. Ibid., p. 207.

57. Barry Lopez, *Arctic Dreams: Imagination and Desire in a Northern Landscape* (London: Macmillan, 1986), p. 277.

58. I am quoting here from Andrew Collier's lucid and accessible book *Critical Realism: An Introduction to Roy Bhaskar's Philosophy* (London: Verso, 1994), p. 76.

59. Philippe Descola, "Constructing Natures: Symbolic Ecology and Social Practice," in Philippe Descola and Gísli Pálsson (eds.), *Nature and Society: Anthropological Perspectives* (London: Routledge, 1996), pp. 82, 84.

60. Collier, *Critical Realism*, p. 86.

61. R. Lidskog, Review of Ulrich Beck's *Risk Society. Acta Sociologica* 36 (1993): 400–403.

62. Macnaghten and Urry, *Contested Natures*, pp. 21, 35.

63. John Hannigan, *Environmental Sociology* (London: Routledge, 1995), p. 31.

64. See, for example, Kate Burningham and Geoff Cooper, "Being Constructive: Social Constructionism and the Environment," *Sociology* 33.2 (1999): 297–316.

65. Sergio Sismondo, "Some Social Constructions," *Social Studies of Science* 23 (1993): 515–53.

66. Hannigan, *Environmental Sociology: A Social Constructionist Perspective*, p. 187.

67. Salvador Minuchin, Bernice Rosman, and Lester Baker, *Psychosomatic Families: Anorexia Nervosa in Context* (Cambridge, Mass.: Harvard University Press, 1978).

68. Richard Shweder, *Thinking Through Cultures* (Cambridge, Mass.: Harvard University Press, 1991); Kleinman, *Rethinking Psychiatry: From Cultural Category to Personal Experience.*

69. Reiss, *The Discourse of Modernism.*

70. Gregory Bateson and Mary Catherine Bateson, *Angels Fear: Towards an Epistemology of the Sacred* (New York: Macmillan, 1987), p. 195.

71. Noam Chomsky, *Keeping the Rabble in Line: Interviews with David Barsamian.* (Monroe, Maine: Common Courage Press, 1994), pp. 163–64.

72. Guntrip, *Schizoid Phenomena, Object Relations, and the Self*, p. 73.

73. Ian Craib, "Social Constructionism as a Social Psychosis," *Sociology* 31 (1997): pp. 1–15.

74. See David Kidner, "Fabricating Nature," *Environmental Ethics* (forthcoming), for a more thorough critique.

75. Kleinman, " 'Everything That Really Matters': Social Suffering, Subjectivity, and the Remaking of Human Experience in a Disordered World," p. 333.

76. Edward Sampson, "Cognitive Psychology as Ideology," *American Psychologist* 36 (1981): 730–43.

77. Sampson, "Cognitive Psychology as Ideology," p. 738.

78. Diagnostic and Statistical Manual IV (Washington D.C.: American Psychiatric Association, 1994) (my italics).

79. See, for example, George W. Brown and Tirrell Harris, *The Social Origins of Depression* (London: Tavistock, 1978).

80. James Hillman, for example, has emphasized the constructive role of cultural structures in diffusing and contextualizing individual trauma. Considering parents' brutality toward children, he suggests that, in the absence of such structures, "I remain a victim in my memory. My memory continues to make me a victim.

"[However], these wounds that he caused have done something to me to make me understand punishment, make me understand vengeance, make me understand submission, make me understand the depth of rage between fathers and sons, which is a universal theme—and *I* took part in that. And so I've moved the memory, somehow, from just being a child victim of a mean father. I've entered fairy tales and I've entered myths, literature, movies. With my suffering I've entered an imaginal, not just a traumatic, world." (James Hillman and Michael Ventura, *We've Had a Hundred Years of Psychotherapy and the World's Getting Worse* [San Francisco: HarperCollins, 1992], p. 26).

81. For a lucid discussion of this point, see David Smail, *The Origins of Unhappiness* (London: HarperCollins, 1993).

82. Ibid.

83. Russell Jacoby, *Social Amnesia: A Critique of Conformist Psychology from Adler to Laing* (Boston: Beacon Press, 1977); Kleinman, *Rethinking Psychiatry.*

84. Sarnoff Mednick, "Breakdown in Individuals at High Risk for Schizophrenia: Possible Predispositional Perinatal Factors," *Mental Hygiene* 54 (1970): 50–63.

85. Mary Douglas, "The Lele of Kasai," in Daryll Forde (ed.), *African Worlds: Studies in the Cosmological Ideas and Values of African Peoples* (London: Oxford University Press, 1954).

86. Sarnoff Mednick, "An Associational Interpretation of the Creative Process," in Calvin Taylor (ed.), *Widening Horizons in Creativity* (New York: Wiley, 1964).

87. Anita Barrows, "The Ecopsychology of Child Development," in Roszak et al. (eds.), *Ecopsychology: Restoring the Earth, Healing the Mind*, p. 103.

88. Barry Lopez, *Of Wolves and Men* (London: Dent, 1978). See also Larry Arnhart, *Darwinian Natural Right* (Albany: State University of New York Press, 1998).

89. Gregory Bateson, *Steps to an Ecology of Mind* (St. Albans, U.K.: Paladin, 1973), p. 284.

90. For a clear statement of how family therapy contextualises the symptomatic individual, see Minuchin, Rosman, and Baker, *Psychosomatic Families: Anorexia Nervosa in Context.*

91. Michael P. Nichols, *The Self in the System: Expanding the Limits of Family Therapy* (New York: Brunner/Mazel, 1987).

92. Minuchin, Rosman, and Baker, *Psychosomatic Families.*

93. Sigmund Freud, "Civilised Sexual Morality and Modern Nervous Illness," in J. Strachey (ed.), *Complete Psychological Works of Sigmund Freud* (London: Hogarth, 1957), vol. 9.

94. Sandor Ferenczi, *Thalassa* (London: Maresfield, 1990), p. 47.

95. Erich Fromm, *The Sane Society* (London: Routledge and Kegan Paul, 1963), p. 166.

96. Carl Jung, "Psychological Aspects of the Mother Archetype," in Herbert Read, Michael Fordham, and Gerhard Adler (eds.), *The Collected Works of C. G. Jung*, 2nd ed. London: Routledge and Kegan Paul, 1968), vol. 9, p. 81.

97. Stephen Jay Gould, *The Mismeasure of Man* (Harmondsworth, U.K.: Penguin, 1981); Russell Marks, *The Idea of IQ* (Washington, D.C.: University Press of America, 1981).

98. Jacqueline Goodnow, "The Nature of Intelligent Behavior: Questions Raised by Cross-Cultural Studies," in L. Resnick (ed.), *The Nature of Intelligence,* (Hillsdale, N.J.: Lawrence Erlbaum, 1968).

99. As, for example, by Robert Sternberg, in his *Beyond IQ: A Triarchic Theory of Human Intelligence* (Cambridge: Cambridge University Press, 1985), p. 43.

100. Rene Descartes, *The Meditations, and Selections from the Principles* (La Salle, Ill.: Open Court Publishing, 1950), p. 194.

101. Herman Witkin, *Manual for the Embedded Figures Test* (New York: Consulting Psychologists Press, 1989).

102. Richard A. Shweder and Edmund J. Bourne, "Does the Concept of the Person Vary Cross-Culturally?" in Richard A. Shweder and Robert A. LeVine (eds.), *Culture Theory: Essays on Mind, Self, and Emotion* (Cambridge: Cambridge University Press, 1984).

103. Hans Kummer and Jane Goodall, "Conditions of Innovative Behaviour in Primates," in L. Weiskrantz (ed.), *Animal Intelligence* (Oxford: Oxford University Press, 1985), p. 203.

104. Casey, *Getting Back into Place: Toward a Renewed Understanding of the Place-World*, p. 36.

105. Aleksandr Luria, *Cognitive Development: Its Social and Cultural Foundations* (Cambridge, Mass.: Harvard University Press, 1976), pp. 130–31.

106. Ibid., pp. 108–9.

107. Reported by Jacqueline Goodnow, "The Nature of Intelligent Behaviour." There are all too many examples in the psychological literature of the way that dissociation of intellect from the world is often regarded by psychologists as a *desirable* quality rather than a problem, in keeping with the preference for "pure" cognition that we noted above. For example, in Sylvia Scribner's (1977) research with unschooled Vai people of Liberia, one of the problems used was: "All women who live in Monrovia are married. Kemu is not married. Does she live in Monrovia?" Respondents "working from . . . the known fact that there *are* unmarried women in Monrovia . . . could arrive at an incorrect answer" because they abandoned the premise that "all women who live in Monrovia are married." Scribner's characterization of such answers as "incorrect" seems to reflect the preference of many experimenters for an abstract, logical world that is only tenuously connected with knowledge gained through direct experience. She goes on to argue that her findings represent "the strongest evidence to date that traditional people can and do engage in valid deductive reasoning . . . provided they put brackets about what they know to be true and confine their reasoning to the terms of the problems." Among "uneducated" people, Scribner continues, "performance is rarely free from the intrusion of real-world knowledge." From an ecological viewpoint, however, it could be argued that such "intrusions of real-world knowledge" are essential in aligning conceptual functioning with material, social, and ecological realities; and that it is their *absence* rather than their presence that should be regarded as problematic.

108. Thomas Gladwin, *East Is a Big Bird: Navigation and Logic on Puluwat Atoll* (Cambridge, Mass.: Harvard University Press, 1970).

109. Edmund Carpenter, *Eskimo Realities* (New York: Holt, Rinehart, and Winston, 1973), pp. 44–45.

110. Martin Heidegger, "What Is Called Thinking?" In *The Question Concerning Technology and Other Essays* (New York: Harper & Row, 1968), p. 14.

111. Carpenter, *Eskimo Realities*, p. 59.

112. Ibid., p. 137.

113. Ibid., p. 180.

114. The statistical convergence toward a central factor is found even in models such as that of Howard Gardner that are not primarily derived from factor analyses of performance on IQ-style tests, as Messick (1992) has pointed out. See Howard Gardner, *Frames of Mind: The Theory of Multiple Intelligences* (London: Heinemann, 1983); S. Messick, "Multiple Intelligences or Multilevel Intelligence? Selective Emphasis on Distinctive Properties of Hierarchy: On Gardner's *Frames of Mind* and Sternberg's *Beyond IQ* in the Context of Theory and Research on the Structure of Human Abilities," *Journal of Psychological Inquiry* 1.3 (1992): 305–84.

115. See, for example, Steven Jay Gould, *The Mismeasure of Man*. But Gould exaggerates the purity of Binet's intentions, later to be "dismantled" in America. The Frenchman clearly envisaged the social applications of the concept of "intelligence," arguing that "without doubt, one could conceive many possible applications of the process [of intelligence testing], in dreaming of a future where the social sphere would be better organised than ours; where everyone would work according to his known aptitudes in such a way that no particle of psychic force should be lost for society." See Alfred Binet and T. Simon, *The Development of Intelligence in Children*, trans. Elizabeth S. Kite (Baltimore: Williams and Wilkins, 1916).

116. Although there are embarrassing exceptions. See, for example, J. Philippe Rushton, "Race Differences, r/K theory, and a Reply to Flynn," *The Psychologist: Bulletin of the British Psychological Society* 5 (May 1990): 195–98.

117. Paul Rozin, "The Selection of Food by Rats, Humans, and Other Animals," in Jay S. Rosenblatt, Robert A. Hinde, Evelyn Shaw, and Colin Beer (eds.), *Advances in the Study of Behavior*, vol. 6 (New York: Academic Press, 1976), pp. 21–76.

118. Margaret Wilson, *Descartes* (London: Routledge and Kegan Paul, 1982), p. 184.

119. Tim Ingold, "The Optimal Forager and Economic Man," in Philippe Descola and Gísli Pálsson (eds.), *Nature and Society: Anthropological Perspectives* (London: Routledge, 1996), p. 26.

120. Interviewed on *Horizon: Look Who's Talking Now*, BBC 2, December 13, 1993.

121. Ibid.

122. See note 1.

123. See, for example, John S. Kennedy, *The New Anthropomorphism* (New York: Cambridge University Press, 1992).

124. Jonathan Schull, "Are Species Intelligent?" *Behavioral and Brain Sciences* 13 (1990): 63–108.

125. Sigmund Freud, *Civilisation and its Discontents* (London: Hogarth, 1949), p. 13.

126. Ibid., p. 30.

127. Erich Fromm, *Escape from Freedom* (New York: Holt, Rinehart and Winston, 1941), pp. 50–51.

128. Erik H. Erikson, *Childhood and Society,* 3rd ed., (New York: Norton, 1963), p. 149.

129. Denis de Rougement, *Love in the Western World* (Princeton, N.J.: Princeton University Press, 1956), p. 59.

130. Joel Kovel, *History and Spirit: An Inquiry into the Philosophy of Liberation* (Boston: Beacon Press, 1991).

131. W. Ronald D. Fairbairn, *Psychoanalytic Studies of the Personality* (London: Routledge and Kegan Paul, 1952); Guntrip, *Schizoid Phenomena, Object Relations, and the Self.*

132. Levin, *The Opening of Vision: Nihilism and the Postmodern Situation,* p. 218.

133. Hugh Gunnison, "The Uniqueness of Similarities: Parallels of Milton H. Erickson and Carl Rogers," *Journal of Counseling and Development* 63 (1985): 561–64.

134. Carl Rogers, "Reaction to Gunnison's Article on the Similarities between Erickson and Rogers," *Journal of Counseling and Development* 63 (1985): 565–66.

135. John (Fire) Lame Deer and Richard Erdoes, *Lame Deer, Seeker of Visions* (New York: Simon & Schuster, 1972).

136. Hillman and Ventura, *We've Had a Hundred Years of Psychotherapy and the World's Getting Worse,* p. 26.

137. Sigmund Koch, "Psychology and Emerging Conceptions of Knowledge as Unitary," in T. W. Wann (ed.), *Behaviorism and Phenomenology* (Chicago: University of Chicago Press, 1964).

Chapter 3. The Colonization of the Psyche

1. Aldo Leopold, *A Sand County Almanac, with Essays on Conservation from Round River* (New York: Ballantine, 1990; Oxford University Press, 1949), p. 138.

2. Julian Jaynes, *The Origin of Consciousness in the Breakdown of the Bicameral Mind* (Boston: Houghton Mifflin, 1976).

3. Bruno Snell, *The Discovery of the Mind: The Greek Origins of European Thought,* trans. T. G. Rosenmeyer (Oxford: Blackwell, 1953), p. 30.

4. Ibid., pp. 28–29.

5. Ibid., pp. 32–33.

6. Ruth Padel, *In and Out of the Mind: Greek Images of the Tragic Self* (Princeton: Princeton University Press, 1992), p. 11.

7. Ibid., p. 34.

8. Ibid., p. 35.

9. George Lakoff, *Women, Fire, and Dangerous Things: What Categories Reveal about the Mind* (Chicago: University of Chicago Press, 1987).

10. Plumwood, *Feminism and the Mastery of Nature.*

11. Colin Morris, *The Discovery of the Individual 1050–1200* (London: Society for the Propagation of Christian Knowledge, 1972).

12. Marie-Dominique Chenu, *Nature, Man, and Society in the Twelfth Century* (Chicago: University of Chicago Press, 1968), p. 25.

13. Lee Patterson, *Chaucer and the Subject of History*, (London: Routledge, 1991), p. 18.

14. Owen Barfield, *Saving the Appearances: A Study in Idolatry* (New York: Harcourt, Brace, Jovanovich, n.d.), p. 78.

15. Lopez, *Of Wolves and Men*, p. 233.

16. Margaret Thatcher in a BBC interview, quoted by Marilyn Strathern in *After Nature: English Kinship in the Twentieth Century* (Cambridge: Cambridge University Press, 1992), p. 144.

17. Samuel Edgerton, *The Renaissance Rediscovery of Linear Perspective* (New York: Harper & Row, 1975), p. 5.

18. Colin Turnbull, *The Forest People* (New York: Simon & Schuster, 1961), pp. 252–53.

19. Edmund Carpenter, *Oh, What a Blow That Phantom Gave Me!* (St. Albans, U.K.: Paladin, 1976), p. 34.

20. Bateson and Bateson, *Angels Fear: Toward an Epistemology of the Sacred*, p. 161.

21. Romanyshyn, *Technology as Symptom and Dream*, p. 82.

22. See, for example, James Cowan, *Mysteries of the Dreamtime: The Spiritual Life of Australian Aborigines*, revised ed. (Bridport, U.K.: Prism Press, 1992).

23. Of course, TV does in one sense bring events closer to us; but in a similar manner to Galileo's telescope, it also derealizes them. For example, the juxtaposition of news items and ads trivializes the former.

24. John Dewey, *Democracy and Education* (New York: Macmillan, 1916), p. 393.

25. Romanyshyn, *Technology as Symptom and Dream*, p. 77.

26. Ibid., p. 82.

27. Sigmund Freud, "Fixation to Traumas: The Unconscious," in *The Standard Edition of the Complete Psychological Works of Sigmund Freud* (London: Hogarth Press, 1957), 17:355.

28. Romanyshyn, *Technology as Symptom and Dream*, p. 47.

29. Stanley Aronowitz, *Dead Artists, Live Theories, and Other Cultural Problems* (New York: Routledge, 1994), p. 21.

30. Quoted by Carpenter, *Oh, What a Blow That Phantom Gave Me!* p. 29.

31. Casey, *Getting Back into Place*, pp. 229–30.

32. Kenneth Coutts-Smith, "Some General Observations on the Problem of Cultural Colonialism," in Susan Hiller (ed.), *The Myth of Primitivism: Perspectives on Art* (London: Routledge, 1991), pp. 20–21.

33. Coutts-Smith, "Some General Observations," p. 24.

34. Morris Berman, *Coming to Our Senses: Body and Spirit in the Hidden History of the West* (New York: Bantam, 1990).

35. Edgerton, *The Renaissance Rediscovery of Linear Perspective*, p. 162.

36. Coutts-Smith, "Some General Observations," pp. 26–27.

37. Today, the part played by art in the colonization of the life-world by an economistic ideology has become largely obsolete, art itself having been sidelined by that same ideology that it helped to establish—unless, that is, one sees television commercials as the ideological descendants of post-Renaissance painting.

38. Oliver Wendell Holmes, "The Stereoscope and the Stereograph," *The Atlantic Monthly* 3 (June 1859), reprinted in Beaumont Newhall (ed.), *Photography: Essays and Images* (London: Secker and Warburg, 1981).

39. Oliver Wendell Holmes, quoted by Stuart Ewen, *All Consuming Images* (New York: Basic Books, 1988), p. 25.

40. Ewen, *All Consuming Images*, p. 25.

41. Holmes, "The Stereoscope and the Stereograph," p. 60.

42. Ansel Adams, "A Personal Credo," reprinted in Newhall (ed.), *Photography*, pp. 257–261.

43. Charles Baudelaire, "Photography," reprinted in Newhall, *Photography*, p. 113.

44. Ansel Adams, "A Personal Credo," reprinted in Newhall, *Photography*.

45. The term "*f* 64" refers to the ratio of the aperture of a lens to its focal length. Since a setting of *f* 64 reflects a very small aperture, the resulting photograph will have a large depth of field; and this was considered by members of the *f* 64 group to be an essential feature of an approach that was intended to express the natural world as truthfully and completely as possible, rather than imposing the individual artist's conception upon the world.

46. Quoted by John Paul Edwards, "Group *f* 64," reprinted in Newhall, *Photography*, p. 252. Italics in original.

47. Barry Lopez, *Arctic Dreams: Imagination and Desire in a Northern Landscape* (London: Macmillan, 1986).

48. Ibid., p. 245.

49. Ibid., p. 247.

50. Carolyn Merchant, *The Death of Nature: Women, Ecology, and the Scientific Revolution* (New York: Harper & Row, 1980).

51. C. F. Hockett, *Man's Place in Nature* (New York: McGraw-Hill, 1973), p. 612.

52. David Michael Levin, *The Body's Recollection of Being* (London: Routledge and Kegan Paul, 1985), p. 128.

53. Jack Turner, *The Abstract Wild* (Tucson: University of Arizona Press, 1996), p. 8.

54. Jay Bernstein, *The Fate of Art: Aesthetic Alienation from Kant to Derrida and Adorno* (Cambridge: Polity Press, 1992) (my italics).

55. This, of course, is the theme of much science fiction such as Michael Crichton's *Jurassic Park.*

56. Alexander J. Argyros, *A Blessed Rage for Order: Deconstruction, Evolution, and Chaos* (Ann Arbor: University of Michigan Press, 1991), p. 2.

57. T. G. H. Strehlow, *Songs of Central Australia* (Sydney: Angus and Robertson, 1971), p. 9.

58. Guntrip, *Schizoid Phenomena, Object Relations, and the Self,* pp. 237–38.

59. Kovel, *History and Spirit,* p. 54.

60. Sigmund Freud, "Thoughts for the Times on War and Death," *Complete Works,* 14:287.

61. Sigmund Freud, "The Question of a Weltanschauuing," *Complete Works* 22:170.

62. Sandor Ferenczi, *Final Contributions to the Problems and Methods of Psychoanalysis* (London: Maresfield Reprints, 1955), p. 245.

63. Ibid., p. 246 (italics in original).

64. Lewis Mumford, *Technics and Civilisation* (London: Burlingame, 1963).

65. Isaac Prilleltensky, "Psychology and the Status Quo," *American Psychologist* 44 (1989): 795–802; Edward Sampson, *Justice and the Critique of Pure Psychology* (New York: Plenum, 1983).

66. Cushman, "Why the Self is Empty; Toward a Historically Located Psychology," 599–611.

67. Guntrip, *Schizoid Phenomena.*

68. See Anthony Storr, *The Dynamics of Creation* (London; Secker and Warburg, 1972), for a full discussion of this point.

69. Charles Darwin, *The Autobiography of Charles Darwin* (London: Collins, 1958), p. 91.

70. Ibid., p. 139.

71. Guntrip, *Schizoid Phenomena,* p. 56.

72. Evelyn Fox Keller, *Reflections on Gender and Science* (New Haven and London: Yale University Press, 1985), p. 148.

73. John Dewey, *The Quest for Certainty: A Study of the Relation between Knowledge and Action* (London: Allen and Unwin, 1930), p. 210.

74. Victor Ferkiss, *Nature, Technology, and Society: Cultural Roots of the Current Environmental Crisis* (New York: New York University Press, 1993), p. 26. (my emphasis).

75. See, for example, Carolyn Merchant, *The Death of Nature*; Karen J. Warren, "The Power and the Promise of Ecological Feminism," *Environmental Ethics* 12 (1990): 125–46.

76. Plumwood, *Feminism and the Mastery of Nature.*

77. Pascal, *Pensées* (Paris: Garnier, 1964; originally published 1670).

78. Jay M. Bernstein, "Introduction," in Theodor W. Adorno, *The Culture Industry: Selected Essays on Mass Culture*, ed. J. M. Bernstein (London: Routledge, 1991), p. 4.

79. Murray L. Wax and Rosalie H. Wax, "Cultural Deprivation as an Educational Ideology," in Eleanor B. Leacock (ed.), *The Culture of Poverty: A Critique* (New York: Simon & Schuster, 1971).

80. Francis La Flesche, *The Middle Five: Indian Schoolboys of the Omaha Tribe* (Madison: University of Wisconsin Press, 1963), p. xx. Quoted by Wax and Wax, "Cultural Deprivation as an Educational Ideology."

81. See, for example, Michael Maccoby and Nancy Modiano, "On Culture and Equivalence: 1," in J. S. Bruner et al. (eds.), *Studies in Cognitive Growth.* (New York: Wiley, 1965).

82. Veronica Strang, *Uncommon Ground: Cultural Landscapes and Environmental Values* (Oxford: Berg, 1997), pp. 182–83.

83. Strehlow, *Songs of Central Australia*, p. 313.

84. Goodnow, "The Nature of Intelligent Behaviour: Questions Raised by Cross-Cultural Studies."

85. Barfield, *Saving the Appearances: A Study in Idolatry.*

86. Gladwin, *East Is a Big Bird: Navigation and Logic on Puluwat Atoll,* pp. 220–22.

87. Cushman, "Why the Self Is Empty: Toward a Historically Situated Psychology," pp. 599–611.

88. Jean Piaget, *The Psychology of Intelligence* (London: Routledge and Kegan Paul, 1950), p. 7. More recent theories also normalize the modern self and its course of development. See Philip Cushman, "Ideology Obscured: Political Uses of the Self in Daniel Stern's Infant," *American Psychologist* 46 (1991): 206–19.

89. Barfield, *Saving the Appearances*, p. 67.

90. Richard A. Shweder and Edmund J. Bourne, "Does the Concept of the Person Vary Cross-Culturally?" in Richard A. Shweder, *Thinking Through Cultures: Expeditions in Cultural Psychology* (Cambridge, Mass.: Harvard University Press, 1991).

91. Paul Shepard, *Nature and Madness*, (San Francisco: Sierra Club, 1982), p. 102.

92. See, for example, W. Rostow, *The Stages of Economic Growth* (Cambridge: Cambridge University Press, 1960).

93. John Dewey, *The Quest for Certainty: A Study of the Relation between Knowledge and Action* (London: Allen and Unwin, 1930), p. 276.

94. Giordano Bruno, *The Heroic Frenzies*, translated by Paul E. Memmo (Chapel Hill: University of North Carolina Press, 1964).

95. Piaget, *The Psychology of Intelligence*, pp. 8–9.

96. Ibid., p. 151 (my italics).

97. Susan Buck-Morss, "Socio-economic Bias in Piaget's Theory and Its Implications for Cross-Cultural Studies," *Human Development* 18 (1975): 40.

98. Buck-Morss, " Socioeconomic bias in Piaget's Theory," p. 39.

99. Teresa Brennan, *History after Lacan* (London: Routledge, 1993), p. 11.

100. Michael Garfield "Possible Worlds or Real Worlds?" in S. Modgil, C. Modgil, and G. Brown (eds.), *Jean Piaget: An Interdisciplinary Critique.* (London: Routledge and Kegan Paul, 1983.), p. 187 (my italics).

101. Angus Gellatly "The Myth of Cognitive Diagnostics," in A. Gellatly, D. Rogers, and J. A. Sloboda, (eds.), *Cognition and Social Worlds* (Oxford: Clarendon Press, 1989), p. 129. See also R. W. Byrne, and A. Whiten (eds.), *Machiavellian Intelligence: Social Expertise and the Evolution of Intellect in Monkeys, Apes, and Humans* (Oxford: Clarendon Press, 1988); and L. B. Resnick et al. (eds.), *Perspectives on Socially Shared Cognition* (Washington, D.C.: American Psychological Association, 1991).

102. Bateson, *Steps to an Ecology of Mind*, p. 469.

103. This fertile but unjustifiably forgotten distinction was introduced by Donald O. Hebb in *The Organisation of Behaviour* (New York: Wiley, 1949). See also Raymond B. Cattell, *Abilities: Their Structure, Growth, and Action* (Boston: Houghton Mifflin, 1971) for a factor analytic understanding.

104. Paul Shepard, "Nature and Madness," in Theodore Roszak, Mary E. Gomes, and Allen Kanner (eds.), *Ecopsychology: Restoring the Earth, Healing the Mind* (San Francisco: Sierra Club, 1995), p. 30.

105. Margaret Donaldson, *Children's Minds* (New York: Norton, 1978).

106. Buck-Morss, "Socio-economic Bias in Piaget's Theory," pp. 35–49.

107. Leopold, *A Sand County Almanac*, p. 168.

Chapter 4. Natural Cultures, Psychic Landscapes

1. Hillman and Ventura, *We've Had 100 Years of Psychotherapy and the World's Getting Worse*, p. 70.

2. Clifford Geertz, *The Interpretation of Cultures* (New York: Basic Books, 1973); Shweder, *Thinking through Cultures: Explorations in Cultural Psychology*; Edward Sampson, *Justice and the Critique of Pure Psychology* (New York: Plenum, 1983).

3. Geertz, *The Interpretation of Cultures*, p. 49.

4. Ibid., p. 89.

5. Don Cupitt, interviewed in the Channel 4 series *The Wisdom of the Dream.*

6. Sam Gill, "The Shadow of a Vision Yonder," in Walter Holden Capps (ed.), *Seeing with a Native Eye* (New York: Harper & Row, 1976).

7. Lauriston Sharp, "Steel Axes for Stone Age Australians," in Edward H. Spicer (ed.), *Human Problems in Technological Change* (New York: Wiley, 1952), pp. 77–78, 80.

8. Ibid., pp. 85–86.

9. For a discussion, see Strang, *Uncommon Ground: Cultural Landscapes and Environmental Values.*

10. Ibid., p. 130.

11. C. G. Jung, "The Spiritual Problem of Modern Man." *Collected Works* 10.

12. Strang, *Uncommon Ground,* p. 231.

13. Ibid., p. 252.

14. Ibid., p. 253.

15. Ibid., p. 179.

16. Strehlow, *Songs of Central Australia,* pp. 706–8. I have not attempted to reproduce here the complex notation that Strehlow uses in printing Aranda speech.

17. David Lewis, "Observations on Route Finding and Spatial Orientation among the Aboriginal Peoples of the Western Desert Region of Central Australia," *Oceania* 46.4 (June 1976): 249–82.

18. Kovel, *History and Spirit: An Inquiry into the Philosophy of Liberation,* p. 24.

19. J Donald Hughes, *American Indian Ecology* (El Paso: Texas Western Press, 1983), p. 14. Some readers will probably object that this much too brief and rather grumpy account of Christianity and its role as cultural mediator fails to do justice to the potential that writers such as Susan Bratton and Max Oelschlaeger have perceived. I hope that they are right. The fact remains, however, that Christianity's record in the environmental debate has been a sorry one, with the attitude of many Christian theologians being at best indifferent toward environmental issues, and at worst awe-inspiringly naive or even actively supporting the destruction of the natural world. Those few Christian writers who *have* courageously spoken out for the natural world, such as Matthew Fox and Thomas Berry, have often been marginalized or expelled from their churches. For a summary of this topic, see Joseph K. Sheldon, *The Rediscovery of Creation: A Bibliographical Study of the Church's Response to the Environmental Crisis* (Metuchen, N.J.: American Theological Library Association, 1992).

20. Vine Deloria Jr., *God Is Red* (New York: Grosset and Dunlap, 1973), p. 11.

21. Elizabeth Rees, *Christian Symbols, Ancient Roots* (London: Jessica Kingsley, 1992), pp. 14–21.

22. Deloria, *God Is Red*, p. 91.

23. Turner, *Beyond Geography: The Western Spirit against the Wilderness*, p. 23.

24. W. E. H. Stanner, "Some Aspects of Aboriginal Religion," in *The Australian and New Zealand Theological Review* 76 (1976): 31. Quoted by Strang, *Uncommon Ground*, p. 159.

25. Casey, *Getting Back into Place: Toward a Renewed Understanding of the Place-World*, pp. 36–37.

26. Ibid., p. 35.

27. Kleinman, *Rethinking Psychiatry*, p. 56.

28. Colin Turnbull, *The Mountain People* (London: Picador, 1974), pp. 24–25.

29. Turnbull's ethnography has been challenged in an ill-tempered series of critiques in the pages of *Current Anthropology*. See, for example, Fredrik Barth, "On Responsibility and Humanity: Calling a Colleague to Account," *Current Anthropology* 15.1 (1974): 99–103.

30. Turnbull, *The Mountain People*, pp. 238–39.

31. Joanna Macy, "Working through Environmental Despair," in Roszak et al. (eds.), *Ecopsychology: Restoring the Earth, Healing the Mind*, p. 244.

32. Smail, *The Origins of Unhappiness*, p. 93.

33. See Roszak et al (eds.), *Ecopsychology*.

34. Elizabeth Rees, in her *Christian Symbols, Ancient Roots*, discusses tree symbolism at length.

35. Quoted by Turner, *Beyond Geography*, p. 129.

36. Turner, *Beyond Geography*, pp. 130–31.

37. Bartolomé de Las Casas, *History of the Indies*, Trans. André Collard (New York: Harper & Row, 1971).

38. Turner, *Beyond Geography*, p. 142.

39. Michel de Montaigne, *Essays* (London, 1910; reprinted, 1965), 3:144.

40. Robert Pogue Harrison, *Forests: The Shadow of Civilisation* (Chicago: University of Chicago Press, 1992), p. 225.

41. Las Casas, *History of the Indies*, documents many incidents that confirm this.

42. Turner, *Beyond Geography*, p. 68.

43. Annie Heloise Abel, *Tabeau's Narrative of Loisel's Expedition to the Upper Missouri* (Norman: University of Oklahoma Press, 1939), pp. 134–35.

44. Turner, *Beyond Geography*, p. 287.

45. Frank Waters, *The Colorado* (New York: Holt, Rinehart, and Winston, 1974), p. 257.

46. Michael Taussig, *Shamanism, Colonialism, and the Wild Man: A Study in Terror and Healing* (Chicago: University of Chicago Press, 1987), p. 56.

47. Jules Crévaulx, "Exploración del Inzá y del Yapura," in *América pintoresca: Descripción de viajes al nuevo continente* (Barcelona: Montaner y Simon, 1884). Quoted by Taussig, *Shamanism, Colonialism, and the Wild Man*, p. 56.

48. Roy Harvey Pearce, *Savagism and Civilisation: A Study of the Indian and the American Mind* (Baltimore: John Hopkins Press, 1965), p. 85.

49. Ibid., p. 118.

50. Ibid., p. 139.

51. Hector St. John de Crevecoeur, *Letters from an American Farmer* (Everyman's Library Edition, p. 11); quoted by Pearce, *Savagism and Civilisation*, p. 140.

52. William Bartram, *The Travels of William Bartram*, ed. Mark van Doren (New York, 1928); quoted by Pearce, *Savagism and Civilisation*, pp. 142–43.

53. James Hillman, *The Essential James Hillman: A Blue Fire* (Routledge, 1990), p. 28.

54. John M. Ellis, *Literature Lost: Social Agendas and the Corruption of the Humanities* (New Haven: Yale University Press, 1997), p. 47.

55. For an account of this continuing oppression and poverty, see M. Annette Jaimes (ed.), *The State of Native America* (Boston: South End Press, 1992).

56. For example, Michael Shanks and Christopher Tilley, in *Social Theory and Archeology* (Cambridge: Polity Press, 1987), argue that "[t]he past . . . is gone; it can't be recaptured in itself, relived as object. It exists now in its connection with the present, in the present's practice of interpretation" (p. 26). According to this view, then, the past only has validity in terms of current interpretations—interpretations that are detached, intellectual, unemotional, that exist only in the world of the academy. What is repressed by this approach is the *empathic* connection that relates us to the past, the geographically distant, the generalized Other, that acknowledges the emotional and spiritual validity of an other that is not materially present. In denying these connective realms, an approach based on the study of competing interpretations renders historical and cultural others as unreal as the feelings of Claude Bernard's vivisected dogs.

57. Mason, *Reconstructing America: Representations of the Other*, p. 2.

58. Ibid., p. 24.

59. Ibid., p. 20.

60. Ibid., p. 15.

61. George E. Marcus and Michael M. J. Fischer, *Anthropology as Cultural Critique: An Experimental Moment in the Human Sciences* (Chicago: University of Chicago Press, 1976), p. 161.

62. Robert B. Edgerton, *Sick Societies: Challenging the Myth of Primitive Harmony* (New York: Free Press, 1992), p. 7 (my emphasis).

63. The defensive appeal to obfuscating qualification and relativization, and to pleas of "insufficient evidence," need for "more research," and the denial of realities other than the rational as excuses for inaction are by now, of course, ploys familiar to environmental activists.

64. Marcus and Fischer, *Anthropology as Cultural Critique*, p. 112.

65. Ibid., p. 113.

66. Ibid., p. 115.

67. Ibid., p. 131.

68. S. Hecht, "Tropical Deforestation in Latin America: Myths, Dilemmas, and Reality," paper presented at the Systemwide Workshop on Environment and Development Issues in Latin America, University of California, Berkeley, October 16, 1990. Quoted by Gómez-Pompa and Kaus, "Taming the Wilderness Myth," p. 274.

69. Marcus and Fischer, *Anthropology as Cultural Critique*, p. 133.

70. Ibid., p. 134.

71. Ibid., p. 113.

72. Ibid., p. 116.

73. Ibid., p. 134.

74. Ibid., p. 167.

75. Ibid., p. 166–67.

76. Stanley Diamond, *In Search of the Primitive* (New Brunswick, N.J.: Transaction, 1974), pp. 109–10.

77. For another, equally courageous attempt to challenge the taboo on involvement in fieldwork, see Don Kulick and Margaret Willson (eds.), *Taboo: Sex, Identity, and Erotic Subjectivity in Anthropological Fieldwork* (London: Routledge, 1995).

78. Johannes Fabian, *Time and the Other: How Anthropology Makes Its Object* (New York: Columbia University Press, 1983), p. 121.

79. Ulrich Beck, *Risk Society: Towards a New Modernity* (London: Sage, 1992).

80. Edgerton, *Sick Societies.*

81. Edgerton, *Sick Societies*, p. 186.

82. Bateson and Bateson, *Angels Fear,* p. 72. Bateson is quoting from W. L. Thomas Jr. (ed.), *Man's Role in Changing the Face of the Earth: Symposium of the Wenner-Gren Foundation* (Chicago: University of Chicago Press, 1956), p. 953.

83. Bateson and Bateson, *Angels Fear,* p. 73.

84. Turnbull, *The Mountain People*, p. 243. It is no coincidence which views which affirm the importance of culture in relation to the psyche are invariably those of anthropologists, that is, researchers who have lived in nonindustrialized cultures. Those who have not ventured beyond the boundaries of academia are typically unable to transcend the epistemological limits of this universe.

85. Turnbull, *The Mountain People*, p. 240.

86. Cisco Lassiter, "Relocation and Illness: The Plight of the Navaho," in David M. Levin (ed.), *Pathologies of the Modern Self: Postmodern Studies on Narcissism, Schizophrenia, and Depression* (New York: New York University Press, 1987), p. 229.

87. Robert Desjarlais, Leon Eisenberg, Byron Good, and Arthur Kleinman, *World Mental Health: Problems and Priorities among Low Income Countries* (New York: Oxford University Press, 1995).

88. Guntrip, *Schizoid Phenomena, Object Relations, and the Self*, p. 237. Italics in original.

89. See Kleinman, *Rethinking Psychiatry*.

90. Ibid., p. 39.

91. R. Warner, *Recovery from Schizophrenia: Psychiatry and Political Economy* (New York, Routledge and Kegan Paul, 1985).

92. Nancy Scheper-Hughes, *Death without Weeping: The Violence of Everyday Life in Brazil* (Berkeley: University of California Press, 1982), p. 36.

93. Mark Cohen, *Health and the Rise of Civilisation* (New Haven: Yale University Press, 1989), p. 61.

94. Scheper-Hughes, *Death without Weeping*, chapter 5.

95. Ibid., p. 207.

96. Ibid., p. 195.

97. Ibid., p. 131.

98. Ibid., p. 132. The experiential realities that Turnbull revealed in *The Mountain People* were not only those of the Ik with whom he lived; for he is quite open about his own emotional reactions to the often extreme situations to which he was exposed. Perhaps what was most disturbing to his critics was this emotional interplay between the anthropologist and his informants, contrasting as it does with the partial, intellectualized form of relationship that most anthropologists feel comfortable with. See the reviews of *The Mountain People*, and Turnbull's response, in *Current Anthropology*, vols. 15 and 16.

99. Scheper-Hughes, *Death without Weeping*, p. 195.

100. Michael Taussig, *The Devil and Commodity Fetishism in South America* (Chapel Hill: University of North Carolina Press, 1980).

101. Ibid., pp. 86–87.

102. Ibid., pp. 155–58.

103. Ibid., p. 161.

104. Ibid., p. 162.

105. Ibid., pp. 136–38.

106. Ibid., p. 158.

107. Ibid., p. 74.

108. Ibid., p. 3.

109. Taussig describes this extensively not only in *The Devil and Commodity Fetishism*, but also in his later *Shamanism, Colonialism, and the Wild Man*.

110. Bronislaw Malinowski, *Coral Gardens and Their Magic*, 2 vols. (Bloomington: Indiana University Press, 1965), 1:19–20. Quoted by Taussig, *The Devil and Commodity Fetishism*, p. 19.

111. Taussig, *The Devil and Commodity Fetishism*, p. 30.

112. Ibid., p. 93.

113. Ibid., pp. 120–21.

Chapter 5. The Psychodynamnics of Self-World Relations

1. Quoted in *Secret Lives: The Young Freud*, Channel 4 Television, 2/3/1995.

2. Garrett Hardin, "The Tragedy of the Commons," *Science* 162 (1968):1243–48.

3. Sigmund Freud, *Civilisation and Its Discontents* (New York: Norton, 1961), p. 33.

4. Ibid., p. 24.

5. Ibid., p. 39.

6. Quoted by Ferkiss, Nature, *Technology, and Society: Cultural Roots of the Current Environmental Crisis*, p. 45.

7. Clifford Geertz, " 'From the Native's Point of View': On the Nature of Anthropological Understanding," in Richard Shweder and Robert LeVine (eds.), *Culture Theory: Essays on Mind, Self, and Emotion* (Cambridge: Cambridge University Press, 1984), p. 126.

8. Freud, *Civilisation and Its Discontents*, p. 91.

9. Ibid., p. 34.

10. Ibid., pp. 65–66.

11. Sigmund Freud, "Anxiety and Instinctual Life," in *The Standard Edition of the Complete Psychological Works of Sigmund Freud*, 22:106.

12. Freud, *Civilisation and Its Discontents*, p. 65. See also his "On Narcissism: An Introduction," *The Standard Edition of the Complete Psychological Works of Sigmund Freud*, vol. 14: 67–102.

13. Freud, *Civilisation and Its Discontents*, p. 68.

14. Frank Sulloway, *Freud: Biologist of the Mind* (London: Burnett, 1979), p. 4.

15. Erich Fromm, *Escape from Freedom* (New York: Avon, 1969; originally published 1941), p. 51.

16. Erich Fromm, *The Sane Society* (London: Routledge and Kegan Paul, 1963), p. 25.

17. Ibid., p. 24.

18. Ibid., p. 49.

19. Margaret Mead, *Sex and Temperament in Three Primitive Societies* (New York: New American Library, 1952), p. 100; quoted by Herbert Marcuse, *Eros and Civilisation* (Boston: Beacon Press, 1977 [1955]), p. 216.

20. Marcuse, *Eros and Civilisation*, p. 216.

21. Ibid., p. 201.

22. Ibid., p. 202.

23. Ibid.

24. Rodman, "The Liberation of Nature?" p. 105.

25. Ibid., p. 104.

26. *Black's Medical Dictionary*, 36th ed. (London: A. and C. Black, 1990).

27. William James, *The Varieties of Religious Experience* (London: Longman's, 1935), p. 416.

28. Lasch, *The Culture of Narcissism*; *The Minimal Self: Psychic Survival in Troubled Times* (New York: Norton, 1984); Freud, "On Narcissism: An Introduction."

29. Lasch, *The Culture of Narcissism*, p. 36.

30. Freud, "On Narcissism: An Introduction," p. 88.

31. Lasch, *The Minimal Self*, p. 59.

32. Paul Shepard, *Traces of an Omnivore* (Washington, D.C.: Shearwater Press, 1996), p. 68.

33. Shepard, *Traces of an Omnivore*, p. 31.

34. Charlene Spretnak, *The Spiritual Dimension of Green Politics* (Santa Fe: Bear and Co., 1986), p. 15.

35. Freud, *Civilisation and Its Discontents*, p. 52.

36. Ibid., p. 52.

37. Michel Foucault, *The History of Sexuality, Vol. 1: An Introduction* (New York: Random House, 1978).

38. Lasch, *The Culture of Narcissism*, p. 191.

39. Norman O. Brown, *Life against Death: The Psychoanalytic Meaning of History* (Middletown, Conn.: Weslyan University Press, 1959), p. 172.

40. Marcuse, *Eros and Civilisation*, p. 207.

41. David E. Scharff and Ellinor Fairbairn Birtles (eds.), *From Instinct to Self: Selected Papers of W. R. D. Fairbairn. Vol. 1: Clinical and Theoretical Papers* (Northants, N.J.: Jason Aronson, 1994), pp. 133–36.

42. Freud, *Civilisation and Its Discontents*, p. 55.

43. Marcuse, *Eros and Civilisation*, p. 228.

44 Shepard, *Nature and Madness*, p. 28.

45. Colin Turnbull, *The Human Cycle*, (London: Jonathan Cape, 1984), pp. 31–32.

46. Turnbull, *The Human Cycle* p. 116.

47. Marcuse, *Eros and Civilisation*, p. 210.

48. Morris Berman, *Coming to Our Senses: Body and Spirit in the Hidden History of the West* (New York: Simon & Schuster, 1989).

49. Hillman, *Re-Visioning Psychology* (New York: Harper & Row, 1975), p. 245.

50. Romanyshyn, *Psychological Life: From Science to Metaphor*, p. 96.

51. Hillman and Ventura, *We've Had 100 Years of Psychotherapy and the World's Getting Worse*, p. 183.

52. W. Ronald D. Fairbairn, *Psychoanalytic Studies of the Personality* (London: Routledge and Kegan Paul, 1952), p. 139.

53. Ibid., p. 140.

54. Ibid., p. 139.

55. Joel Kovel, *The Age of Desire: Reflections of a Radical Psychoanalyst* (New York: Pantheon, 1981), p. 233.

56. Carl Jung, quoted by Dolores LaChapelle, *Sacred Land, Sacred Sex* (Durango: Kivaki Press, 1988), p. 74.

57. Eugene Gendlin, "A Philosophical Critique of the Concept of Narcissism," in David M. Levin (ed.), *Pathologies of the Modern Self: Postmodern Studies on Narcissism, Schizophrenia, and Depression* (New York: New York University Press, 1987), p. 257.

58. Joel Kovel, *The Radical Spirit: Essays on Psychoanalysis and Society* (London: Free Association, 1988), p. 295.

59. Marcuse, *Eros and Civilisation*, p. 86.

60. Donald Winnicott, *Playing and Reality* (Harmondsworth, U.K.: Penguin, 1974), p. 123.

61. Ibid., p. 3.

62. Ibid., p. 6 (my italics).

63. Dorothy Lee, *Freedom and Culture* (New York: Prentice Hall, 1959), p. 34.

64. Ibid.

65. Guntrip, *Schizoid Phenomena, Object Relations, and the Self*, p. 419.

66. Winnicott, *Playing and Reality*, p. 76.

67. Ibid., p. 3.

68. Ibid., p. 118.

69. Ibid., p. 128.

70. Ibid., p. 76.

71. Fairbairn, *Psychoanalytic Studies of the Personality*, p. 8.

72. Guntrip, *Schizoid Phenomena*, pp. 237–38.

73. Ronald. D. Laing, *The Politics of Experience* (Harmondsworth, U.K.: Penguin, 1969), p. 22.

74. Abraham H. Maslow, *Toward a Psychology of Being* (New York: Van Nostrand, 1968), p. 16.

75. Evelyn Fox Keller, *Reflections on Gender and Science* (New Haven and London: Yale University Press, 1985), p. 148.

76. Roszak et al. (eds). *Ecopsychology: Restoring the Earth, Healing the Mind.*

77. David Ingleby, "Ideology and the Human Sciences: Some Comments on the Role of Reification in Psychology and Psychiatry," *The Human Context* 11 (1970): 159–87.

Chapter 6. Resymbolizing Nature

1. Warwick Fox, *Toward a Transpersonal Ecology* (Boston: Shambhala, 1990), p. 252.

2. Arne Naess, *Ecology, Community, and Lifestyle: Outline of an Ecosophy* (Cambridge: Cambridge University Press, 1989).

3. Fox, *Toward a Transpersonal Ecology*, p. 231.

4. Sigmund Freud, "Mourning and Melancholia," in *The Standard Edition of the Complete Psychological Works of Sigmund Freud*, 14:249–50.

5. Fox, *Toward a Transpersonal Ecology*, p. 258.

6. Ibid., p. 252.

7. See, for example, William M. Shaffer, "Stretching and Folding in Lynx Fur Returns: Evidence for a Strange Attractor in Nature?" *American Naturalist* 124 (1984): 798–820; William M. Shaffer and M. Kot, "Do Strange Attractors Govern Ecological Systems?" *BioScience* 35 (1985): 342–50. For an excellent introduction to this field, see James Gleick, *Chaos: The Making of a New Science* (London: Heinemann, 1988).

8. Jack Turner, " 'In Wildness Is the Preservation of the World,' " in George Sessions (ed.), *Deep Ecology for the Twenty-First Century* (Boston: Shambhala, 1995).

9. Naess, *Ecology, Community, and Lifestyle: Outline of an Ecosophy*, p. 165.

10. Fox, *Toward a Transpersonal Ecology*, p. 261.

11. Loren Eiseley, *Night Country* (London: Garnstone Press, 1974). Quoted by Linda Hogan, *Dwellings: A Spiritual History of the Living World* (New York: Norton, 1995), pp. 96–97.

12. Naess, *Ecology, Community, and Lifestyle*, p. 173.

13. Michael A. Wallach and Lise Wallach, *Psychology's Sanction for Selfishness: The Error of Egoism in Theory and Therapy* (New York: Freeman, 1983).

14. Liam Hudson and Bernadine Jacot, *The Way We Think: Intellect, Intimacy, and the Erotic Imagination* (New Haven: Yale University Press, 1991), p. 13.

15. Charles Taylor, *The Ethics of Authenticity* (Cambridge, Mass.: Harvard University Press, 1992), pp. 40–41.

16. Ibid., pp. 60–61.

17. Shan Guisinger and Sidney J. Blatt, "Individuality and Relatedness: Evolution of a Fundamental Dialectic," *American Psychologist* 49.2 (1994): 104–11.

18. Naess, *Ecology, Community, and Lifestyle: Outline of an Ecosophy*, p. 165.

19. As Jacqueline Rose has put it, "The unconscious is the only defence against a language frozen into pure, fixed, or institutionalised meaning." See her *Sexuality in the Field of Vision* (London: Verso, 1986).

20. Hillman, *Re-Visioning Psychology*, p. 150.

21. Fox, *Toward a Transpersonal Ecology*, p. 261.

22. Theodore Roszak, *Where the Wasteland Ends: Politics and Transcendence in Postindustrial Society* (New York: Doubleday, 1972), p. 382.

23. Philip Cushman, "History, Psychology, and the Abyss," *Psychohistory Review* 15 (1987): 41.

24. Sigmund Freud, "The Question of a Weltanschauung," *The Standard Edition of the Complete Psychological Works of Sigmund Freud*, 22:171.

25. David M. Levin, *The Body's Recollection of Being: Phenomenological Psychology and the Deconstruction of Nihilism* (London: Routledge and Kegan Paul, 1985).

26. Edward Abbey, *Desert Solitaire: A Season in the Wilderness* (New York: Ballantine Books, 1968).

27. Sigmund Freud, "A Case of Obsessional Neurosis," *The Standard Edition of the Complete Psychological Works of Sigmund Freud*, vol. 3.

28. See: Goldsmith, "Gaia: Some Implications for Theoretical Ecology," Also, Neil Evernden, *The Natural Alien* (Toronto: University of Toronto Press, 1985); McIntosh, *The Background of Ecology*; Peters, *A Critique for Ecology*.

29. Carlos Castaneda, *A Separate Reality* (Harmondsworth, U.K.: Penguin, 1973).

30. See: J. Baird Callicott, *Earth's Insights* (Berkeley: University of California Press, 1994), p. 131, for a discussion of the "Seattle Affair."

31. Richard De Mille, *Castaneda's Journey: The Power and the Allegory* (London: Abacus, 1978).

32. Christopher Bollas, *The Shadow of the Object: Psychoanalysis of the Unthought Known* (New York: Columbia University Press, 1987).

33. James Hillman, *Healing Fictions* (Barrytown, N.Y.: Station Hill, 1983), p. 93.

34. Alan Sokal and Jean Bricmont, *Intellectual Impostures* (London: Profile, 1998).

35. Robert Elliot, "Why Preserve Species?" in Don S. Mannison et al. (eds.), *Environmental Philosophy* (Canberra: Australian National University, 1980).

36. Robert Pogue Harrison, *Forests: The Shadow of Civilisation*, p. 199.

37. Erich Neumann, *The Origins and History of Consciousness* (London: Routledge and Kegan Paul, 1954), p. 365.

38. Rees, *Christian Symbols, Ancient Roots*, p. 21.

39. Shepard, *Traces of an Omnivore*, p. 28.

40. Luigi Zoja, *Growth and Guilt: Psychology and the Limits of Development* (London: Routledge, 1995).

41. Zoja, *Growth and Guilt*, p. 11.

42. Lopez, *Of Wolves and Men*, p. 104.

43. Zoja, *Growth and Guilt*, p. 161.

44. Jacques Derrida, *Writing and Difference* (Chicago: University of Chicago Press, 1978).

45. Geertz, *The Interpretation of Cultures*, p. 52.

46. Robert Bellah, *Habits of the Heart: Individualism and Commitment in American Life* (New York: Harper & Row, 1986), p. 150.

47. Diamond, *In Search of the Primitive*, p. 172.

48. Lasch, *The Culture of Narcissism*, p. 11.

49. Ibid.

50. Carl G. Jung, "Approaching the Unconscious," in C. G. Jung (ed.), *Man and His Symbols* (Garden City, N.Y.: Doubleday, 1964), pp. 94-95.

51. Gendlin, "A Philosophical Critique of the Concept of Narcissism," pp. 265–66.

52. Carl Jung, "The Structure and Dynamics of the Psyche," in *Complete Works* 8 (London: Routledge, 1969), pp. 132–33.

53. Luther Standing Bear, *Land of the Spotted Eagle* (Lincoln: University of Nebraska Press, 1978), p. 212.

54. Rodman, "The Liberation of Nature?" p. 112.

55. Carl Jung, "Mind and Earth," in *Complete Works* 10 (London: Routledge and Kegan Paul, 1970), p. 31.

56. Carl Jung, "Archetypes of the Collective Unconscious," in *Complete Works* 9.

57. Victor Turner, *The Forest of Symbols* (Ithaca: Cornell University Press, 1967).

58. Turner, *The Abstract Wild*, p. 15.

59. Karl Stern points out that the words for "mother" and "matter" (Latin *mater* and *materia*) are etymologically related in several languages. See his *The Flight from Woman* (London: George Allen and Unwin, 1966), pp. 77–78.

60. K. A. Busia, "The Ashanti," in Darryl Forde (ed.), *African Worlds: Studies in the Cosmological Ideas and Social Values of African Peoples* (London: Oxford University Press, 1954).

61. Cf. Romanyshyn, *Technology as Symptom and Dream*, chapter 5.

62. Bateson and Bateson, *Angels Fear: Towards an Epistemology of the Sacred*, p. 147.

63. Carl Jung, *Man and His Symbols*, p. 9.

64. Bateson and Bateson, *Angels Fear*, p. 188.

65. Roderick Nash, *Wilderness and the American Mind* (New Haven: Yale University Press, 1967), p. 1.

66. Gary Snyder, *The Practice of the Wild* (Berkeley, Calif.: North Point Press, 1990), p. 6.

67. Bateson and Bateson, *Angels Fear*, p. 30.

68. Ibid., p. 28.

69. Ibid., p. 27.

70. Loren J. Chapman and Jean P. Chapman, *Disordered Thought in Schizophrenia* (Englewood Cliffs, N.J.: Prentice Hall, 1973).

71. Bateson and Bateson, *Angels Fear*, pp. 27–30.

72. "Running with Wolves" (Joel Sartore interviewed by Graeme Fordyce). *Outdoor Photographer*, September 1999.

73. Quoted by Lasch, *The Culture of Narcissism*, p. 21.

74. Cushman, "Why the Self Is Empty: Toward a Historically Located Psychology."

75. Cushman, "Why the Self Is Empty," p. 600.

76. Romanyshyn, *Psychological Life: From Science to Metaphor*, p. 103.

77. Ibid., p. 109.

78. Ibid., p. 127.

79. See, for example, Peter Marin, "The New Narcissism," *Harper's Magazine*, October 1975, pp. 45–56.

80. Romanyshyn, *Psychological Life*, p. 129.

81. Aldo Leopold, *A Sand County Almanac*, p. 139.

82. Lopez, *Of Wolves and Men*.

83. Romand Coles, "Ecotones and Environmental Ethics: Adorno and Lopez," In Bennett and Chaloupka (eds.), *In the Nature of Things: Language, Politics, and the Environment*, p. 241.

84. Theodore Roszak, "Where Psyche Meets Gaia," in Roszak et al. (eds.), *Ecopsychology: Restoring the Earth, Healing the Mind*.

Chapter 7. Healing the World of Wounds

1. Octavio Paz, quoted by Jamake Highwater, *The Primal Mind: Vision and Reality in Indian America* (New York: Harper & Row, 1981).

2. Joanna Macy, "Working through Environmental Despair," in Roszak et al. (eds.), *Ecopsychology: Restoring the Earth, Healing the Mind*. See also her *Despair and Personal Power in the Nuclear Age* (Philadelphia: New Society, 1983).

3. Erik H. Erikson, *Identity: Youth and Crisis* (New York: Norton, 1968).

4. Carl R. Rogers, *On Becoming a Person: A Therapist's View of Psychotherapy* (London: Constable, 1967), p. 114.

5. Carl R. Rogers, *A Way of Being* (Boston: Houghton Mifflin, 1980), p. 351.

6. Rogers, *On Becoming a Person*, p. 105.

7. Eugene Gendlin, "The Client's Client: The Edge of Awareness," in Ronald F. Levant and John H. Shlien, *Client-Centered Therapy and the Person-Centered Approach* (New York: Praeger, 1984). All subsequent Gendlin quotes in this section are taken from this paper.

8. Ibid., pp. 81–82.

9. Casey, *Getting Back into Place: Toward a Renewed Understanding of the Place-World*, p. 186.

10. David J. Grove and B. I. Panzer, *Resolving Traumatic Memories* (New York: Irvington, 1991), p. 19.

11. Quoted by John Clay, *R. D. Laing: A Divided Self* (London: Hodder and Stoughton, 1996), p. 240.

12. Turner, *Beyond Geography: The Western Spirit against the Wilderness*, p. 9.

13. Bateson, *Steps to an Ecology of Mind*, p. 286 (italics in original).

14. Extracted from *Defenders of the Wild: Paul Watson* (Channel 4 Television).

15. See, for example, Richard Rabkin, "Critique of the Clinical Use of the Double-Bind Hypothesis"; and Gregory Bateson, "Double Bind, 1969"; both in Carlos E. Sluzki (ed.), *Double Bind: The Foundation of the Communicational Approach to the Family* (New York: Grune and Stratton, 1976).

16. *Defenders of the Wild: Paul Watson.*

17. Bruce Chatwin, *The Songlines* (New York: Penguin, 1987), p. 70.

18. Gary Nabhan, *Cultures of Habitat: On Nature, Culture, and Story* (Washington D.C.: Counterpoint, 1997), pp. 183, 275.

19. Carpenter, *Eskimo Realities*, p. 59.

20. Fairbairn, *Psychoanalytic Studies of the Personality*, p. 139.

21. See Alf Hornborg, "Ecology as Semiotics: Outlines of a Contextualist Paradigm for Human Ecology," in Philippe Descola and Gísli Pálsson, *Nature and Society: Anthropological Perspectives* (London: Routledge, 1996).

22. Geertz, *The Interpretation of Cultures*, pp. 134–135.

23. Taussig, *The Devil and Commodity Fetishism*, pp. 167–168.

24. James Cowan, *Mysteries of the Dreamtime: The Spiritual Life of Australian Aborigines.* (Revised edition, Bridport, U.K.: Prism Press, 1992), p. 98.

25. Ted Strehlow, *Aranda Traditions.* (Melbourne: Melbourne University Press, 1947). See also Chatwin, *The Songlines.*

26. Steven Feld, *Sound and Sentiment: Birds, Weeping, Poetics, and Song in Kaluli Expression* (Philadelphia: University of Pennsylvania Press, 1982).

27. Ibid., p. 164.

28. Ibid., p. 162.

29. Ibid., p. 41.

30. Ibid., p. 84.

31. Gina Abeles, "Researching the Unresearchable: Experimentation on the Double Bind," in Sluzki, *Double Bind: The Foundation of the Communicational Approach to the Family*.

32. Wallace Stegner, *This Is Dinosaur* (New York: Knopf, 1955), p. 15.

33. Geertz, *The Interpretation of Cultures*, pp. 404–405.

34. Ibid., p. 130.

35. Ibid., p. 127.

36. Ibid., p. 131.

37. Claude Lévi-Strauss, *Tristes Tropiques*. Quoted by Clifford Geertz, "The Cerebral Savage: On the Work of Claude Lévi-Strauss," *Encounter* 28.4 (1967): 25–32.

38. Romanyshyn, *Technology as Symptom and Dream*, chapter 3.

39. See Kidner, "Fabricating Nature."

40. Geertz, *The Interpretation of Cultures*, p. 405.

41. Dorothy Lee, *Freedom and Culture* (Englewood Cliffs, N.J.: Prentice Hall, 1959), p. 17.

42. Eugene C. Hargrove, *Foundations of Environmental Ethics* (Englewood Cliffs, N.J.: Prentice-Hall, 1989), chapter 5.

43. John Dewey, *Psychology*, 3rd revised ed. (New York: American Book Company, 1891), pp. 199–200.

44. See Collier, *Critical Realism*.

45. Romand Coles, "Ecotones and Environmental Ethics: Adorno and Lopez," in Bennett and Chaloupka, *In the Nature of Things*, p. 236.

46. Lopez, *Arctic Dreams: Imagination and Desire in a Northern Landscape*, p. 277. See also David Abram, *The Spell of the Sensuous: Perception and Language in a More-than-Human World* (New York: Pantheon, 1996), chapter 3.

Bibliogrpahy

Abbey, Edward. *Desert Solitaire: A Season in the Wilderness* (New York: Ballantine Books, 1968).

Abel, Annie Heloise. *Tabeau's Narrative of Loisel's Expedition to the Upper Missouri* (Norman: University of Oklahoma Press, 1939).

Abeles, Gina. "Researching the Unresearchable: Experimentation on the Double Bind." In Carlos Sluzki and Donald C. Ransom (eds.), *Double Bind: The Foundation of the Communicational Approach to the Family* (New York: Grune and Stratton, 1976).

Abram, David. *The Spell of the Sensuous: Perception and Language in a More-than-Human World* (New York: Pantheon, 1996).

Adams, Ansel. "A Personal Credo." Reprinted in Beaumont Newhall (ed.), *Photography: Essays and Images* (London: Secker and Warburg, 1981).

Argyros, Alexander J. *A Blessed Rage for Order: Deconstruction, Evolution, and Chaos* (Ann Arbor: University of Michigan Press, 1991).

Arnhart, Larry. *Darwinian Natural Right* (Albany: State University of New York Press, 1998).

Aronowitz, Stanley. *Science as Power: Discourse and Ideology in Modern Society* (Minneapolis: University of Minnesota Press, 1988).

———. *Dead Artists, Live Theories, and Other Cultural Problems* (New York: Routledge, 1994).

Atkinson, Rita L, Richard C. Atkinson, Edward E. Smith, and Daryl J. Bem. *Introduction to Psychology*. 10th ed. (San Diego: Harcourt Brace Jovanovich, 1990).

Bannister, Don. "Psychology as an Exercise in Paradox." *Bulletin of the British Psychological Society* 19. 63 (1966): 21–26.

Barfield, Owen. *Saving the Appearances: A Study in Idolatry* (New York: Harcourt, Brace, Jovanovich, n.d.).

Barrows, Anita. "The Ecopsychology of Child Development." In Theodore Roszak et al. (eds.), *Ecopsychology: Restoring the Earth, Healing the Mind* (San Francisco: Sierra Club, 1995).

Barth, Fredrik. "On Responsibility and Humanity: Calling a Colleague to Account." *Current Anthropology* 15 (1974): 99–103.

Bartram, William. *The Travels of William Bartram,* ed. Mark van Doren (New York: Publisher unknown, 1928).

Bateson, Gregory. *Steps to an Ecology of Mind* (St. Albans, U.K.: Paladin, 1973).

———. "Double Bind, 1969," in Carlos E. Sluzki (ed.), *Double Bind: The Foundation of the Communicational Approach to the Family* (New York: Grune and Stratton, 1976).

Bateson, Gregory, and Mary Catherine Bateson. *Angels Fear: Towards an Epistemology of the Sacred* (New York: Macmillan, 1987).

Baudelaire, Charles. "Photography." Reprinted in Beaumont Newhall (ed.), *Photography: Essays and Images* (London: Secker and Warburg, 1981).

Beck, Ulrich. *Risk Society: Towards a New Modernity* (London: Sage, 1992).

Bellah, Robert. *Habits of the Heart: Individualism and Commitment in American Life* (New York: Harper & Row, 1986).

Bennett, Jane, and William Chaloupka (eds.). *In the Nature of Things: Language, Politics, and the Environment* (Minneapolis: University of Minnesota Press, 1993).

Berman, Morris. *Coming to Our Senses: Body and Spirit in the Hidden History of the West* (New York: Simon & Schuster, 1989).

Bernstein, Jay M. *The Fate of Art: Aesthetic Alienation from Kant to Derrida and Adorno.* (Cambridge: Polity Press, 1992).

———. "Introduction." In Theodor W. Adorno, *The Culture Industry: Selected Essays on Mass Culture,* ed. J. M. Bernstein (London: Routledge, 1991).

Bhaskar, Roy. *Reclaiming Reality* (London: Verso, 1989).

———. *The Possibility of Naturalism.* 2nd ed. (Hemel Hempstead, U.K.: Harvester Wheatsheaf, 1989).

Billig, Michael. *Arguing and Thinking: A Rhetorical Approach to Social Psychology* (Cambridge: Cambridge University Press, 1987).

Binet, Alfred, and T. Simon. *The Development of Intelligence in Children,* trans Elizabeth S. Kite (Baltimore: Williams and Wilkins, 1916).

Bird, Elizabeth A. R. "The Social Construction of Nature: Theoretical Approaches to the History of Environmental Problems." *Environmental Review* 11 (1987): 255–64.

Black's Medical Dictionary. 36th ed. (London: A. and C. Black, 1990).

Bollas, Christopher. *The Shadow of the Object: Psychoanalysis of the Unthought Known* (New York: Columbia University Press, 1987).

Brennan, Teresa. *History after Lacan* (London: Routledge, 1993).

Brown, George W., and Tirrell Harris. *The Social Origins of Depression* (London: Tavistock, 1978).

Brown, Norman O. *Life against Death: The Psychoanalytic Meaning of History* (Middletown, Conn.: Weslyan University Press, 1959).

Bruno, Giordano. *The Heroic Frenzies*, trans. Paul E. Memmo (Chapel Hill: University of North Carolina Press, 1964).

Buck-Morss, Susan. "Socio-economic Bias in Piaget's Theory and Its Implications for Cross-Cultural Studies." *Human Development* 18 (1975): 40.

Burningham, Kate, and Geoff Cooper. "Being Constructive: Social Constructionism and the Environment." *Sociology* 33 (1999): 297–316.

Burr, Vivien. *An Introduction to Social Constructionism* (London: Routledge, 1995).

Burtt, Edwin A. *The Metaphysical Foundations of Modern Physical Science* (London: Routledge and Kegan Paul, 1959; originally published 1924).

Busia, K. A. "The Ashanti." In Darryl Forde (ed.), *African Worlds: Studies in the Cosmological Ideas and Social Values of African Peoples* (London: Oxford University Press, 1954).

Byrne, R. W., and Whiten, A. (eds.). *Machiavellian Intelligence: Social Expertise and the Evolution of Intellect in Monkeys, Apes, and Humans* (Oxford: Clarendon Press, 1988).

Callicott, J. Baird. *Earth's Insights* (Berkeley: University of California Press, 1994).

Carpenter, Edmund. *Eskimo Realities* (New York: Holt, Rinehart, and Winston, 1973).

———. *Oh, What a Blow That Phantom Gave Me!* (St. Albans, U.K.: Paladin, 1976).

Casey, Edward S. *Getting Back into Place* (Bloomington: Indiana University Press, 1993).

Castaneda, Carlos. *A Separate Reality* (Harmondsworth, U.K.: Penguin, 1973).

Cattell, Raymond B. *Abilities: Their Structure, Growth, and Action* (Boston: Houghton Mifflin, 1971).

Chapman, Loren J., and Jean P. Chapman, *Disordered Thought in Schizophrenia* (Englewood Cliffs, N.J.: Prentice Hall, 1973).

Chatwin, Bruce. *The Songlines* (New York: Penguin, 1987).

Cheney, Jim. "Ecofeminism and Deep Ecology," *Environmental Ethics* 9 (1987): 115–45.

Chenu, Marie-Dominique. *Nature, Man, and Society in the Twelfth Century* (Chicago: University of Chicago Press, 1968).

Chomsky, Noam. *Keeping the Rabble in Line: Interviews with David Barsamian* (Monroe, Maine: Common Courage Press, 1994).

Clay, John. *R. D. Laing: A Divided Self* (London: Hodder and Stoughton, 1996).

Cohen, Mark. *Health and the Rise of Civilisation* (New Haven: Yale University Press, 1989).

Coles, Romand. "Ecotones and Environmental Ethics: Adorno and Lopez." In Jane Bennett and William Chaloupka (eds.), *In the Nature of Things: Language, Politics, and the Environment* (Minneapolis: University of Minnesota Press, 1993).

Collier, Andrew. *Critical Realism: An Introduction to Roy Bhaskar's Philosophy* (London: Verso, 1994).

Coutts-Smith, Kenneth. "Some General Observations on the Problem of Cultural Colonialism." In Susan Hiller (ed.), *The Myth of Primitivism: Perspectives on Art* (London: Routledge, 1991).

Cowan, James. *Mysteries of the Dreamtime: The Spiritual Life of Australian Aborigines*. Revised ed. (Bridport, U.K.: Prism Press, 1992).

Craib, Ian. "Social Constructionism as a Social Psychosis." *Sociology* 31 (1997): 1–15.

Crévaulx, Jules. "Exploración del Inzá y del Yapura." *In América pintoresca: Descripción de viajes al nuevo continente* (Barcelona: Montaner y Simon, 1884).

Crevecoeur, Hector St. John de. *Letters from an American Farmer* (New York: Everyman's Library, n.d.).

Cronon, William (ed.). *Uncommon Ground: Rethinking the Human Place in Nature* (New York: Norton, 1996).

Cushman, Philip. "History, Psychology, and the Abyss." *Psychohistory Review* 15 (1987): 29-45.

———. "Why the Self Is Empty: Toward a Historically Situated Psychology." *American Psychologist* 45 (1990): 599–611.

———. "Ideology Obscured: Political Uses of the Self in Daniel Stern's Infant." *American Psychologist* 46 (1991): 206–19.

Danziger, Kurt. *Constructing the Subject: Historical Origins of Psychological Research* (Cambridge: Cambridge University Press, 1990).

Darwin, Charles. *The Autobiography of Charles Darwin* (London: Collins, 1958).

Davis, Susan G. "Touch the Magic." In William Cronon (ed.), *Uncommon Ground: Rethinking the Human Place in Nature* (New York: Norton, 1996).

Deloria, Vine. *God Is Red* (New York: Grosset and Dunlap, 1973).

De Mille, Richard. *Castaneda's Journey: The Power and the Allegory* (London: Abacus, 1978).

De Rougement, Denis. *Love in the Western World* (Princeton, N.J.: Princeton University Press, 1956).

Derrida, Jacques. *Writing and Difference* (Chicago: University of Chicago Press, 1978).

Descartes, Rene. *The Meditations, and Selections from the Principles* (La Salle, Ill.: Open Court Publishing, 1950; originally published 1641).

Descola, Philippe. "Constructing Natures: Symbolic Ecology and Social Practice." In Philippe Descola and Gísli Pálsson (eds.), *Nature and Society: Anthropological Perspectives* (London: Routledge, 1996).

Desjarlais, Robert, Leon Eisenberg, Byron Good, and Arthur Kleinman. *World Mental Health: Problems and Priorities among Low Income Countries* (New York: Oxford University Press, 1995).

Devall, Bill. *Simple in Means, Rich in Ends: Practicing Deep Ecology* (Salt Lake City: Peregrine Smith Books, 1988).

Dewey, John. *Psychology*. 3rd revised ed. (New York: American Book Company, 1891).

———. *Democracy and Education* (New York: Macmillan, 1916).

———. *The Quest for Certainty: A Study of the Relation between Knowledge and Action* (London: Allen and Unwin, 1930).

Diagnostic and Statistical Manual IV. (Washington D.C.: American Psychiatric Association, 1994).

Diamond, Stanley. *In Search of the Primitive* (New Brunswick, N.J.: Transaction, 1974).

Donaldson, Margaret. *Children's Minds* (New York: Norton, 1978).

Douglas, Mary. "The Lele of Kasai." In Daryll Forde (ed.), *African Worlds: Studies in the Cosmological Ideas and Values of African Peoples* (London: Oxford University Press, 1954).

Eagleton, Terry. *Literary Theory: An Introduction* (Minneapolis: University of Minnesota Press, 1983).

Edgerton, Robert B. *Sick Societies: Challenging the Myth of Primitive Harmony* (New York: Free Press, 1992).

Edgerton, Samuel. *The Renaissance Rediscovery of Linear Perspective* (New York: Harper & Row, 1975).

Eiseley, Loren. *The Night Country: Reflections of a Bone-Hunting Man* (London: Garnstone Press, 1974).

Eliot, T. S. *The Cocktail Party* (London: Faber and Faber, 1950).

Elliot, Robert. "Why Preserve Species?" In Don S. Mannison et al. (eds.), *Environmental Philosophy* (Canberra: Australian National University, 1980).

Ellis, John M. *Literature Lost: Social Agendas and the Corruption of the Humanities* (New Haven: Yale University Press, 1997).

Erikson, Erik H. *Childhood and Society*. 3rd ed. (New York: Norton, 1963).

———. Identity: *Youth and Crisis* (New York: Norton, 1968).

Evernden, Neil. *The Natural Alien* (Toronto: University of Toronto Press, 1985).

Ewen, Stuart. *All Consuming Images* (New York: Basic Books, 1988).

Fabian, Johannes. *Time and the Other: How Anthropology Makes Its Object* (New York: Columbia University Press, 1983).

Fairbairn, W. Ronald D. *Psychoanalytic Studies of the Personality* (London: Routledge and Kegan Paul, 1952).

Feld, Steven. *Sound and Sentiment: Birds, Weeping, Poetics, and Song in Kaluli Expression* (Philadelphia: University of Pennsylvania Press, 1982).

Ferenczi, Sandor. *Thalassa* (London: Maresfield, 1990).

———. *Final Contributions to the Problems and Methods of Psychoanalysis* (London: Maresfield Reprints, 1955).

Ferkiss, Victor. *Nature, Technology, and Society: Cultural Roots of the Current Environmental Crisis* (New York: New York University Press, 1993).

Fordyce, Graeme. "Running with Wolves" (An interview with Joel Sartore). *Outdoor Photographer*, September 1999.

Foucault, Michel. *The Birth of the Clinic: An Archeology of Medical Perception* (New York: Vintage Books, 1975).

———. *The History of Sexuality, Vol. 1: An Introduction* (New York: Random House, 1978).

Fox, Warwick. *Toward a Transpersonal Ecology* (Boston: Shambhala, 1990).

Freud, Sigmund. "Civilised Sexual Morality and Modern Nervous Illness." In J. Strachey (ed.), *Complete Psychological Works of Sigmund Freud* (London: Hogarth Press, 1957), vol. 9.

———. "Fixation to Traumas: The Unconscious." In *The Standard Edition of the Complete Psychological Works of Sigmund Freud*, vol. 17.

———. "Thoughts for the Times on War and Death." In *The Standard Edition of the Complete Psychological Works of Sigmund Freud*, vol. 14.

———. *Civilisation and Its Discontents* (Norton, 1961).

———. "The Question of a Weltanschauuing." In *The Standard Edition of the Complete Psychological Works of Sigmund Freud*, vol. 22.

———. "Anxiety and Instinctual Life." In *The Standard Edition of the Complete Psychological Works of Sigmund Freud*, vol. 22.

———. "On Narcissism: An Introduction." In *The Standard Edition of the Complete Psychological Works of Sigmund Freud*, vol. 14.

———. "Mourning and Melancholia." In *The Standard Edition of the Complete Psychological Works of Sigmund Freud*, vol. 14.

———. "A Case of Obsessional Neurosis." In *The Standard Edition of the Complete Psychological Works of Sigmund Freud*, vol. 3.

Fromm, Erich. *Escape from Freedom* (New York: Holt, Rinehart and Winston, 1941).

———. *The Sane Society* (London: Routledge and Kegan Paul, 1963).

Gardner, Howard. *Frames of Mind: The Theory of Multiple Intelligences* (London: Heinemann, 1983).

Garfield, Michael. "Possible Worlds or Real Worlds?" In S. Modgil, C. Modgil, and G. Brown (eds.), *Jean Piaget: An Interdisciplinary Critique* (London: Routledge and Kegan Paul, 1983).

Geertz, Clifford. "The Cerebral Savage: On the Work of Claude Lévi-Strauss." *Encounter* 28.4 (1967): 25–32.

———. *The Interpretation of Cultures* (New York: Basic Books, 1973).

Gellatly, Angus. "The Myth of Cognitive Diagnostics." In A. Gellatly, D. Rogers, and J. A. Sloboda, (eds.), *Cognition and Social Worlds* (Oxford: Clarendon Press, 1989).

Gendlin, Eugene. "A Philosophical Critique of the Concept of Narcissism." In David M. Levin (ed.), *Pathologies of the Modern Self: Postmodern Studies on Narcissism, Schizophrenia, and Depression* (New York: New York University Press, 1987).

———. "The Client's Client: The Edge of Awareness." In Ronald F. Levant and John H. Shlien (eds.), *Client-Centered Therapy and the Person-Centered Approach* (New York: Praeger, 1984).

Gergen, Kenneth. "Social Psychology as History." *Journal of Personality and Social Psychology* 26 (1973): 309–20.

———. "Experimentation in Social Psychology: A Reappraisal." *European Journal of Social Psychology* 8 (1978): 507–27.

———. "The Social Constructionist Movement in Modern Psychology." *American Psychologist* 40 (1985): 266–75.

Giddens, Anthony. *The Consequences of Modernity* (Stanford, Calif.: Stanford University Press, 1990).

Gill, Sam. "The Shadow of a Vision Yonder." In Walter Holden Capps (ed.), *Seeing with a Native Eye* (New York: Harper & Row, 1976).

Gladwin, Thomas. *East Is a Big Bird: Navigation and Logic on Puluwat Atoll* (Cambridge, Mass.: Harvard University Press, 1970).

Gleick, James. *Chaos: The Making of a New Science* (London: Heinemann, 1988).

Goldsmith, Edward. "Gaia: Some Implications for Theoretical Ecology." In P. Bunyard and Edward Goldsmith (eds.), *Gaia: The Thesis, the Mechanisms, and the Implications* (Wadebridge, U.K.: Wadebridge Ecological Centre, 1989).

Goodnow, Jacqueline. "The Nature of Intelligent Behaviour: Questions Raised by Cross-cultural Studies." In L. Resnick (ed.), *The Nature of Intelligence* (Hillsdale, N.J.: Lawrence Erlbaum, 1968).

Gorz, André. *Critique of Economic Reason* (London: Verso, 1989).

Gould, Stephen Jay. *The Mismeasure of Man* (Harmondsworth, U.K.: Penguin, 1981).

Grove, David J., and B. I. Panzer. *Resolving Traumatic Memories* (New York: Irvington, 1991).

Grumbine, R. Edward. "Wildness, Wise Use, and Sustainable Development." *Environmental Ethics* 16 (1994): 227–249.

Guisinger, Shan, and Sidney J. Blatt. "Individuality and Relatedness: Evolution of a Fundamental Dialectic." *American Psychologist* 49.2 (1994): 104–11.

Gunnison, Hugh. "The Uniqueness of Similarities: Parallels of Milton H. Erickson and Carl Rogers." *Journal of Counseling and Development* 63 (1985): 561–64.

Guntrip, Harry. *Schizoid Phenomena, Object Relations, and the Self* (London: Hogarth, 1980).

Hannigan, John. *Environmental Sociology: A Social Constructionist Perspective* (London: Routledge, 1995).

Haraway, Donna. "A Cyborg Manifesto." In *Simians, Cyborgs, and Women: The Reinvention of Nature* (London: Free Association, 1991).

Hardin, Garrett. "The Tragedy of the Commons." *Science* 162 (1968): 1243–48.

Hargrove, Eugene. *Foundations of Environmental Ethics* (Englewood Cliffs, N.J.: Prentice Hall, 1989).

Harrison, Robert Pogue. *Forests: The Shadow of Civilization* (Chicago: University of Chicago Press, 1992).

Heat-Moon, William Least. *Blue Highways: A Journey into America* (London: Secker and Warburg, 1983).

Hebb, Donald O. *The Organisation of Behavior* (New York: Wiley, 1949).

Hecht, S. "Tropical Deforestation in Latin America: Myths, Dilemmas, and Reality." Paper presented at the Systemwide Workshop on Environment and Development Issues in Latin America, University of California, Berkeley, October 16, 1990.

Heidegger, Martin. "What Is Called Thinking?" In *The Question Concerning Technology and Other Essays* (New York: Harper and Row, 1968).

Held, David. *Introduction to Critical Theory: From Horkheimer to Habermas* (London: Hutchinson, 1980).

Highwater, Jamake. *The Primal Mind: Vision and Reality in Indian America* (New York: Harper & Row, 1981).

Hillman, James. *Re-Visioning Psychology* (New York: Harper & Row, 1975).

———. *The Essential James Hillman: A Blue Fire*, ed. Thomas Moore, (London: Routledge, 1990).

Hillman, James, and Michael Ventura. *We've Had a Hundred Years of Psychotherapy and the World's Getting Worse* (San Francisco: HarperCollins, 1992).

Hockett, C. F. *Man's Place in Nature* (New York: McGraw-Hill, 1973).

Hogan, Linda. *Dwellings: A Spiritual History of the Living World* (New York: Norton, 1995).

Holmes, Oliver Wendell. "The Stereoscope and the Stereograph." *The Atlantic Monthly* 3 (June 1859). Reprinted in Beaumont Newhall (ed.), *Photography: Essays and Images* (London: Secker and Warburg, 1981).

Horkheimer, Max, and Theodor Adorno. *Dialectic of Enlightenment* (New York: Continuum, 1991).

Hornborg, Alf. "Ecology as Semiotics: Outlines of a Contextualist Paradigm for Human Ecology." In Philippe Descola and Gísli Pálsson, *Nature and Society: Anthropological Perspectives* (London: Routledge, 1996).

Hudson, Liam, and Bernadine Jacot. *The Way We Think: Intellect, Intimacy, and the Erotic Imagination* (New Haven: Yale University Press, 1991).

Hughes, J. Donald. *American Indian Ecology* (El Paso: Texas Western Press, 1983).

Ihde, Don. *Technology and the Lifeworld* (Bloomington: Indiana University Press, 1990).

Ingleby, David. "Ideology and the Human Sciences: Some Comments on the Role of Reification in Psychology and Psychiatry." *The Human Context* 11 (1970), 159–87.

Ingold, Tim. "The Optimal Forager and Economic Man." In Philippe Descola and Gísli Pálsson (eds.), *Nature and Society: Anthropological Perspectives* (London: Routledge, 1996).

Jacoby, Russell. *Social Amnesia: A Critique of Conformist Psychology from Adler to Laing* (Boston: Beacon Press, 1977).

Jaimes, M. Annette (ed.). *The State of Native America* (Boston: South End Press, 1992).

James, William. *The Varieties of Religious Experience* (London: Longman's, 1935).

Jameson, Fredric. *The Prison-House of Language: A Critical Account of Structuralism and Russian Formalism* (Princeton, N.J.: Princeton University Press, 1972).

Jaynes, Julian. *The Origin of Consciousness in the Breakdown of the Bicameral Mind.* (Boston: Houghton Mifflin, 1976).

Jung, Carl G. "Psychological Aspects of the Mother Archetype." In Herbert Read, Michael Fordham, and Gerhard Adler (eds.), *The Collected Works of C. G. Jung*, 2nd ed. (London: Routledge and Kegan Paul, 1968), vol. 9.

———. "The Spiritual Problem of Modern Man." *Collected Works*, vol. 10.

———. "The Psychology of the Transference." *Collected Works*, vol. 16.

———. "Approaching the Unconscious." In C. G. Jung (ed.), *Man and His Symbols* (Garden City, N.Y.: Doubleday, 1964).

———. "The Structure and Dynamics of the Psyche." *Collected Works*, vol. 8.

———. "Mind and Earth." *Collected Works*, vol. 10.

———. "Archetypes of the Collective Unconscious." *Collected Works*, vol. 9.

Kaplan, Rachel, and Stephen Kaplan. *The Experience of Nature: A Psychological Perspective* (Cambridge: Cambridge University Press, 1989).

Keller, Evelyn Fox. *Reflections on Gender and Science* (New Haven and London: Yale University Press, 1985).

Kennedy, John S. *The New Anthropomorphism* (New York: Cambridge University Press, 1992).

Kidner, David W. "Why Psychology Is Mute about the Environmental Crisis." *Environmental Ethics* 16.4 (Winter 1994): 359–76.

———. "Fabricating Nature." *Environmental Ethics* (forthcoming).

Kleinman, Arthur. " 'Everything That Really Matters': Social Suffering, Subjectivity, and the Remaking of Human Experience in a Disordered World." *Harvard Theological Review* 90 (1997): 315-35.

———. *Rethinking Psychiatry: From Cultural Category to Personal Experience* (New York: Free Press, 1988).

Kleinman, Arthur, and Joan Kleinman. "The Appeal of Experience, the Dismay of Images: Cultural Appropriations of Suffering in Our Times." *Daedalus* 125 (1996), part 1, 1–23.

Koch, Sigmund. "The Nature and Limits of Psychological Knowledge: Lessons of a Century *qua* 'Science'." In Sigmund Koch and David E. Leary (eds.), *A Century of Psychology as a Science* (New York: McGraw-Hill, 1985).

———. "Psychology and Emerging Conceptions of Knowledge as Unitary." In T. W. Wann (ed.), *Behaviorism and Phenomenology* (Chicago: University of Chicago Press, 1964).

Korzybski, Alfred. *Science and Sanity* (New York: Science Press, 1941).

Kovel, Joel. *The Age of Desire: Reflections of a Radical Psychoanalyst* (New York: Pantheon, 1981).

———. *The Radical Spirit: Essays on Psychoanalysis and Society* (London: Free Association, 1988).

———. *History and Spirit: An Inquiry into the Philosophy of Liberation* (Boston: Beacon Press, 1991).

Kulick, Don, and Margaret Willson (eds.). *Taboo: Sex, Identity, and Erotic Subjectivity in Anthropological Fieldwork* (London: Routledge, 1995).

Kummer, Hans, and Jane Goodall. "Conditions of Innovative Behaviour in Primates." In L. Weiskrantz (ed.), *Animal Intelligence* (Oxford: Oxford University Press, 1985).

Laing, Ronald D. *The Divided Self; An Existential Study in Sanity and Madness* (Harmondsworth, U.K.: Penguin, 1965).

———. *The Politics of Experience* (Harmondsworth, U.K.: Penguin, 1969).

La Flesche, Francis. *The Middle Five: Indian Schoolboys of the Omaha Tribe* (Madison: University of Wisconsin Press, 1963).

Lakoff, George. *Women, Fire, and Dangerous Things: What Categories Reveal about the Mind* (Chicago: University of Chicago Press, 1987).

Lame Deer, John (Fire), and Richard Erdoes. *Lame Deer, Seeker of Visions* (New York: Simon & Schuster, 1972).

Las Casas, Bartolomé de. *History of the Indies*, trans. André Collard (New York: Harper & Row, 1971).

Lasch, Christopher. *The Culture of Narcissism: American Life in an Age of Diminishing Expectations* (New York: Norton, 1979).

———. *The Minimal Self: Psychic Survival in Troubled Times* (New York: Norton, 1984).

Lassiter, Cisco. "Relocation and Illness: The Plight of the Navaho." In David M. Levin (ed.), *Pathologies of the Modern Self: Postmodern Studies on Narcissism, Schizophrenia, and Depression* (New York: New York University Press, 1987).

Lee, Dorothy. *Freedom and Culture* (New York: Prentice Hall, 1959).

Leopold, Aldo. *A Sand County Almanac, with Essays on Conservation from Round River* (New York: Ballantine, 1990; originally published 1949).

Levin, David Michael. *The Body's Recollection of Being* (London: Routledge and Kegan Paul, 1985).

———. *The Opening of Vision: Nihilism and the Postmodern Situation* (New York: Routledge, 1988).

Lévi-Strauss, Claude. *Tristes Tropiques*, trans. John and Doreen Weightman (Harmondsworth, U.K.: Penguin, 1976).

Lewis, David. "Observations on Route Finding and Spatial Orientation among the Aboriginal Peoples of the Western Desert Region of Central Australia." *Oceania* 46.4 (June 1976): 249–82.

Lidskog, R. Review of Ulrich Beck's *Risk Society. Acta Sociologica* 36 (1993): 400–403.

Lopez, Barry. *Of Wolves and Men* (London: Dent, 1978).

———. *Arctic Dreams: Imagination and Desire in a Northern Landscape* (London: Macmillan, 1986).

Luke, Tim. *Ecocritique: Contesting the Politics of Nature, Economy, and Culture* (Minneapolis: University of Minnesota Press, 1997).

Luria, Aleksandr. *Cognitive Development: Its Social and Cultural Foundations* (Cambridge, Mass.: Harvard University Press, 1976).

Maccoby, Michael, and Nancy Modiano. "On Culture and Equivalence: 1." In J. S. Bruner et al. (eds.), *Studies in Cognitive Growth* (New York: Wiley, 1965).

Macnaghten, Phil, and John Urry. *Contested Natures* (London: Sage, 1998).

Macy, Joanna. "Working through Environmental Despair." In Theodore Roszak et al. (eds.), *Ecopsychology: Restoring the Earth, Healing the Mind* (San Francisco: Sierra Club Books, 1995).

———. *Despair and Personal Power in the Nuclear Age* (Philadelphia: New Society, 1983).

Malinowski, Bronislaw. *Coral Gardens and Their Magic.* 2 vols. (Bloomington: Indiana University Press, 1965).

Manes, Christopher. *Green Rage: Radical Environmentalism and the Unmaking of Civilisation* (Boston: Little, Brown, 1990).

Marcus, George E., and Michael M. J. Fischer. *Anthropology as Cultural Critique: An Experimental Moment in the Human Sciences* (Chicago: University of Chicago Press, 1976).

Marcuse, Herbert. *Eros and Civilisation* (Boston: Beacon Press, 1977; originally published 1955).

Marin, Peter. "The New Narcissism." *Harper's Magazine,* October 1975, 45–56.

Marks, Russell. *The Idea of IQ* (Washington, D.C.: University Press of America, 1981).

Maslow, Abraham H. *Toward a Psychology of Being* (New York: Van Nostrand, 1968).

Mason, Peter. *Deconstructing America: Representations of the Other* (London: Routledge, 1990).

McCrane, Bernard. *Beyond Anthropology: Society and the Other* (New York: Columbia University Press, 1989).

McIntosh, Robert P. *The Background of Ecology* (Cambridge: Cambridge University Press, 1985).

Mead, Margaret. *Sex and Temperament in Three Primitive Societies* (New York: New American Library, 1952).

Mednick, Sarnoff. "An Associational Interpretation of the Creative Process." In Calvin Taylor (ed.), *Widening Horizons in Creativity* (New York: Wiley, 1964).

———. "Breakdown in Individuals at High Risk for Schizophrenia: Possible Predispositional Perinatal Factors." *Mental Hygiene* 54 (1970): 50–63.

Merchant, Carolyn. *The Death of Nature: Women, Ecology, and the Scientific Revolution* (San Francisco: Harper & Row, 1983).

Messick, S. "Multiple Intelligences or Multilevel Intelligence? Selective Emphasis on Distinctive Properties of Hierarchy: On Gardner's *Frames of Mind* and Sternberg's *Beyond IQ* in the Context of Theory and Research on the Structure of Human Abilities." *Journal of Psychological Inquiry* 1.3 (1992): 305–84.

Minuchin, Salvador, Bernice L. Rosman, and Lester Baker. *Psychosomatic Families: Anorexia Nervosa in Context* (Cambridge, Mass.: Harvard University Press, 1978).

Mitchell, Sandra D. et al. "The Whys and Hows of Interdisciplinarity." In P. Weingart et al. (eds.), *Human by Nature: Between Biology and the Social Sciences* (Mahwah, N.J.: Lawrence Erlbaum, 1997).

Montaigne, Michel de. *Essays* (Harmondsworth, U.K.: Penguin, 1958), vol. 3.

Morris, Colin. *The Discovery of the Individual 1050-1200* (London: Society for the Propagation of Christian Knowledge, 1972).

Mumford, Lewis. *Technics and Civilisation* (London: Burlingame, 1963).

Nabhan, Gary. *Cultures of Habitat: On Nature, Culture, and Story* (Washington D.C.: Counterpoint, 1997).

Naess, Arne. *Ecology, Community, and Lifestyle: Outline of an Ecosophy* (Cambridge: Cambridge University Press, 1989).

Nash, Roderick. *Wilderness and the American Mind* (New Haven: Yale University Press, 1967).

Neumann, Erich. *The Origins and History of Consciousness* (London: Routledge and Kegan Paul, 1954).

Nichols, Michael P. *The Self in the System: Expanding the Limits of Family Therapy* (New York: Brunner/Mazel, 1987).

Ong, Walter J. *Interfaces of the Word: Studies in the Evolution of Consciousness and Culture* (Ithaca: Cornell University Press, 1977).

Padel, Ruth. *In and Out of the Mind: Greek Images of the Tragic Self* (Princeton: Princeton University Press, 1992).

Parker, Ian, and John Shotter (eds.). *Deconstructing Social Psychology* (London: Routledge, 1990).

Pascal, Blaise. *Pensées* (Paris: Garnier, 1964; originally published 1670).

Pearce, Roy Harvey. *Savagism and Civilisation: A Study of the Indian and the American Mind* (Baltimore: John Hopkins University Press, 1965).

Peters, Robert H. *A Critique for Ecology* (Cambridge: Cambridge University Press, 1991).

Piaget, Jean. *The Psychology of Intelligence* (London: Routledge and Kegan Paul, 1950).

Plumwood, Val. *Feminism and the Mastery of Nature* (London: Routledge, 1993).

Popper, Karl. *The Open Society and Its Enemies,* vol. 1 (London: Routledge and Kegan Paul, 1966).

Prilleltensky, Isaac. "Psychology and the Status Quo." *American Psychologist* 44 (1989): 795–802.

Rabkin, Richard. "Critique of the Clinical Use of the Double-Bind Hypothesis," in Carlos E. Sluzki (ed.), *Double Bind: The Foundation of the Communicational Approach to the Family* (New York: Grune and Stratton, 1976).

Rees, Elizabeth. *Christian Symbols, Ancient Roots* (London: Jessica Kingsley, 1992).

Reiss, Timothy. *The Discourse of Modernism* (Ithaca: Cornell University Press, 1982).

Resnick, Lauren B. et al. (eds.), *Perspectives on Socially Shared Cognition* (Washington D.C.: American Psychological Association, 1991).

Rodman, John. "The Liberation of Nature?" *Inquiry* 20 (1977): 83-145.

Rogers, Carl R. *On Becoming a Person: A Therapist's View of Psychotherapy* (London: Constable, 1967).

———. *A Way of Being* (Boston: Houghton Mifflin, 1980).

———. "Reaction to Gunnison's Article on the Similarities between Erickson and Rogers." *Journal of Counseling and Development* 63 (1985): 565–66.

Rolston, Holmes III. "Nature for Real: Is Nature a Social Construct?" In: Timothy D. J. Chappell (ed.), *The Philosophy of the Environment* (Edinburgh: Edinburgh University Press, 1997).

Romanyshyn, Robert D. *Psychological Life: From Science to Metaphor* (Milton Keynes, U.K.: Open University Press, 1982).

———. *Technology as Symptom and Dream* (London: Routledge, 1989).

Rose, Jacqueline. *Sexuality in the Field of Vision* (London: Verso, 1986).

Rostow, W. *The Stages of Economic Growth* (Cambridge: Cambridge University Press, 1960).

Roszak, Theodore. *Where the Wasteland Ends: Politics and Transcendence in Postindustrial Society* (New York: Doubleday, 1972).

———. "Where Psyche Meets Gaia." In Theodore Roszak, Mary E. Gomes, Allen D. Kanner (eds.), *Ecopsychology: Restoring the Earth, Healing the Mind* (San Francisco: Sierra Club, 1995).

Roszak, Theodore, Mary E. Gomes, and Allen Kanner (eds.). *Ecopsychology: Restoring the Earth, Healing the Mind* (San Francisco: Sierra Club, 1995).

Rozin, Paul. "The Selection of Food by Rats, Humans, and Other Animals." In D. Lehrman, R. A. Hinde, and E. Shaw (eds.), *Advances in the Study of Behavior,* vol. 6 (New York: Academic Press, 1976).

Rushton, J. Philippe. "Race Differences, r/K Theory, and a Reply to Flynn." *The Psychologist: Bulletin of the British Psychological Society* 5 (May 1990): 195–98.

Sampson, Edward. "Cognitive Psychology as Ideology." *American Psychologist* 36 (1981): 730–43.

———. *Justice and the Critique of Pure Psychology* (New York: Plenum, 1983).

Sarason, Seymour. *Psychology Misdirected* (New York: Free Press, 1981).

Scharff, David E., and Ellinor Fairbairn Birtles (eds.), *From Instinct to Self: Selected Papers of W. R. D. Fairbairn. Vol. 1: Clinical and Theoretical Papers* (Northvale, N.J.: Jason Aronson, 1994).

Scheper-Hughes, Nancy. *Death without Weeping: The Violence of Everyday Life in Brazil* (Berkeley: University of California Press, 1982).

Schull, Jonathan. "Are Species Intelligent?" *Behavioral and Brain Sciences* 13 (1990): 63–108.

Scribner, Sylvia. "Modes of Thinking and Ways of Speaking: Culture and Logic Reconsidered." In P. N. Johnson-Laird and P. C. Wason (eds.), *Thinking: Readings in Cognitive Science* (Cambridge: Cambridge University Press, 1977).

Shaffer, William M. "Stretching and Folding in Lynx Fur Returns: Evidence for a Strange Attractor in Nature?" *American Naturalist* 124 (1984): 798-820.

Shaffer, William M., and M. Knot, "Do Strange Attractors Govern Ecological Systems?" *BioScience* 35 (1985): 342–50.

Shanks, Michael, and Christopher Tilley. *Social Theory and Archeology* (Cambridge: Polity Press, 1987).

Sharp, Lauriston. "Steel Axes for Stone Age Australians." In Edward H. Spicer (ed.), *Human Problems in Technological Change* (New York: Wiley, 1952).

Sheldon, Joseph K. *The Rediscovery of Creation: A Bibliographical Study of the Church's Response to the Environmental Crisis* (Metuchen, N.J.: American Theological Library Association, 1992).

Shepard, Paul. *Nature and Madness* (San Francisco: Sierra Club, 1982).

———. "Nature and Madness." In Theodore Roszak, Mary E. Gomes, and Allen Kanner (eds.), *Ecopsychology: Restoring the Earth, Healing the Mind* (San Francisco: Sierra Club, 1995).

———. *Traces of an Omnivore* (Washington, D.C.: Shearwater Press, 1996).

Shotter, John. *Images of Man in Psychological Research* (London: Methuen, 1975).

Shweder, Richard A. *Thinking through Cultures* (Cambridge, Mass.: Harvard University Press, 1991).

Shweder, Richard A., and Edmund J. Bourne. "Does the Concept of the Person Vary Cross-Culturally?" In Richard A. Shweder and Robert A. LeVine (eds.), *Culture Theory: Essays on Mind, Self, and Emotion* (Cambridge: Cambridge University Press, 1984).

Sismondo, Sergio. "Some Social Constructions." *Social Studies of Science* 23 (1993): 515–53.

Smail, David. *The Origins of Unhappiness* (London: HarperCollins, 1993).

Snell, Bruno. *The Discovery of the Mind: The Greek Origins of European Thought*, trans. T. G. Rosenmeyer (Oxford: Blackwell, 1953).

Snyder, Gary. *The Practice of the Wild* (Berkeley, Calif.: North Point Press, 1990).

Sokal, Alan and Jean Bricmont. *Intellectual Impostures* (London: Profile, 1998).

Spretnak, Charlene. *The Spiritual Dimension of Green Politics* (Santa Fe: Bear and Co., 1986).

Standing Bear, Luther. *Land of the Spotted Eagle* (Lincoln: University of Nebraska Press, 1978).

Stanner, W. E. H. "Some Aspects of Aboriginal Religion." *The Australian and New Zealand Theological Review* 76 (1976).

Stegner, Wallace. *This Is Dinosaur* (New York: Knopf, 1955).

Stern, Karl. *The Flight from Woman* (London: George Allen and Unwin, 1966).

Sternberg, Robert. *Beyond IQ: A Triarchic Theory of Human Intelligence* (Cambridge: Cambridge University Press, 1985).

Storr, Anthony. *The Dynamics of Creation* (London: Secker and Warburg, 1972).

Strang, Veronica. *Uncommon Ground: Cultural Landscapes and Environmental Values* (Oxford: Berg, 1997).

Strathern, Marilyn. *After Nature: English Kinship in the Twentieth Century* (Cambridge: Cambridge University Press, 1992).

Strehlow, Ted G. H. *Aranda Traditions* (Melbourne: Melbourne University Press, 1947).

———. *Songs of Central Australia* (Sydney: Angus and Robertson, 1971).

Sulloway, Frank. *Freud: Biologist of the Mind* (London: Burnett, 1979).

Taussig, Michael. *The Devil and Commodity Fetishism in South America* (Chapel Hill: University of North Carolina Press, 1980).

———. *Shamanism, Colonialism, and the Wild Man: A Study in Terror and Healing* (Chicago: University of Chicago Press, 1987).

Taylor, Charles. *The Ethics of Authenticity* (Cambridge, Mass.: Harvard University Press, 1992).

Thomas, W. L. Jr. (ed.). *Man's Role in Changing the Face of the Earth: Symposium of the Wenner-Gren Foundation* (Chicago: University of Chicago Press, 1956).

Todorov, Tzvetan. *The Conquest of America: The Question of the Other* (Cambridge: Harper & Row, 1984).

Turnbull, Colin. *The Forest People* (New York: Simon & Schuster, 1961).

———. *The Mountain People* (London: Picador, 1974).

———. *The Human Cycle* (London: Jonathan Cape, 1984).

Turner, Frederick. *Beyond Geography: The Western Spirit against the Wilderness* (New York: Viking Press, 1980).

Turner, Jack. " 'In Wildness Is the Preservation of the World.' " In George Sessions (ed.), *Deep Ecology for the Twenty-First Century* (Boston: Shambhala, 1995).

———. *The Abstract Wild* (Tucson: University of Arizona Press, 1996).

Turner, Victor. *The Forest of Symbols* (Ithaca: Cornell University Press, 1967).

Velody, Irving, and Robin Williams (eds.). *The Politics of Constructionism* (London: Sage, 1998).

Vygotsky, Lev S. *Mind in Society* (Cambridge, Mass.: Harvard University Press, 1978).

Wallach, Michael A., and Lise Wallach, *Psychology's Sanction for Selfishness: The Error of Egoism in Theory and Therapy* (New York: Freeman, 1983).

Warner, Richard. *Recovery from Schizophrenia: Psychiatry and Political Economy* (New York, Routledge and Kegan Paul, 1985).

Warren, Karen J. "The Power and the Promise of Ecological Feminism." *Environmental Ethics* 12 (1990): 125–46.

Waters, Frank. *The Colorado* (New York: Holt, Rinehart, and Winston, 1974).

Wax, Murray L., and Rosalie H. Wax. "Cultural Deprivation as an Educational Ideology." In Eleanor B. Leacock (ed.), *The Culture of Poverty: A Critique* (New York: Simon & Schuster, 1971).

Weston, Edward. *Photography* (Esto Publishing, 1934). Reprinted in Beaumont Newhall (ed.), *Photography: Essays and Images* (London: Secker and Warburg, 1981).

Wilson, Margaret. *Descartes* (London: Routledge and Kegan Paul, 1982).

Winnicott, Donald W. *Playing and Reality* (Harmondsworth, U.K.: Penguin, 1974).

Witkin, Herman. *Manual for the Embedded Figures Test* (New York: Consulting Psychologists Press, 1989).

Zoja, Luigi. *Growth and Guilt: Psychology and the Limits of Development* (London: Routledge, 1995).

Index